T0304480

Water Supply and Distribution Systems

Second edition

Water Supply and Distribution Systems

Second edition

Dragan A. Savić and John K. Banyard

Published by Emerald Publishing Limited, Floor 5, Northspring, 21-23 Wellington Street, Leeds LS1 4DL.

ICE Publishing is an imprint of Emerald Publishing Limited

Other ICE Publishing titles:
ICE Handbook of Urban Drainage
Richard Ashley, Brian Smith, Paul Shaffer and Issy Caffoor.
ISBN 9780727741783
Sustainable Water
Charles Ainger and Richard Fenner. ISBN 9780727757739
Land Drainage and Flood Defence Responsibilities, Fifth edition
Institution of Civil Engineers. ISBN 9780727760630

A catalogue record for this book is available from the British Library

ISBN 978-1-83549-847-7

Cover photo: Caroline Ericson / Alamy Stock Photo

Commissioning Editor: Viktoria Hartl-Vida
Content Development Editor: Cathy Sellars
Books Production Lead: Emma Sudderick

Typeset by KnowledgeWorks Global Ltd.
Index created by David Gaskell

Printed and bound by CPI Group (UK) Ltd, Croydon, CR0 4YY

Contents

Joby Boxall, Neil Dewis, John Machell, Ken Gedman, Adrian Saul, Frank van der Kleij, Adam Smith and Nathan Sunderland

Paul Jowitt, Adrian Johnson and Kees van Leeuwen

Preface to the second edition

The first edition of this book was published in 2011; since then, the field has undergone significant developments, driven by advancements in technology, changing environmental concerns and evolving regulatory frameworks. Much of what was viewed in 2011 as leading edge is now in many cases normal practice.

At a time when there is much criticism of the UK Water Industry it worth remarking that current compliance with UK standards (incorporating EU standards) stands at 99.97% overall. The quality of UK tap water is among the best in the world. As Editors, we have meticulously curated this updated edition to reflect the dynamic nature of water management in the belief that by doing so, we can help the UK and Europe to maintain their world leading position and provide the tools for others to meet similar standards.

All chapters have been thoroughly reviewed, and where appropriate the latest research findings, case studies and practical insights have been incorporated. Our expert contributors have delved deeper into critical topics, ensuring accuracy and relevance.

This second edition introduces fresh perspectives through revised chapters, addressing various design, operational, financial and sustainability considerations for water supply and distribution system management. Entirely new chapters, dealing with smart water systems and digitalisation, explore the integration of smart water technologies and predictive analytics in water infrastructure management, and, while previously in one chapter, water resource management and water treatment are now separated, giving these two topics their own identities.

This book would not be possible without the generosity of contributors, reviewers and practitioners who have shaped this edition and to them we extend our heartfelt gratitude. Their expertise and dedication have elevated the quality of this work.

As they embark on this journey through the ever-expanding intricacies of water supply and distribution, we invite our readers to engage with the evolving landscape. It is our hope that like the first edition this book will serve as a valuable resource for researchers, engineers, policymakers and students committed to ensuring a sustainable water future.

Dragan A. Savić FICE, FCIWEM
KWR Water Research Institute, The Netherlands and
University of Exeter, UK

John K. Banyard OBE, F.R.Eng, FCGI, FICE, FCIWEM
Independent consultant, Warwick, UK

Preface to the first edition

There are already a large number of textbooks covering hydraulics and water engineering, so why do we need yet another one to fill our bookshelves?

While the above is certainly a valid question, we, the editors, with long experience in both academia and with Water Service Providers were aware that there was a gap. With privatisation of the UK Water Industry in the late eighties, which brought changes to the organisation of the industry including a high level of regulation, a reduction in in-house expertise and a decline in research investment, there is a need to provide a useful reference book for practising engineers, particularly to provide information on up to date practice in today's increasingly complex Water Industry. This is important as practitioners in the developed (and the less developed) parts of the world face not only classical design and management problems, but also ever increasing environmental and sustainability requirements and concerns and at the same time few engineers can hope to keep pace with the vast amount of material presented at conferences and seminars. We also felt that there needed to be a book, which would provide a suitable guide for final year undergraduates and MSc students of water and environmental engineering courses, but which at the same time could be a useful reference after graduation when they enter employment with the Water Industry, environmental protection agencies or consultancies.

We have not tried to cover the whole of the Water Industry, but rather have focused on Water Distribution, where there have been major advances in the engineer's ability to optimise solutions, and obtain levels of understanding that have been denied to previous generations of practitioners.

To achieve this aim, we have assembled authors from academia, consultancy and the Water Service Providers to ensure that each chapter provides a balanced view of not only what is theoretically possible, but also what is practical both in the design office and in the world of water distribution system operation.

We hope that our readers will find this book helpful in their working lives, and that it will not become yet another tome that gathers dust on the bookshelves.

Dragan A. Savić FICE, FCIWEM
KWR Water Research Institute, The Netherlands and
University of Exeter, UK

John K. Banyard OBE, F.R.Eng, FCGI, FICE, FCIWEM
Independent consultant, Warwick, UK

About the editors

Dragan A. Savić FREng, CEng, FICE, FCIWEM

Dragan Savić is a Global Advisor on Digital Sciences and former Chief Executive Officer at KWR Water Research Institute based in The Netherlands and Professor of Hydroinformatics at the University of Exeter in the UK.

Dragan is an internationally recognised leading water engineering and pioneering hydroinformatics expert with over 40 years of experience working in engineering technology, academia and consultancy. He has influenced the water sector through academic research, mentoring of future water leaders, undertaking leading roles in international organisations, visiting/distinguished professor roles at various universities worldwide (e.g. China, Malaysia, Italy, Saudi Arabia, Serbia), and serving on advisory boards of water technology companies and government bodies. In addition to innovation and leadership skills, he is known for believing in bringing science into practice in the broader water sector and utilities in general.

Dragan is a founder and former director of the Centre for Water Systems at the University of Exeter, an internationally recognised group for excellence in water and environmental science research. He is a Chartered Civil and Water Engineer and an elected Fellow of the Royal Academy of Engineering, Member of the European Academy of Sciences (EURASC), Fellow of the Institution of Civil Engineers (ICE), Fellow of the Chartered Institution of Water and Environmental Management (CIWEM) and Fellow of the International Water Association (IWA).

John K. Banyard OBE, FREng, FCGI, BSc(Eng), CEng, FICE, FCIWEM

John joined Severn Trent Water on its foundation in 1974. In January 1990 he became Director of Engineering and in 1997 Asset Management for Severn Trent Water Ltd and was appointed to the main Board of Severn Trent Plc in January 1998. He was also a non-executive director of several subsidiary and joint venture companies from 1990 onwards including an appointment to the Board of the US subsidiary in April 2000.

On retirement from Severn Trent he became a board member of the Water Industry Commission for Scotland July 2005–June 2011. He was Chairman of the West Midlands Innovation and Technology Council 2006–2011. He was also Chairman of the Standing Joint Committee for the Infrastructure Conditions of Contract and Chairman of the Civil Engineering Standard Method of Management Panel for many years. Additionally, he works as an independent consultant.

He is a Chartered Civil Engineer and was elected a Fellow of the Royal Academy of Engineering in 1997. He became a Fellow of the City & Guilds of London Institute in 2000 and was awarded an OBE for services to Engineering and The Water Industry in December 2004.

John is the author of over 20 published papers on topics ranging from Project Appraisal and Computer Aided Drafting to the Collapse of Carsington Dam and Water Privatisation. In 1999, he was awarded the Frederick Palmer Prize by the Institution of Civil Engineers for his work on asset management. In 2004, he was invited to deliver the 5th International Brunel lecture for the Institution of Civil Engineers, involving over 30 presentations in 17 countries around the world.

He was a Royal Academy of Engineering Visiting Professor at the University of Loughborough 2001–2012. In 2014, he became Chairman of the Water Informatics Science and Engineering Doctoral Training Centre Advisory Board, a joint venture between water engineering departments at Bath, Bristol, Cardiff and Exeter Universities; retiring in 2023. He has been a mentor for recipients of the R A Eng leadership and Entrepreneurs Awards for many years.

John is a past President of the Pipeline Industries Guild, a past Master of the Worshipful Company of Engineers, and is also a member of the Water Conservators Livery Company; both being City of London Livery Companies.

Contributors

Peter Boden	*Edge Analytics Ltd, Leeds, UK*
Joby Boxall	*University of Sheffield, UK*
Rob Casey	*Thames Water, Reading, UK*
Adrian Cashman	*Independent Water Management Consultant and University of the West Indies, Barbados*
Emile Cornelissen	*KWR Water Research Institute, The Netherlands*
Geoff Darch	*Anglian Water, Peterborough, UK*
Neil Dewis	*WSP, UK*
Ken Gedman	*Stantec UK, UK*
Julien J. Harou	*University of Manchester, UK*
Adrian Johnson	*Stantec, Edinburgh, UK*
Paul Jowitt	*Heriot-Watt University, Edinburgh, UK*
Zoran Kapelan	*Delft University of Technology, The Netherlands*
Myles Key	*South West Water Ltd, Exeter, UK*
Chris Lambert	*Thames Water, UK*
John Machell	*University of Sheffield, UK*
Mohamed Mansoor PHD	*Jacobs, UK*
Adrian McDonald	*University of Leeds, UK*
Iain McGuffog	*South West Water, Exeter, UK*
Harrison Mutikanga	*Uganda Electricity Generation Company, Uganda*
Mark Randall-Smith	*Independent Consultant, RS Analytical Solutions Ltd, Bournemouth, UK*

Kees Roest	*KWR Water Research Institute, The Netherlands*
Adrian Saul	*Emeritus Professor, University of Sheffield, UK*
Adam Smith	*Yorkshire Water, UK*
Nathan Sunderland	*Yorkshire Water, UK*
Tiku Tanyimboh	*University of the Witwatersrand, Johannesburg*
Seneshaw Tsegaye	*Florida Gulf Coast University, USA*
Kalanithy Vairavamoorthy	*International Water Association, UK*
Frank van der Kleij	*Stantec UK, Leeds, UK*
Kees van Leeuwen	*University of Utrecht, The Netherlands*
Peter van Thienen	*KWR Water Research Institute, The Netherlands*
Howard S. Wheater	*Imperial College London, UK*

Abbreviations

ABC activity-based costing

ACO ant colony optimisation

AEX anion exchange

AI artificial intelligence

AM area meter

AMI advanced metering infrastructure

AMP asset management plan

AMR automatic meter reading

AoS appraisal of sustainability

ASR aquifer storage and recovery

AZNP average zone night pressure

BABE bursts and background estimates

BAC biological activated carbon

BCI Blue City Index

BMV burst main valve

BPSO best practicable sustainable option

CAPEX capital expenditure

CAMS catchment management strategy

CARE-W Computer Aided Rehabilitation of Water Networks

CBA cost–benefit analysis

CBA City Blueprint Approach

CBF City Blueprint Framework

CEA cost-effectiveness analysis

CFD computational fluid dynamics

CRI compliance risk index

DA decision analysis

DAF dissolved air flotation

DAFF dissolved air flotation over filters

DBP disinfection by-products

DCF discounted cashflow

DCM domestic consumption monitor

DDA demand-driven analysis

Defra Department for Environment, Food and Rural Affairs

DEM digital elevation model

DG Director General

DLF deflection lag factor

DMA distribution management areas

DMA district metered area

DOMS distribution operation and maintenance strategies

DPC direct procurement for customers

DPM discolouration propensity model

DRIP data rich, information poor

DRM discolouration risk model

DWDS drinking water distribution systems

DWI Drinking Water Inspectorate

DWQR Drinking Water Quality Regulations for Scotland

DWSP drinking water safety plan

EA Environment Agency

ECF electro-coagulation–flotation

EGL energy grade line

EIA economic impact assessment

EIA environmental impact assessment

ELL economic level of leakage

EPA Environmental Protection Agency

EPR evolutionary polynomial regression

EPS extended period simulation

ESG environmental, social, and governance

ESP extended period simulation

ET evapotranspiration

FBC flat-bottomed clarifier

FDO flexible design option

FEX fluidized ion exchange

FOSM first order second moment model

FSD fixed speed drive

GA genetic algorithm

GAC granular activated carbon

GCF Governance Capacity Framework

GDP gross domestic product

GGA global gradient algorithm

GHG greenhouse gas

GIS geographic information system

GPS global positioning system

GUI graphical user interface

GSS Guaranteed Standards Scheme

HDA head-driven analysis

HDN heuristic derived from nature

HDPE high-density polyethylene

HGL hydraulic grade line

HPPE high-performance polyethylene

IMF International Monetary Fund

IPCC Intergovernmental Panel on Climate Change

IRR internal rate of return

ISO International Organization for Standardization

IWA International Water Association

IWRM integrated water resources management

IWS intermittent water supplies

lcd litres per capita per day

MAIDE monitor, analysis, intervention, decision and evaluation

MCA multi-criteria analysis

MCS Monte Carlo simulation

MDPE medium-density polyethylene

MF microfiltration

MIEX magnetic ion exchange

MOGA multi-objective genetic algorithm

NF nanofiltration

NFW National Framework for Water

NHPP non-homogeneous Poisson process

NIC National Infrastructure Commission

NOM natural organic matter

NPSH net positive suction head

NPV net present value

NRV non-return valve

NRW Natural Resources Wales

NSO National Statistics Office

NTU nephelometric turbidity units

OECD Organization for Economic Co-operation and

Development

OEP Office of Environmental Protection

Ofwat Office of Water Services

Ofwat Water Services Regulation Authority

OPA overall performance assessment

OPEX operational expenditure

OPI operational performance index

OPI operational performance indicator

OSEC on-site electrolytic chlorination

PAH polycyclic aromatic hydrocarbons

PALMM prevention, awareness, location, mitigation and mend

PCC per capita consumption

PDF probability density function

PE polyethylene

PFAS polyfluoroalkyl substances

PFI public finance initiative

PI performance indicator

PIC Public Interest Commitment

PLC programmable logic controller

PMA pressure managed area

PODDS Prediction of Discolouration in Distribution Systems

PPRA pre- and post-rehabilitation assessment

PRV pressure-reducing valve

PSBR public sector borrowing requirement

PSV pressure-sustaining valve

PU polyurethane

PVC polyvinyl chloride

RCP rapid crack propagation

RGF rapid gravity filter

RO reverse osmosis

ROA real options analysis

RTC real-time control

SA simulated annealing

SaaS software as a service

SBTI Science-based Targets Initiative

SCADA supervisory control and data acquisition

SEA strategic environmental assessment

SELL sustainable level of leakage

SIC Standard Industrial Classification

SIX suspended ion exchange

SoSI security of supply index

SOSM second order second moment model

SPEA Strength Pareto Evolutionary Algorithm

SROL social return on investment

SSF slow sand filtration

STPR social time preference rate

SWAN Smart Water Network Forum

TCFD Taskforce on Climate-related Financial Disclosure

TPF trends and pressures framework

TPI trends and pressures index

UF ultrafiltration

UKAS UK Accreditation Service

UKWIR UK Water Industry Research

UNCED UN Conference on Environment and Development

uPVC unplasticised polyvinyl chloride

VOC volatile organic compounds

VSD variable speed drive

WCED World Commission on Environment and Development

WDS water distribution system

WFD Water Framework Directive

WHO World Health Organization

WINEP Water Industry National Environmental Programme

WIS Water Industry Standard

WIS water into supply

WLC whole-life costing

WRAS The Water Regulations Advisory Scheme

WRc Water Research Centre

WRMP water resources management plan

WSP water safety plan

WSP water service provider

WTW water treatment works

Dragan A. Savić and John K. Banyard
ISBN 978-1-83549-847-7
https://doi.org/10.1108/978-1-83549-846-020242001
Emerald Publishing Limited: All rights reserved

Chapter 1
Historical development of water distribution practice

Dragan A. Savić
KWR Water Research Institute, the Netherlands, and University of Exeter, UK

John K. Banyard
Independent Consultant, Warwick, UK

1.1. Introduction

One of the problems facing the water distribution engineer is the longevity of the assets that they are responsible for. Today those fortunate enough to live in the developed nations of the world have come to take the supply of potable water for granted. Yet it is certainly within living memory of some of the population that there was a time when many houses had a single cold-water tap, with no bathroom, and certainly no washing machine or dishwasher. The very significant increase in volume of water consumed, not to mention far higher quality standards, are often overlooked and the technology is viewed as 'the water flows through pipes as it did in my grandfather's day' with the result that many believe that the industry has ossified. Even more sadly, it is possible to find practitioners who also share the same beliefs, with a view that what was satisfactory for their predecessors must be satisfactory today. For that reason alone it is worth taking a short space within this book to review the historical development of water treatment and distribution. However, it is also important to provide at least a basic foundation of what has gone before to help better understand today's technology and good practice. In doing so we shall hopefully prepare the way for future developments to be introduced to enhance today's practices.

In reality the history of water supply and distribution is one of over 200 years of constant innovation and development. Where there have been lulls, it has not been through lack of effort, but rather waiting for science and technology to develop and, sometimes, waiting for those concepts to be capable of being deployed. The advent of the digital age has facilitated huge strides forward, with a result that the technical sophistication is now beyond the wildest imaginings of those who worked in the industry only 60 years ago.

1.2. History of water treatment and supply

It is impossible to say precisely when the first installations of artificial water supply were introduced. We know that man must always have needed access to clean water for survival. For nomadic peoples this was simply a matter of finding a clean river or spring.

Early conurbations were sited near to water sources, and there is evidence of early civilisations going back to at least the fourth millennium BC. The earliest form of water engineering appears to have been the construction of irrigation canals, but at some stage wells must have been constructed. Rather than get involved in lengthy discussions as to where or when the first water supply system was constructed, it will meet our needs in understanding the historical developments of water supply if we rely on Roman sources, and, in particular, the work of Frontinus (35–103 AD). Frontinus was certainly not the first Roman to write about water supply; approximately 100 years earlier, the architect Vitruvius (c.75–15 BC) had produced a large work on architecture, which included among many other topics the construction of aqueducts. However, Frontinus is a more helpful source because he was appointed manager of Rome's water supply in 97 AD. Furthermore, he left a report on his work, which has survived as a text book, which, well beyond his intentions, has been used to instruct (willingly or otherwise) generations of Latin students.

Frontinus was not an engineer; he was an extremely successful professional soldier who, in 76 AD, was appointed governor of Britain. At the end of his term as governor he returned to Rome having already written a book outlining military stratagems. He was faced by a new challenge, one for which he was not wholly prepared, and having tackled it, he produced 'De aquis urbis Romae'. In this he sets out his understanding of the history of Rome's water supply, saying that for the first 441 years of Rome's existence it was supplied by wells, springs and, of course, the River Tiber. However, around 312 BC the first aqueduct was brought into use, known as the Appian Aqueduct, after Appius Claudius Crassus, who was also responsible for the Appian Way. The aqueduct was around five miles long and much of it was constructed underground. He goes on to detail a further ten aqueducts, all constructed before he took office. Today's engineers might care to ponder the engineering feat of building such a structure, which Frontinus would describe as still being in use some 400 years after its initial construction.

It is interesting that Frontinus condemns the construction of the Alsietian (or Augusta) aqueduct by the emperor Augustus since the water is described as unwholesome and not used for consumption by the people. We have clear evidence, therefore, that there was at this time clear understanding of the link between water and illness, indeed it would be surprising if this were not the case.

The book goes on to reveal that, in modern parlance, Frontinus inherited a mess. He explains how he had all of the aqueducts surveyed and drawings produced, so that he did not have to waste his time going out to view problems personally, his subordinates could explain the issue to him with the help of the appropriate drawing. He had the aqueducts relined with lead to prevent leakage and vigorously pursued the owners of villas along the route of the aqueducts who had tapped into them to provide a free water supply to their properties. There was no water treatment as we would recognise it, but water discharged at the end of the aqueducts into tanks where impurities could settle out. Water was generally distributed around Rome by water sellers, and Frontinus has some harsh words for them. Overall, Frontinus applied his military background for standards and discipline very successfully to the management of the Roman aqueducts and, in doing so, gave a good indication of the tasks required of the asset manager, which would be recognised some 1900 years later.

Although Frontinus was only concerned with Rome's aqueducts, the provision of water supplies was extremely important to all of Rome's cities. Perhaps the most famous is the spectacular Pont du Gard near the French city of Nimes but it is by no means unique, and the ruins of the aqueduct that brought water into the city of Barcelona are still visible near to the Gothic cathedral.

Unfortunately, with the decline and, eventually, the fall of the Roman Empire, the aqueducts fell into disrepair and the population returned to the methods that had served Rome for the first 440 years of its existence (if Frontinus is indeed correct). It appears that in medieval times, monasteries started to pipe water as an addition to the supplies from wells on which they were frequently founded, and it is possible that some of this water found its way to the local population.

In 1589 Sir Francis Drake was instrumental in providing the city of Plymouth with a new water supply known as Drake's Leat, dug by hand (although not in the single day ascribed by legend), and in 1613 the New River was constructed to bring fresh water to the ever expanding city of London from the River Lea some 20 miles away. However, none of these schemes can really be compared with the technical achievements of the Romans. Equally, some 1400–1500 years after Frontinus, civilisation was indeed catching up with the Romans of his era.

The lack of water treatment and clean water supplies began to manifest themselves as the Industrial Revolution took place. Farm workers flocked to the new industrial cities to better their existence, but this placed huge strains on both water supplies and sanitation, both of which were extremely basic, and outbreaks of both typhoid and cholera became common place, albeit by no means continuous. The state of the working classes was exposed in a report by Edwin Chadwick, published in 1842, and quotations from these reports adequately demonstrate the misery of those days.

The various forms of epidemic, endemic and other disease caused, or aggravated or propagated chiefly among the labouring classes by atmospheric impurities produced by decomposing animal and vegetable substances, by damp and filth and close and overcrowded dwellings prevail among the population in every part of the kingdom…

That such disease wherever its attacks are frequent is always founding connection with the above circumstances…

The formation of all habits of cleanliness is obstructed by defective supplies of water.

The poet Shelley went further in one of his works, stating that '*Hell is a city rather like London*'.

Although Chadwick's reports did start the movement to provide better living conditions, particularly in terms of sanitation, the lack of scientific understanding about the cause of disease was a major hurdle.

It is difficult now to fully understand why the link between impure water and disease was not appreciated, particularly as there is ample evidence that there was a desire for clean wholesome water. In 1852, the Metropolis Water Act required all water derived from the Thames and supplied in London within 5 miles of St Paul's Cathedral to be at first filtered, but, even so, the accepted medical theory for much of the nineteenth century was that typhoid and cholera were airborne diseases, spread by the breathing in of miasmas (foul air) and had nothing to do with water, which was considered 'clean' as long as it was clear and not turbid.

The germ theory of disease did, indeed, exist, but it was not accepted by the majority of the population nor scientific opinion. It had its proponents, one of whom was Dr William Budd of Bristol who recognised that clean sources of water were required to avoid the spread of cholera. In the early 1840s he had used a microscope to examine rice water, the term used for the stools produced by cholera patients; he found organisms that he had also found in drinking water and concluded (incorrectly) that the organism was a fungus, and was responsible for cholera. Dr Budd became one of the people responsible for establishing the Bristol Waterworks Company in 1846, which brought clean water to the city from the Mendip Hills through an aqueduct.

Budd wrote up his work in 1849, and was in correspondence with a Dr John Snow in London (who in 1849 had produced a paper suggesting that cholera was not spread by miasmas), but the paper produced little interest. Interestingly, a careful reading of Chadwick's report shows that he fully subscribed to the miasmic theory.

In 1854, there was an outbreak of cholera in Soho, a district of London loosely defined by what in modern London are Oxford Street, Regent Street, Leicester Square and Charing Cross Road. There was nothing exceptional about this, as London in those days suffered regular outbreaks of both cholera and typhoid, as indeed did most cities and large towns within the UK and, indeed, Europe and the USA. Dr Snow came to realise that all of his patients obtained water from the same pump in Broad Street. He plotted the progress of the outbreak on a map and demonstrated to the public authorities that the outbreak in Soho was indeed linked to the Broad Street pump, and persuaded them to remove the handle of the pump, after which the outbreak rapidly subsided. It is ironic that in the paper that he subsequently wrote in 1855 concerning the outbreak and his work in stopping it, he identified a number of patients who had used the Broad Street pump in preference to alternative sources closer to their homes. Subsequent investigation showed that the well which supplied the Broad Street pump was located very close to a cess pit. Despite the success in stopping this particular outbreak of cholera, the pump handle was in fact reinstated by the authorities once the epidemic had passed.

Snow's work was not accepted by the medical establishment. His case was not helped by the fact that he had subjected the water to chemical and microscopic examination and failed to identify any agent that he could categorically state was the cause of the cholera outbreak. Today this seems to us almost incredible, particularly as it was well known from microscopic examination that there were organisms in the water. However, it was to be another 30 years before the 'germ theory of disease' was to be established by the French scientist Louis Pasteur. It is impossible to put a precise date on Pasteur's work, as he published a number of papers from 1865 on, but by 1880 the case for germ theory was established as correct. Pasteur did not invent the germ theory, it had existed as a hypothesis for many years; what he did was to demonstrate convincingly that the germ theory was correct and the more widely accepted miasmic theory was invalid.

Looking back at historical actions is always difficult, and today it seems unbelievable that the perceived wisdom of the medical establishment supported the miasmic theory. It was certainly well known that drinking dirty water led to illness and that had been the case for centuries, it was one of the reasons given for medieval monasteries brewing beer, although there may also have been other incentives. Nonetheless, it was undoubtedly the case that until Pasteur's work, the majority opinion was that serious illnesses such as typhoid, cholera and even the plague were transmitted by miasmas. Although it easy to be critical of such failure to accept what is now obvious, history has many such examples: the failure of the Church to accept the work of Copernicus is one example, a far more recent one is the opposing views about the nature of the universe. Einstein's theory of relativity leads to the conclusion that the universe is constantly expanding as proposed by Hubble (the Big Bang theory) but there are others like Hoyle and Bondi who proposed a 'steady state' model; Hubble's ideas are now generally accepted, but there still persist a number of other models with their proponents which may eventually be proved to have some validity. Scientific progress tends to be incremental and it is against that background that we now need to view the development of the principal water treatment processes.

The first recorded instance of use of a filtration system for water treatment was Paisley, Scotland, in 1808. However, this did not reflect some farsighted public health concerns by the city fathers, rather it was installed by John Gibb to improve his cloth bleaching business. The town is sited on the side of the River Cart, which was notorious for becoming turbid in times of storm. This variability in water quality affected the colouring of the cloth being bleached. It is reported that he was so successful he was able to sell surplus water to those who wished to pay for it.

The first municipal installation of water filters (slow filters) was at Chelsea by James Simpson in 1829, some 23 years before the Metropolis Water Act referred to above.

The first installation of filters in the USA was in Richmond, Virginia in 1833, and there was a large installation of 'English' filters installed at Poughkeepsie in New York by James P Kirkwood in 1872. Thereafter the efficacy of water filtration as a means of avoiding outbreaks of disease became more readily demonstrable, however, although a well-designed filter made a tremendous improvement to the quality of water, it could not guarantee purity. By 1880 Pasteur's demonstration of the germ theory was becoming accepted and water professionals could start to concentrate on removing the offending organisms.

In 1895, George W Fuller, working at Louisville in the USA, was attempting to find the most appropriate way of treating the waters of the Ohio River and successfully combined the addition of chemical coagulants and water filtration to produce the now classical two stage process that is at the heart of many water treatment plants around the world. He formed his own consultancy practice subsequently, and was responsible for a much larger installation at Little Falls, New Jersey, in 1902.

Fuller also worked on the development of so called rapid gravity filters, which are commonplace today, being cleaned mechanically as part of the operational cycle, rather than depending on manual excavation of sand associated with the original slow sand filters. Although the basic concept of chemical coagulation followed by filtration has remained for over 120 years, that does not mean that there has been no progress. Doubtless if Fuller were to visit a modern treatment works he would recognise his basic process sheet, but he would also be amazed at the sophistication now deployed. The coagulation process is now carefully controlled by computers, and the separation process is undertaken in a variety of clarifier designs that improve performance beyond anything that he was able to achieve. Even his dependence on gravity for separation has been replaced, at times, by use of dissolved air flotation. More recently, the advent of membrane filtration has started to challenge the traditional flow sheet, but these devices, employing as they do filtration at a truly molecular level, still require protection in the form of roughing filters or the use of chemical coagulants to break the molecular bonds before separation takes place across the membrane with use of a partial vacuum.

Despite the huge advances in coagulation and filtration, there remains one further element in the process of modern water treatment that has a major impact on the work of all water distribution engineers. This is the question of disinfection. While Fuller and his contemporaries were pursuing the removal of dangerous bacteria from the water supplies, an alternative approach was also being developed, that of simply killing the bacteria. Of course, the chemicals used had to be harmless to man and, hence, strong oxidising agents were used.

There are several claims made for the first use of chlorine in treating drinking water around the start of the twentieth century, particularly in Middlekerk and Antwerp. It is probable that its first use was not in the United Kingdom, although there are some references to Maidstone in 1897. What is well established is that, in 1903, there was an outbreak of typhoid at Fulborne Asylum in Cambridgeshire. The authorities sought permission from the House of Lords to introduce chlorine into the drinking water, but the request was refused on the grounds that it would be too dangerous. Twelve months later there was an outbreak of cholera in Lincoln and, one assumes learning from the experience at Fulborne, the city fathers did not seek any approval but went ahead with the use of chlorine and the outbreak rapidly abated. At this point they made a fatal mistake and stopped using chlorine, which in turn resulted in a further outbreak several years later. However, Lincoln is generally credited with the first use of chlorine in water treatment in the UK. From that time on, the use of chlorine became increasingly commonplace in the control of both typhoid and cholera. In 1909, the city of New Jersey became the first municipality in the USA to use chlorine as permanent element of its water purification process. Gradually, disinfection with chlorine (or compounds of chlorine such as hypochlorite) became part of the established water treatment process flow sheet.

In 1935 there was an outbreak of typhoid in Croydon, UK, and again chlorine had not been used in the treatment process. Following that outbreak, the UK legislation was changed and it became a requirement for potable water to contain a residual disinfectant when distributed through the public water supply network. Since that date there has not been a single reported case of waterborne cholera or typhoid within the UK.

The requirement for a residual disinfectant provides a major challenge to the distribution engineer. Initially, it was sufficient for the water to leave the treatment works with a high chlorine residual and for it to simply decay as it passed through the distribution system. However, public opinion turned against the taste of highly chlorinated water and, while the inaugural address of one president of the American Waterworks Association contained the suggestion that 'If you can't taste the chlorine, don't drink the water', this can only have brought comfort to the manufacturers of

Figure 1.1 Effect of water purification on the death rate from typhoid fever in a city drawing water from a clear lake

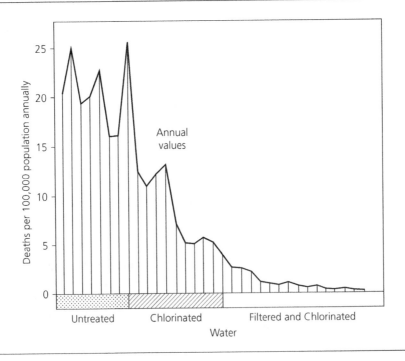

bottled water. Today's distribution engineer has to be able to manage the levels of chlorine within the distribution system to achieve a balance between the competing demands of public acceptability and public health. To do that, it is necessary to be able to predict the flow regimes within pipe networks, which is the subject of the next part of this review of the historical development of water distribution practice.

But before we move on, there is one final piece of this particular 'treatment jigsaw' that needs to be understood. Three processes are required for effective water treatment and each stage is essential, it is not a 'pick and mix' situation. Figure 1.1 shows the efficacy of the standard three stage approach.

Typical improvement in death rate following introduction of Water Treatment Technology (Fair, Geyer and Okun, 1966).

1.3. Evolution of pipeline materials

As we have seen with water treatment technology, we shall also find continuous evolution of pipeline materials over the last 150 years.

We can again start with Rome, where Frontinus makes it clear that lead was used for pipes and also to make discharge devices to control flows from aqueducts and cisterns. It is also interesting to note in passing that 100 years earlier, Vitruvius appears to report (his writings are not always clear because he presumes detailed knowledge of then current practice) problems with what appears to be an inverted siphon, where material technology could not withstand the bursting forces. It may therefore be the case that Roman preference for aqueducts rather than pipes across valleys related to limitations in pipe materials rather than ignorance of the hydraulic phenomena which allow inverted siphons to function.

It is also clear that clay pipes were available to Roman engineers and, indeed, earlier civilisations but they also would have limitations in terms of bursting resistance and were probably used for sewers (where such existed). We should, perhaps, stress that there is no suggestion of Roman houses all having running water. A few very rich individuals had water piped to their houses to a single point of discharge but, as Frontinus makes clear, the vast majority depended on collection themselves from cisterns at the end of aqueducts or the services of water sellers who (supposedly) collected water from cisterns and distributed it around Rome in carts.

Again, with the fall of the Roman empire much of this was lost and we have to wait until the Middle Ages to find any return to public water supply provision.

Early water mains were made from hollowed out tree trunks, joined with a socket and spigot joint which itself was sealed by wrapping in lead or some other material. There are numerous examples of this technology in museums.

1.3.1 Iron pipe

It is not clear when the first cast iron water pipes were introduced, but it is known that cast iron pipe was used to distribute water to the various fountains of the Palace of Versailles in 1672. It is of passing interest that there was, and still is, insufficient water to feed all of the fountains at the same time, and so they had to be switched on or off as the King approached or passed them.

In all probability, there were earlier uses of cast iron pipes but they are less celebrated and, sadly, not recorded; although the American Ductile Iron Pipe Research Association reports their first use as 1455 in Siegerland, Germany. However, it is clear that cast iron pipe began to be used for water supply in the eighteenth and nineteenth centuries. The first recorded use of cast iron pipe for water supplies in the USA appears to be in the early 1800s in the New Jersey and Philadelphia areas, where they replaced traditional wooden pipes. Initially, these pipes were cast horizontally and were of uneven quality, but around 1850 vertical casting was developed, which produced a far more reliable product. There was no standardisation of sizes for these early pipes, and engineers would specify not only the internal diameter of the pipe they required but also its wall thickness.

The next major advance was the invention of the centrifugal spinning process in 1918. In this, instead of a mould which defined the internal and external diameters, there was only a hollow circular mould that fixed the external diameter. This was rotated at speed about its longitudinal axis and molten metal was poured into it. The centrifugal force took the molten metal to the internal surface of the mould and the quantity of liquid iron applied defined the internal diameter. This process provided a far more uniform pipe material with less possibility of air bubbles or casting flaws. This process gradually replaced vertical and horizontal casting, but of course could not be used for pipe fittings such as Tee junctions and bends.

In 1955, a further advance was made with the development of ductile iron pipes. Traditionally, pipes had been made of gray iron, which was strong and durable, but brittle. By modifying the metallurgical composition, a material was developed that was far less brittle and could accept a small amount of deformation. Although being developed in the United States in the mid 1950s, it did not reach the UK until the 1970s. Initially it appeared to be a far superior material, and was manufactured in part from scrap steel (although still technically an iron), however, experience showed that it was more prone to corrosion than the gray iron that it had replaced. There then followed a series of developments aimed at corrosion prevention, starting with on-site wrapping in polythene sheets, through to today's standard of factory applied multicoated protection systems.

A number of other materials were developed and challenged the traditional approach of iron water mains, and we shall look at those shortly, however, before doing so we should also look at jointing of pipes.

The early cast iron pipes were sealed with what was known as a run lead joint. Molten lead was poured into the annulus between the spigot of one pipe and the socket of the other pipe. As the lead cooled, it solidified and shrank, it then had to be compressed by use of a series of chisel like tools and a hammer to form a water tight joint. It was of course necessary to insert caulking before the lead to prevent it running down the length of the spigot. Some pipes also had a small channel cut into the inner circumference of the spigot to assist in making the joint and providing a path and holding channel for the lead. This highly skilled and somewhat dangerous technique was replaced by the invention of the sealing ring. There are several different styles of ring, generally protected by patents, but for the purpose of this introduction they can be viewed as an 'O' ring that is compressed between the spigot and the socket at each joint. Flanged joints and various mechanically restrained joints are also available, but push fit joints are most commonly used for external infrastructure (buried pipes).

1.3.2 Asbestos cement pipe
One of the early rivals to cast iron was the development of asbestos cement pipe. Essentially, this is a cement pipe that gets its tensile strength from the incorporation of 11% asbestos fibres. The pipes were used from the 1920s to the 1980s, and many kilometres are still in use around the world,

including the UK, Europe and the USA. They were lighter and lower cost than conventional cast iron. They suffered some bad press over alleged high burst rates, but an investigation by the UK Drinking Water Inspectorate, in 1985 (DWI0033), showed that this was a fallacy and burst rates were very similar to those for cast iron and plastic water pipes.

A further DWI report in 2002 (DWI0822) suggests that there is little if any health risk from the pipes in terms of water supply, but the hazards involved in manufacture and, particularly, on site cutting of the pipes meant that most countries in the world had ceased installing these pipes.

So although an early contender to cast iron pipe, asbestos is no longer a competitor.

1.3.3 PVC pipes

The next real contender to the supremacy of cast iron was the introduction of unplasticised poly-vinyl chloride pipe (uPVC). Although uPVC as a compound had been discovered in Germany in 1835, it was not used to produce water pipes until approximately 100 years later. Some of these early pipes were installed as water supply pipes in Germany on a trial basis and performed well, but development was interrupted by the war.

Large scale production of uPVC pipes started in the 1950s and the product became increasingly popular in the 1960s, and started to emerge as a major competitor to traditional gray iron. It had many advantages: it was light and easy to handle, it did not corrode and it was competitively priced. Unfortunately, these advantages were overtaken by a number of operational issues, as failures of pipes were reported. Investigations in the UK showed that Class B uPVC pipe (a low pressure pipe), when used as a pumping main, suffered failures due to surge pressures gener-ated on starting and stopping pumps, which, although within the pressure rating, caused fatigue failure of the material. As a result class B pipes were largely shunned by the UK water industry. The second problem came with a number of failures of larger diameter pipes, which this time were linked with poor handling of the material. Pipes made of uPVC deteriorate if left in sunlight and also have to be laid on a bed that does not contain sharp stones – a sand bed and haunch is often specified.

These problems and the emergence of alternative materials led to a significant decline in the popu-larity of the product in the UK water market; the product was reformulated and relaunched as PVCu, which had overcome the early technical problems, but the early experience hindered its adoption in the UK. In contrast, the American Water Industry has embraced PVC pipe and it is one of the most popular water pipe materials in the USA. It is available in a range of sizes up to 600 mm (24 inches).

1.3.4 MDPE water pipe

With the problems associated with both uPVC and asbestos cement becoming apparent, and the desire to find a corrosion-free alternative to the then dominant gray iron pipe, attention turned to polyethylene (PE). PE had been discovered in 1933 by ICI, but major development and use was again post war. The gas industry had been the first major user of PE pipe, but the UK water industry of the 1970s was in need of an alternative material, and began to explore the possibilities. It was believed that medium density polyethylene was more robust, and better able to withstand the less than ideal handling it would receive on site, than the stronger high density (HDPE) material. This may have been a correct assessment, but there were still problems with rapid crack propagation caused by minute levels of site damage that spread once the pipe was put into service.

There were also considerable problems with pipe jointing; the promise of a continuous pipe with perfect, leak-free butt-welded joints soon gave way to a reality of less than perfect joints leaking or even breaking. Gradually, the technology improved becoming more tolerant of site conditions and the industry, itself, grew accustomed to the need for high standards of workmanship compared to the very abuse-tolerant cast and ductile iron pipe that had traditionally been used.

MDPE was very cost effective and as confidence in its performance grew, it largely displaced the traditional cast iron pipe for sizes up to 300 mm diameter. Above this figure, the thickness of the MDPE material is such that the economics start to move against it, but the actual breakeven point varies with time.

In the UK and many parts of Europe, MDPE is probably the market-leading material for water pipe up to at least 300 mm diameter, but this is not the case in the USA where it has trailed behind both PVCu and ductile iron.

1.3.5   Other pipe materials

Having been squeezed out of the smaller diameter market for water pipe by MDPE and PVC, it might be thought that ductile iron would still enjoy a virtual monopoly in terms of both the medium and large diameter market (greater than 450 mm) and the market for fittings for internal pipework in pumping stations and treatment works, but even here its position is under threat.

For large diameter underground pipelines, there is growing use of steel pipe, usually protected with a factory applied epoxy coating. While for internal pipework, there is a growing use of welded stainless steel pipe.

Other alternatives for large diameter pipes include glass reinforced plastic, which is a generic term for a number of different pipe materials, each requiring slightly different laying techniques and materials that can also be used for pipe renovation work. Additionally, prestressed concrete pipes are available for sizes above 600 mm diameter, and they come in two forms: those with a steel tube cast into the concrete before prestressing and those that rely solely on the prestressed concrete.

So, in the same way as we have seen that water treatment has been an evolving technology, the same is true of pipeline material development. Even the traditional cast iron pipe has undergone tremendous development in the last 100 years, but, additionally, there has been continuous innovation and development of alternative materials; and that development will undoubtedly continue into the future.

1.4.   Development of pipe flow calculations

Having traced the development of water treatment science and pipeline technology up to the present day, it is now necessary to go back into history to do the same for hydraulic calculations that are required to enable the design of water distribution systems. The reader should remember that these developments were taking place in parallel with the development of water treatment science.

The starting point is once again Frontinus and his treatise on the aqueducts of Rome. In it he describes how he came to doubt the figures produced by the water sellers who took water from the tanks at the end of the aqueducts, and conveyed it around Rome selling it to the population in small quantities (very much the system that can be found in many emerging countries today). He, therefore, arranged to have measuring vessels deployed at the end of each aqueduct (not simultaneously) and in this way actually measured the flow being delivered by each aqueduct. Not surprisingly, he found that the water sellers were severely under measuring the amount of water available and had been defrauding the city for many years in respect of the amount of water they actually sold. It is interesting that the reputation of water sellers in developing countries is about the same as their Roman predecessors for both quantity and quality.

However, that is a diversion. What this tells us is that although the Romans were able to construct aqueducts to carry water over very long distances, they did not have the mathematical tools to allow them to accurately predict what volume of water would actually be carried by the aqueduct. A treatise on building aqueducts was produced by Vitruvius some 100 years before Frontinus, who credits Vitruvius with developing the Roman unit 'the quinquaria' which was used to define capacity of a pipe or vessel. However, it was not a measure of flow, but rather volume, being 'as much water as would flow through a pipe one and a quarter digits in diameter constantly discharging under pressure (there must have been a time period used, possibly 24 hours). A digit is believed to have been 1/16 part of a foot, but Frontinus tells us that the standard of a digit is not uniform and varies. By now it should be clear that hydraulic calculations were not understood by Frontinus and probably not by those advising him.

The first published formula for the calculation of flow of water in pipes was by Antoine Chezy (1718–1798) c. 1770, although the actual date is uncertain. His initial formula was for flow in open channels, but pipe flow was also covered as a special case. It is worth remembering that the start of the French Revolution was 1789, which may explain why Chezy's work was lost until one of his students, Prony, published a paper referring to it in 1800. Prony subsequently published his own equation for calculating fluid flow a few years later.

Today the Chezy Formula is usually written

$$V = C \times (R \times S)^{0.5}$$

where
V is average velocity in pipe
R is the hydraulic radius (the area of flow divided by the wetted perimeter)
S is the hydraulic gradient of the pipe
C is a dimensionless constant representing the friction for a particular pipe

However, the work of Fourier (1768–1830) introduced, among other things, the concept of 'dimensional analysis', which came to be a major contribution to the practice of building physical models in fluid mechanics (although the major breakthrough in that sphere was the work of Lord Rayleigh at the end of the nineteenth century and Edward Buckingham in the early part of the twentieth century); as far as pipe hydraulics are concerned, this can be simply stated as requiring the fundamental physical units of mass (M), length (L) and time (T) to be the same on both sides of the equation. A quick inspection of the Chezy formula reveals that on the one side of the equation we have a velocity (L/T) while on the other we have the square root of L and no T term. This indicates that the formula as conceived must be incorrect and, indeed, C must have dimensions. However, this was not appreciated for many years and the Chezy formula was (and to some degree still is) used for calculation. The value of C is based on specific experiments for varying hydraulic structures and Chezy himself derived values for it for designs that he was undertaking. The Chezy formula remains as an empirical tool for hydraulic calculations, and this also applies to derivatives from it.

Like all empirical relationships, they are only valid for the range of experiments on which they are based, and extrapolation outside that range can produce very misleading results.

In 1845, Weisbach (1806–1871) published a formula, which was dimensionally correct. It is usually written in the form

$$H = 4f \frac{l}{d} \frac{v^2}{2g}$$

where
H is the headloss
f is a friction factor
l is length of pipe
d is diameter of pipe
v is average velocity of flow

Checking the above with dimensional analysis shows that both sides are consistent and therefore f is, indeed, a dimensionless number.

In 1857, Henry Darcy published work which was based on Prony's formula and which, although he was apparently unaware of Weisbach's work, added to the understanding of the Weisbach equation and the relationship is now commonly referred to as the Darcy-Weisbach equation.

Unfortunately, practical use of the equation rapidly showed that f was not a constant, but indeed varied with pipe geometry (diameter) and roughness of the inside of the pipe, and it was in this area that Darcy made his contribution. But even that did not resolve the problem of a dimensionally consistent equation which failed to predict pipe flows adequately.

Readers should note that in Chapter 2 the term 4f is replaced by λ, which is a more modern notation adopted because f is used in mathematics to denote a function of other parameters.

Around this time the engineering profession started to split into two camps, divided between those who continued to work on the theory of pipe flow, seeking to find a totally rational relationship on which calculations could be based, and more pragmatic engineers who believed that, although theoretically inferior, empirical equations based on extensive experimentation could produce reliable practical tools for practising engineers.

Many empirical relationships were published, these generally were of the form

$$V = K \times R^x \times S^y$$

The values of K, x and y were derived from experiments, with extrapolation and interpolation used to provide missing values.

Two of these empirical formulae became the foundation for practical hydraulics for the first 60–70 years of the twentieth century.

In 1889, the Irish engineer Robert Manning published his formula for open channel flow and pipe flow. Not only was it found to be accurate over a wide range of conditions, it was also simple in structure and lent itself to the publication of tables, so that engineers could simplify the iterative nature of design calculations by use of the tables.

Manning's formula had several iterations but is now generally given as

$$V = M \times R^{\frac{2}{3}} \times S^{\frac{1}{3}}$$

where M is Manning's constant.

It was soon realised that Manning's M was, in fact, the reciprocal of another constant developed earlier by Kutter, given the nomenclature 'n'. As a result, the Manning formula is often expressed as

$$V = \frac{1}{n} \times R^{\frac{2}{3}} \times S^{\frac{1}{3}}$$

In 1905, the Hazen-Williams formula was proposed in the USA

$$V = C \times R^{0.63} \times S^{0.54} \times \left(1.32\right)$$

The factor 1.32 was introduced to make Hazen's C conform with Chezy's C for the case where $R = 1$ ft and $S = 1$ in 1000.

It should be remembered that the fitting of curves to experimental results was carried out without the benefit of computers and was laborious, and also approximate compared to modern computer-based regression analysis.

Additionally, in the absence of even simple electronic calculators, use of even these simplified empirical equations was extremely time consuming, often requiring the use of Log tables.

The equations were frequently available in the form of design tables and/or charts to allow practising engineers iterate to a specific design, having assumed appropriate roughness coefficients.

Because both Manning's equation and that of Hazen Williams are empirical, as well as requiring caution when used outside the range of supporting experimental work (which was seldom referred to in the published tables), great caution was also required in terms of which units were used when listing appropriate coefficients (C and M), as these were not dimensionless and would therefore vary significantly between Imperial, American and Metric systems.

Despite their drawbacks, these two empirical formulae became the most widely used and have remained in (now significantly declining) use for around 100 years.

The emergence of these empirical formulae did not in any way stop the continued search for a more dimensionally correct approach, as represented by the Darcy-Weisbach Equation. Rather, it was the sheer complexity of the search for f within that equation which resulted in the emergence of the empirical formulae.

It had been recognised in the 1830s that there was a difference between slow and rapid flow in pipes. It was, therefore, recognised that the Weisbach equation required different values of 'f' for these two cases. At that time the concepts of laminar and turbulent flow were not understood, although Darcy's contribution was in studying flows in what we would now describe as being in the turbulent range. As knowledge increased, it became clear that 'f' rather than being a simple coefficient was in fact an extremely complex factor. It was a function of the Reynold's number (1883), but the complexity did not stop there. As research continued the concepts of boundary layer theory had to be incorporated, and the complexity of the function increased as it became apparent that relationships only held good within certain ranges of the Reynolds number. However, one major problem was that the transition zone between laminar and turbulent flow proved extremely difficult and, yet, was the area where many engineering installations operated.

Colebrook and White published their research into commercial pipes using the theories that had been developed to explain the behaviour of 'f' in 1939 and demonstrated that the relationship set out below had very good correlation with measured physical experiments over wide ranges of flows.

$$V = -2\sqrt{2gDS} \log(k_s/3.7D + 2.51\lambda\nu/D\sqrt{2gDS})$$

where
S = the hydraulic gradient
ν = the viscosity

Unfortunately, it is immediately apparent that this is not a user-friendly relationship for practising engineers. Apart from the sheer complexity of the equation, including square roots and log functions, the parameter which is sought 'I' the hydraulic gradient cannot be isolated and, therefore, a laborious iterative solution must be employed.

Matters improved slightly in 1944 when Moody published his now famous diagram, although this was in reality a redrawing of a diagram initially produced by Hunter Rouse (1906–1996), a leading American hydraulics engineer. However despite the availability of both Moody's diagram and Rouse's own work, the use of these complex relationships lay well outside the practical design experience of most water distribution designers and operators, who depended on slide rules and log tables to perform their calculations. Undoubtedly, engineers could have used the equations with the help of the Moody and Rouse charts to assist in the calculations, but using Manning or Hazen Williams was far simpler, and there was a wealth of experience in the use of these formulae by the time that the Moody diagram was developed. It remained, therefore, something that was taught at universities to assist in the understanding of pipe flow, but was not generally taken up by practising engineers.

An additional hurdle was that the solution of pipe flow equations was by no means the end of the analysis process for water distribution networks, and that was yet another factor that militated against the introduction of a more complex method for calculating pipeline flows. But before moving on to another engineering challenge, we need to conclude our understanding of pipe flow analysis.

Thus, for around 15 years, the Moody diagram, which should have given access to the Colebrook-White equations, remained of largely academic interest. However, the availability of computers was to revolutionise previous practice.

In the late 1950s and early 1960s, the Hydraulics Research Station at Wallingford used a then large computer to produce, initially, design charts based on the Colebrook-White equation and then a series of design tables. These should have put the Colebrook White equation on an even footing with both Manning and Hazen Williams, but of course there was by that time over 50 years of practical experience of using both of these formulae and, despite the more rigorous scientific basis of Colebrook White, the industry was slow to adopt the use of this 'new entrant'.

As the use of computers became more common in engineering offices through deployment of, at first, time sharing terminals and even a few mini computers, so software began to be developed to assist in hydraulic calculations. Initially, this was based on the tried and trusted Manning or Hazen Williams formulae, but gradually Colebrook-White began to be offered by vendors as an alternative. As the power of computers increased, and dedicated mini computers were replaced by desktop PCs, the use of the more complex but also more rigorous Colebrook-White equations grew in popularity; when feeding parameters into a computer, there was no extra work involved for the designer, whichever formula was adopted. Additionally, technical papers were being published about the continued development of the Colebrook-White equations, and finally for possibly the preceding 20 years graduates had left university having been taught the theoretical benefits of Colebrook-White but having to accept the impracticalities of adopting the equation for use in a design office. Thus the advent of technical computing on PCs finally allowed the practicalities of design office practice to catch up with the technical advances that had been made over the preceding 60 years in theoretical hydraulics. Today there can be no sound reason for the continued use of the old empirical equations in the design of distribution systems.

However as indicated above, there remained a further problem in the design and operation of distribution systems, and this relates to the analysis of pipe networks rather than simple pipelines.

1.5.   Analysis of pipe networks

So far, hydraulics has been restricted to the analysis of a single pipeline; however, distribution networks are not single pipes in isolation but, rather, many pipes joined together at nodes, as shown in Figure 1.2. In fact, the diagram shows two distinct types of network problem, the first (Figure 1.2a) for a branched system and the second (Figure 1.2b) for a looped system. In both cases the problem is to determine how much water flows down each of the pipes, which, of course, are not necessarily of uniform diameter or roughness factor.

The solution to this problem is well beyond the scope of this introduction, and the reader will find a comprehensive treatment in Chapter 2, 'Basic hydraulic principles'. However, in order to understand the evolution of network analysis, it is necessary to distinguish generically between the two

Figure 1.2 Two distinct types of network: (a) a branched system and (b) a looped system.

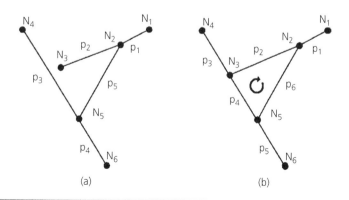

(a)   (b)

idealised problems and to understand that with looped systems there are always more unknowns than available equations, and thus an iterative solution technique has to be adopted.

A number of mathematical techniques were developed over the years to assist in solving these iterative problems, including the use of the Hardy Cross relaxation method, which many engineers have encountered as 'moment distribution' when applied to structural analysis. However, the sketch in Figure 1.2 is a very oversimplified network compared with a normal distribution system. The only way of solving the equations was to simplify the network by restricting the calculations to large pipes only. This resulted in the majority of small pipes never being analysed, and network analysis calculations were only a minor part of the distribution engineer's armoury. They had to be used for new supply calculations and for major reinforcements, but were just too cumbersome to be deployed for day-to-day operational decisions.

The distribution system was therefore managed on the basis of the combined knowledge of operatives and engineers which had been built up from years of practical operational experience. There was a long-established tradition of operators keeping 'little black books' giving details of how their part of the network appeared to operate, and tales of these books being sold to successors at the end of a career to fund retirement parties are not all apocryphal.

By the mid-1970s, some distribution engineers had access to dedicated mini computers such as the PDP-11, which was the size of a kitchen table and had 8 kB of RAM. However, machines such as this could run programs to automate the solving of simple networks. As computers grew more powerful, following Moore's law (the power of computer chips doubles every 2 years), the complexity of networks that could be analysed continued to increase and by the mid-1980s large network models could be solved, although again restricted to 'large pipes' and well removed from the all-pipe models that would have been the ideal.

Even with these simplified models, it was becoming clear that much of the information contained in the little black books was erroneous. However, the programs were the province of expert network modellers, and the outputs were not in graphical form but sheets of numerical output that required to be interpreted by the modeller. It is hardly surprising, therefore, that network modelling remained a tool to be used by experts when considering large capital expenditure, but continued to play almost no part in the day-to-day operation of distribution networks.

By the mid-1990s, the power of computers was such that it was possible to approach the all-pipe-modelling ideal by omitting small pipes, rather than simplifying the principal pipes in the physical network. However, the really big breakthrough was the development of graphical user interfaces. This now meant that the network could be represented on the screen as a network and pipes could be added or isolated on the screen by the operator. The model user no longer had to be a specialist and, by the turn of the millennium, network modelling had finally become an operational tool: at long last the little black books could be seen by all to be of little further value.

Over the last 10 years, further refinements have emerged: all-pipe models are now a reality, links are routinely made to geographic information systems to update models, and new model build has been greatly simplified by these links. The functionality of the models has increased to allow the decay of chlorine to be predicted, thus assisting with control of chlorine levels in the distribution system. Pollution incidents can be tracked and predictions made about the spread of the polluting substance. Isolation of mains can be explored in terms of the pressure reduction experienced by all properties within the model, rather than simply the properties that will not receive water. Links can be made to customer databases to issue warning notices. Chapter 7 is devoted to a detailed explanation of the modern network modelling techniques now available to the distribution engineer.

The advances over the last 10 years have been incredible and will continue, but all of them are reliant on the advances that have been made over the last 220 years of hydraulic engineering.

1.6.   Water distribution engineering in the twenty-first century

Having established that water distribution has an innovative and progressive technical history, it will be helpful to also consider the challenges facing today's engineer and how this book may be of assistance. The history of water supply and distribution is one of over 200 years of innovation and development. Where there have been lulls, it has not been through lack of effort, but rather waiting for science and technology to develop and, sometimes, waiting for those concepts to be capable of deployment. The advent of the digital age has facilitated huge strides forward, with a result that today's technical sophistication is now beyond the wildest imaginings of those who worked in the industry only 50 years ago.

Today, the engineer's ability to calculate pipe flows and predict the consequences of operational decisions is higher than it has ever been: there can be no justification for failing to use these tools for all aspects of water distribution engineering, be it routine isolation of pipes for cleaning or reinforcement of the existing system.

Our modern world is extremely complex and few engineers can expect to work in total isolation. Most engineers are expected to work as part of a team, whether they be employed by consulting engineers or directly by client organisations such as water service providers. These teams will themselves be multidisciplinary and it is essential that the engineer is capable of presenting his or her knowledge and advice to other team members in a clear and understandable form. The idea of a single engineer as an infallible source of technical wisdom is long past. Within the discipline of water engineering there are numerous specialists who have sophisticated and evolving techniques at their disposal. Chapter 8, 'Design of water distribution systems', explores some of the most recent ideas, while Chapter 9 covers modern approaches to operation, maintenance and performance.

The concept of asset management is a relatively recent one, perhaps evolving as a formal discipline over the last 25 years. It is now vital for all water service providers that they have to recognise the disciplines of the financial markets, and Chapter 10, 'Asset planning and management', is devoted to exploring these ideas and associated techniques.

If the engineer is to succeed in this element of their work, they must have at least an understanding of the role of others, as well as a command of their own technology. Chapter 11, covering finance and project appraisal, establishes the basic mechanisms available for financing projects by government, municipalities and private companies; at the very least it is hoped that this chapter will establish that raising finance is a complex business and that no asset owner has access to unlimited funds. It also sets out the various approaches to project appraisal, ranging from those adopted for major international projects, funded by institutions such as the World Bank, to more modest appraisals carried out routinely by engineers requiring to reinforce an existing distribution system.

There is much debate about sustainability and climate change, and Chapter 12 sets out the views of an eminent engineering practitioner. However, this is a rapidly developing field, and one where it is not possible to provide unequivocal advice on all aspects. Issues such as the conflict between 'sustainability' and conventional economic theory can be identified and discussed, but the actual mechanisms will have to be resolved at governmental level before definitive solutions can be given. For example, much increased energy use is associated with meeting higher treatment standards required by EU Directives; however, this conflict between the mutually exclusive need for higher treatment standards and lower carbon dioxide emissions can only be resolved by the lawmakers who imposed both objectives, albeit, one hopes, informed by advice from the scientific community. This chapter differs, therefore, from others in that its purpose is to inform the debate rather than provide well-proven technical and operational solutions.

In concluding this introduction, it is tempting to try to predict which areas of water distribution will be the subject of the next major change, but with such rapid development on so many fronts, it is not possible to provide a comprehensive forecast. However, one issue does appear to be emerging, driven by separate although linked objectives: the question of water metering. This is not a reference to household metering but, rather, measurement of flows at points throughout the distribution system. There are two main drivers for this.

The first is the political focus on water leakage, particularly as electricity usage in the water industry is significant, and, if 30% of water is lost as leakage, then 30% of the energy is also probably wasted. This could make a very significant contribution to the country's Kyoto targets (United Nations, 1997), and with no increased cost or loss of service to customers. The level of metering in the typical water supply and distribution system is well below that found in the hydrocarbons industries, where their products have higher monetary value. It is likely, therefore, that some techniques established in those industries will find their way into the water industry.

The second driver is the 2009 Cave Review (Cave, 2009), which supports the theoretical concept of vertical disaggregation of the water industry. Whether or not this will actually happen is unclear, and Professor Cave himself recognises that the practical barriers are significant, but one of the issues that he uncovers is the paucity of information about what actually happens between water treatment works and the customer. Before decisions can be reached, this gap will have to be filled, and that will, in turn, also require more information on flows.

However, there are other major challenges facing the industry that will also require innovative approaches, not the least of which is the ever-increasing challenge of replacing life-expired infrastructure.

In the early 2020s the political agenda moved from the Cave review recomending disaggregation to wholesale renationalisation of the English and Welsh water companies. This was driven in part by alleged financial mismanagement and also by a political belief that natural monopolies should remain with the state. We do not wish to take sides in such debates but feel that we should point out that in the Drinking Water Inspectorate report for 2022 the average compliance for the Water Industry with EU plus additional GB standards is 99.97%. This is a compliance with international standards that is world leading.

New challenges await water engineers as indicated above and demands for higher performance standards will continue to increase. But irrespective of ownership the quality standards currently achieved by adopting the technical principles set out in this book are currently at the forefront of world class performance.

In closing this introduction, the reader's attention is drawn to the poem published in 1849 by Edgar Alan Poe (Box 1.1). It tells the story of a knight setting forth on a quest to find the fabled city of gold, Eldorado. He starts off full of enthusiasm, but as the poem progresses he ages without fulfilling his quest. He becomes melancholy until, in the final verse, he receives less than clear advice from a mysterious stranger who tells him that he must, 'Ride, boldly ride …If you seek for Eldorado'.

Box 1.1 Eldorado by Edgar Allan Poe, 1849

Gaily bedight,
A gallant knight
In sunshine and in shadow,
Had journeyed long,
Singing a song,
In search of Eldorado.

But he grew old –
This knight so bold –
And o'er his heart a shadow
Fell as he found
No spot of ground
That looked like Eldorado.

And, as his strength
Failed him at length,
He met a pilgrim shadow –
"Shadow," said he,
"Where can it be –
This land of Eldorado?"

"Over the Mountains
Of the Moon,
Down the Valley of the Shadow,
Ride, boldly ride,"
The shade replied –
"If you seek for Eldorado!"

Like Poe's knight, it is most unlikely that engineers will ever achieve their Eldorado of a perfect water distribution system, and the road to further improvement is not always clear, but, nonetheless, that is no reason for failing to pursue the objective with all the tools available, both currently existing and yet to be developed; it is hoped that this book will provide some assistance in that quest.

REFERENCES

Barty-King H (1992) *Water: The Book – An Illustrated History of Water Supply and Wastewater in the United Kingdom*. Quiller Press, London, UK.

Cave M (2009) *Independent Review of Competition and Innovation in Water Markets: Final Report (Cave Review)*. Defra, London, UK.

Chadwick E (1842) *Report on the Sanitary Condition of the Labouring Population of Great Britain*. HMSO, London, UK.

Colebrook CF (1939) Turbulent flow in pipes with particular reference to the transition region between the smooth and rough pipe laws. *Proceedings of the Institution of Civil Engineers* **12**: 393–422.

Council for Science and Technology (2009) *Improving Innovation in the Water Industry: 21st Century Challenges and Opportunities*. The Stationery Office, London, UK.

Darcy H (1857) *Recherches Expériméntales Relatives au Mouvement de l'Eau dans les Tuyaux*. Mallet-Bachelier, Paris, France. (In French.)

DWI (1985) *Usage and Performance of Asbestos Cement Pressure Pipe*, DWI0033. Drinking Water Inspectorate, London, UK.

DWI (2002) *Asbestos Cement Drinking Water Pipes and Possible Health Risks*. Drinking Water Inspectorate, London, UK.

Fair, Geyer and Okun (1966) *Water & Wastewater Engineering*. John Wiley & Sons, Oxford, UK.

Frontinus, Sextus Julius (1980) *Stratagems and Aqueducts*. Heinemann, London, UK.

Manning R (1891) On the flow of water in open channels and pipes. *Transactions of the Institution of Civil Engineers of Ireland* **20**: 161–207.21

Metropolis Water Act 1852 (15 & 16 Victoria c. 84). HMSO, London, UK.

Moody LF (1944) Friction factors for pipe flow. *Transactions of the ASME* **66(8)**: 671–684.

Reynolds O (1883) An experimental investigation of the circumstances which determine whether the motion of water shall be direct or sinuous, and of the law of resistance in parallel channels. *Philosophical Transactions of the Royal Society* 174: 935–982.

Rouse H (1943) Evaluation of boundary roughness. *Proceedings of the 2nd Hydraulics Conference, The University of Iowa Studies in Engineering. Bulletin 27*, pp. 105–116. Wiley, New York, USA.

Shelley PB (1839) *Peter Bell the Third* [publisher unknown].

Snow J (1849) *On the Mode of Communication of Cholera* [publisher unknown].

Snow J (1855) *On the Mode of Communication of Cholera*, 2nd edn [publisher unknown].

United Nations (1997) *Kyoto Protocol*. United Nations Framework Convention on Climate Change website: http://unfccc.int/kyoto_protocol/items/2830.php [accessed 20.03.24].

Vitruvius (1970) *De Architectura*. Heinemann, London, UK.

Dragan A. Savić and John K. Banyard
ISBN 978-1-83549-847-7
https://doi.org/10.1108/978-1-83549-846-020242003
Emerald Publishing Limited: All rights reserved

Chapter 2
Basic hydraulic principles

Dragan A. Savić
KWR Water Research Institute, the Netherlands, and University of Exeter, UK

Rob Casey
Thames Water, UK

Zoran Kapelan
Delft University of Technology, the Netherlands

2.1.   Introduction

This chapter provides a brief review of basic hydraulic principles, including fluid properties and fluid dynamics, necessary for understanding the hydraulics of water distribution systems. The review encompasses fluid properties, the governing equations of conservation of mass and energy, and the framework in which they are solved. Then, network analysis equations are introduced and methods for solving them are briefly described.

This chapter is not intended as a substitute for a course in fluid mechanics or hydraulics. It is rather assumed that the reader already has a fundamental understanding of fluid statics and fluid dynamics. The primary aim of this chapter is to remind and update the reader on the concepts used in the modelling of water distribution systems.

The chapter starts with an introduction to basic fluid principles, before moving on to describe the basic flow equations and losses in pipes. The second part of the chapter covers network analysis methods, including steady and unsteady flow modelling and water quality analysis.

2.2.   Basic fluid properties
2.2.1   Density
The density of a fluid ρ (M/L^3) is defined as its mass m (M) per unit volume V (L^3). The SI unit for density is kg/m^3, and for water at 5°C (reference temperature) its quantum is 1000 kg/m^3 and at 20°C is 998.2 kg/m^3.

2.2.2   Viscosity
Viscosity is an internal measure of resistance to flow in a liquid or, more precisely, it is a measure of a fluid's susceptibility to shear deformation. Informally, we can think of viscosity as 'thickness', but for fluids only. For example, water can be considered 'thin', having a lower viscosity, while honey and treacle are 'thick', having a higher viscosity. Generally, fluids resist the relative motion of immersed objects through them and act like different sets of layers with relative motion between them (Figure 2.1(a) shows how friction between the fluid and the moving boundary causes the fluid to shear). In the case of fluid flowing between stationary plates, the velocity of flow varies from zero for the layer next to the plate to a maximum along the centreline (Figure 2.1(b)). During that process, viscosity opposes the flow of fluid, which attributes to viscous force.

Figure 2.1 Viscosity

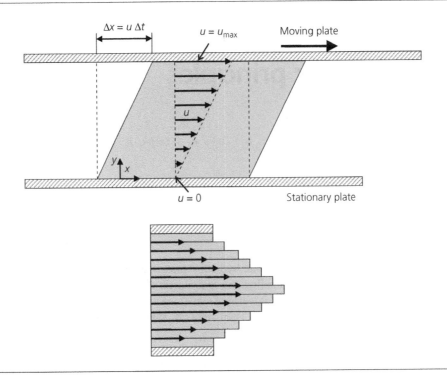

Formally, viscosity (sometimes called dynamic viscosity) is the ratio of the shear stress to the velocity gradient in a fluid. More often, this relationship, also called Newton's equation because of its similarity to Newton's second law of motion, is given as follows

$$\tau = \mu \frac{du}{dy} \tag{2.1}$$

Where τ = shear stress $(M/L/T^2)$
$\qquad \mu$ = absolute (dynamic) viscosity $(M/L/T)$
$\dfrac{du}{dy}$ = velocity gradient perpendicular to the direction of shear $(1/T)$

Water is considered a Newtonian fluid as the shear stress versus velocity gradient relationship is linear. In the SI system the dynamic viscosity units are $N \cdot s/m^2$, $Pa \cdot s$ or $kg/m \cdot s$.

Kinematic viscosity ν is the ratio of dynamic viscosity to the density of the fluid

$$\nu = \frac{\mu}{\rho} \tag{2.2}$$

Note that no force is involved with this quantity. In the SI system, the theoretical unit is (m^2/s) or commonly used Stoke (St) where $1\,St = 10^{-4}\,m^2/s$. The viscosity of a fluid is highly temperature dependent and for either dynamic or kinematic viscosity to be meaningful, the reference temperature must be quoted.

2.3. Basic flow equations

2.3.1 Flow and velocity

Flow through a pipe of uniform bore running completely full with the same velocity at each consecutive cross-section is called *uniform flow*. This is the most basic type of flow in pipes. If, however, the cross-sectional area and velocity of the fluid vary from cross-section to cross-section, but the flow rate does not change over time, the flow is called *steady*. An example is flow through a tapering pipe. The mean velocity V (m/s) at any cross-section of area A (m²) when the volume passing per unit of time is Q (m³/s) is given as

$$V = \frac{Q}{A} \tag{2.3}$$

where Q is also called the discharge. The cross-sectional area of a circular pipe can be expressed directly by using the diameter D (m), thus the velocity equation becomes

$$V = \frac{4Q}{\pi D^2} \tag{2.4}$$

Finally, if both the cross-sectional area and velocity of the flow vary with time, the flow is said to be *unsteady*. A pressure wave travelling along a pipe after a valve has suddenly closed is an example of unsteady flow.

2.3.2 Flow regime

When calculating losses in pipes, it is important to know if the fluid flow is *laminar* or *turbulent* (Box 2.1). Osborne Reynolds (1842–1912), an English scientist, found that the type of flow is determined by the pipe diameter D, the density of the liquid ρ, its dynamic viscosity μ, and the mean velocity V (Reynolds, 1883). The dimensionless parameter, the *Reynolds number* (Re), called after him, gives a measure of the ratio of inertial forces to viscous forces and consequently can be used to distinguish between laminar, transitional and turbulent flow

$$\mathrm{Re} = \frac{VD}{\nu} \tag{2.5}$$

As a general guide, laminar flow of water in pipes occurs for $\mathrm{Re} < 2000$, turbulent flow occurs for $\mathrm{Re} > 4000$ and transitional flow occurs for the values of Re between 2000 and 4000.

Box 2.1 Laminar and turbulent flow

Laminar flow is usually associated with slow moving, viscous fluids in small diameter pipes. Turbulent flow is much faster and chaotic because vortices, eddies and wakes make the flow unpredictable. Laminar flow is relatively rare in real situations while turbulent and *transitional* flow commonly occur. Transitional flow is a mixture of laminar and turbulent flow, with turbulence typically occurring in the centre of the pipe, and laminar flow near the edges.

2.3.3 Mass conservation (continuity) law

The *mass conservation law* states that mass can be neither created nor destroyed – that is, any mass entering a system must either accumulate in it or leave it. Applying this law to a steady flow through a control volume (Figure 2.2) where the stored mass m in it does not change, gives

Figure 2.2 A control volume

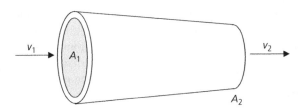

$$m_1 = m_2 \qquad (2.6)$$

where m_1 = mass coming in (M)
 m_2 = mass going out (M)

Assuming that water is an incompressible fluid, the mass conservation law can be applied to volumes or discharge, hence

$$Q_1 = Q_2 \text{ or } v_1 A_1 = v_2 A_2 \qquad (2.7)$$

This is then called the *continuity equation*.

2.3.4 Energy conservation law

Water in a pipe system may possess three forms of internal energy: *potential, pressure* and *kinetic*. If we consider a system such as the one given in Figure 2.3, the *potential energy* arises due to the position of the fluid above some reference (datum) level, thus, if a weight W of liquid is at a height z above datum, then

$$\text{Potential energy} = W_z \qquad (2.8)$$

Note that at points 1 and 2 along the pipe in Figure 2.3 the potential energy changes as the elevation changes (i.e. $z_1 > z_2$). If we now consider the *pressure energy* that arises due to hydrostatic pressure (the internal energy of a fluid due to the pressure exerted on the pipe), thus in travelling through a distance, L, the liquid can do work. Knowing that force due to pressure, p, on the cross-sectional area, A, is pA, then the work done is

$$\text{Pressure energy} = pAL \qquad (2.9)$$

If we express the volume of fluid AL in terms of weight W as $W/\rho g$ (where g is the gravitational acceleration), we get

$$\text{Pressure energy} = p\frac{W}{\rho g} \qquad (2.10)$$

The *kinetic energy* is the energy possessed by the liquid of weight W moving at a velocity v, then

$$\text{Kinetic energy} = \frac{1}{2}\frac{W}{g}v^2 \qquad (2.11)$$

It is normal for pipe systems to express energy per unit weight, thus resulting in

Potential energy per unit weight (or elevation head) $= z$

Pressure energy per unit weight (or pressure head) $= \dfrac{p}{\rho g}$ (2.12)

Kinetic energy per unit weight (or velocity head) $= \dfrac{v^2}{2g}$

Note that all three types of energy in Equation 2.12 are given as the head in terms of metres of fluid column. If we add the three equations (2.12) together to get the total energy (per unit weight), the following expression for the total energy head is obtained

$$\text{total energy per unit weight} = z + \frac{p}{\rho g} + \frac{v^2}{2g}$$ (2.13)

The three components of the total energy head shown in Equation 2.13 are the main building blocks of the *energy (Bernoulli) equation*, which is a statement of the conservation of energy principle. It states that the total energy of each particle of a body of fluid is the same provided that no energy enters or leaves the system at any point. The division of this energy between potential, pressure and kinetic may vary, but the total remains constant, as shown in the following Bernoulli equation

$$z + \frac{p}{\rho g} + \frac{v^2}{2g} = \text{constant}$$ (2.14)

For a static open water system, such as the one in Figure 2.3, the total head (comprising only two types of energy, potential and pressure, also called the piezometric head) is the same at any point along the pipeline. However, the pressure will change as the elevation of the pipeline changes. The piezometric head represents the height to which the water will rise in a piezometer (or a stand-pipe in a pipeline).

Equation 2.14 can be written for two cross-sections of a pipe as follows

$$z_1 + \frac{p_1}{\rho g} + \frac{v_1^2}{2g} = z_2 + \frac{p_2}{\rho g} + \frac{v_2^2}{2g}$$ (2.15)

Figure 2.3 A static pipeline water system

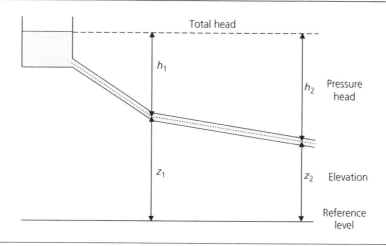

The equation assumes that the flow is one-dimensional (i.e. assumes constant velocity and pressure across a cross-section), steady (i.e. velocity does not change over time), frictionless (i.e. no losses) and ideal (i.e. no energy losses), and that the fluid is incompressible (i.e. ρ is constant). If we now write the energy equation for a real fluid with energy losses (e.g. water) between any two sections of a pipe, the equation becomes

$$z_1 + \frac{p_1}{\rho g} + \frac{v_1^2}{2g} = z_2 + \frac{p_2}{\rho g} + \frac{v_2^2}{2g} + h_f \qquad (2.16)$$

where, $h_f =$ energy head losses (L).

The equation can be used to evaluate energy head losses, which can be caused either by *friction* (resistance) along the pipe walls or by *geometry or shape configuration changes* (minor or local losses). There will be more discussion about these two types of energy losses later.

A schematic in Figure 2.4 illustrates the relationship between velocity, pressure and losses in a simple system consisting of a reservoir and a pipeline comprising two segments.

Figure 2.4 A static pipeline water system

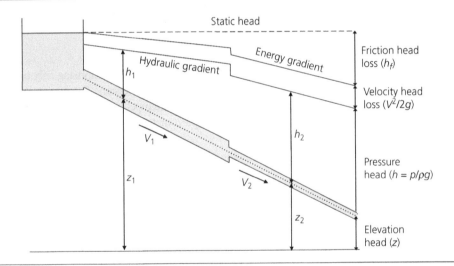

Box 2.2 Static head, energy grade line and hydraulic grade line

The *static head*, which also represents the total energy line for an ideal liquid with no energy losses, is always horizontal. The *energy grade line* (EGL) or *energy gradient* represents the total energy if head losses are taken into consideration. The energy grade line always falls in the direction of flow (unless there is a pump along the pipeline), because there must always be a head loss due to friction and minor losses. The *hydraulic gradient* or *hydraulic grade line* (HGL) is obtained when the piezometric levels along the pipe sections are joined together by a straight line. The hydraulic grade line generally falls in the direction of flow, but it can also go up (for example, when at a sudden expansion in pipe cross-section). The sudden vertical step downward in the hydraulic grade line and the energy grade line (Figure 2.4) at the sudden contraction of the pipeline represents a local loss in energy.

It is worth pointing out that in real-life water distribution systems, flow velocities rarely go above 1-2 m/s, hence velocity heads are typically very small (0.05–0.20 m), i.e. negligible when compared to pressure heads (can be anything from 10 to 100+ metres, usually 20-50 m). This is the reason why many practitioners often equalise the EGL with HGL by referring to HGL as the 'total head' (Box 2.2). Another interesting point to note is that in real-life water distribution systems, the elevation of the pipe centre line z is often not too far from the ground elevation as the top of the pipe is typically buried less than 1 m (perhaps 2 m in very cold climates) below ground level. This is the reason why practitioners often approximate the pressure head as the distance between the 'total head' and the ground elevation (which is often easier to obtain than the pipe centreline elevation).

2.4.   Losses in pipes

The energy head loss term in Equation 2.16 results from two different phenomena. The head loss due to friction between the moving water and the pipe wall is called *friction loss* and is always present throughout the length of a pipe. Additional head loss due to local disruption of the fluid flow (e.g. due to valves and junctions) is called *local loss*. Friction losses are often dominant in long pipes, whereas minor losses may be significant in short pipes. It is commonly considered that in pipe networks local losses do not contribute significantly to overall losses and could be neglected.

2.4.1   Friction losses

The two most commonly used equations for calculating friction losses in pressurized pipes are the Darcy-Weisbach and Hazen-Williams equations.

2.4.1.1   Darcy-Weisbach (Colebrook-White) equation

The Darcy-Weisbach equation is named after the French engineer Henry Darcy (1803–1858) and the German engineer Julius Weisbach (1806–1871), who independently discovered it around 1850, and relates the energy head loss due to friction along a pipe to the average velocity of the fluid flow

$$h_f = \lambda \frac{L}{D} \frac{V^2}{2g} \qquad\qquad (2.17)$$

where, λ is the non-dimensional friction factor.

The equation can be used for all pipe flow categories (i.e. laminar, turbulent and transitional) and, as such, it is a function of the Reynolds number, Re, and relative pipe roughness, k/D, where k is the absolute pipe roughness height or also called Nikuradse's equivalent sand-grain roughness, expressed in mm (Nikuradse, 1933). The functional behaviour of λ is shown in the Moody diagram (Moody, 1944) in Figure 2.5.

On the left of the diagram, the value of λ for laminar flow conditions is

$$\lambda = \frac{64}{\text{Re}} \qquad\qquad (2.18)$$

This shows that λ is a function not only on the pipe material but also of the Reynolds number – that is, depends on viscosity, density and flow velocity. The laminar flow equation holds for Re < 2000. For Reynolds numbers above 2000, the flow changes from laminar to weakly turbulent flow and beyond, approximately, 4000 it becomes turbulent, but it is characterised by three characteristic flow zones.

(a) Smooth turbulent flow (the lower envelope line in Figure 2.5), which represents flow in a hydraulically *smooth pipe*. The flow in PVC and copper pipes is described by this line.

(b) Rough turbulent flow (where the curves become horizontal), for which the friction factor λ is a function only of the relative roughness k/D and not of Re.

(c) Between the smooth and rough turbulent flow zones there is a transitional turbulent region for which λ decreases as Re increases.

One equation that is applicable for Re > 4000 (i.e. over the whole turbulent flow region) is the Colebrook-White formula (Colebrook and White, 1937), which is used to solve for the Darcy-Weisbach friction factor λ.

$$\frac{1}{\sqrt{\lambda}} = -2 \cdot \log_{10} \left(\frac{k}{3.7D} + \frac{2.51}{Re\sqrt{\lambda}} \right)$$
(2.19)

It relates the friction factor, λ, implicitly to the roughness, k, and the Reynolds number, Re. Because of its implicit nature (i.e. λ appears on both sides of the equation) iterative numerical methods have to be used to determine λ from Equation 2.19. This is not always practicable, especially within today's modelling tools when λ needs to be determined for a large number of pipes. The other important feature of this equation is that it uses the base 10 logarithm, whereas in many computer languages the computation is based on the natural logarithm, thus requiring a base change. Therefore, the equation is often written in the natural base as

$$\frac{1}{\sqrt{\lambda}} = -0.8686 \cdot \ln \left(\frac{k}{3.7D} + \frac{2.51}{Re\sqrt{\lambda}} \right)$$
(2.20)

Figure 2.5 Example Moody diagram for the Darcy-Weisbach friction factor λ. (Reproduced with permission from Purcell (2003))

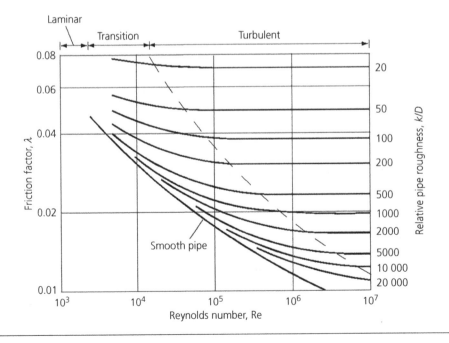

Swamee and Jain (1978) came up with an explicit approximation of the implicit Colebrook-White equation

$$\lambda = \frac{0.25}{\left[\log_{10} \left(\dfrac{k}{3.71D} + \dfrac{5.74}{Re^{0.9}} \right) \right]^2} \tag{2.21}$$

The relative simplicity and accuracy (typically it provides friction factor estimate only 1–2% different from the Colebrook-White equation) of the explicit Swamee-Jain equation has influenced software developers to implement this equation in a number of modelling tools for water distribution system analysis. For example, the widely available hydraulic analysis software EPANET (Rossman, 2000) uses this equation to solve for the Darcy-Weisbach friction factor.

2.4.1.2 Hazen-Williams equation

The Hazen-Williams equation, originally introduced in 1902 (Liou, 1998) is still widely used by water supply engineers to calculate the headloss along a pipe based on the flow in the pipe and the physical properties of the pipe. The equation is empirical and in SI units could be expressed as:

$$h_f = 10.67 \frac{Q^{1.85}}{C^{1.85}D^{4.87}} L \tag{2.22}$$

where C is the Hazen-Williams friction coefficient, which indicates the roughness of the interior surface of a pipe. Note that Equation 2.22 implies that the coefficient is dimensional and it would change for different units of hf, Q and D. However, the C-value is considered as a pipe constant and a numerical conversion coefficient needs to be introduced if different units for headloss, flow and diameter are used in the equation. However, being empirical, the equation is not dimensionally homogeneous and its range of application is limited. Note also that lower C values means higher headlosses in the pipe (opposite to relative roughness used in the Darcy-Weisbach equation).

Both Darcy-Weisbach and Hazen-William equations can be expressed using the following single headloss equation

$$h_f = RQ^n \tag{2.23}$$

Where R is the pipe/flow resistance coefficient and n is the headloss equation exponent. Note that for the Darcy-Weisbach equation $R = 0.8106\lambda L/(gD^5)$ and $n = 2.0$ and for the Hazen-Williams equation $R = 10.67L/(C^{1.85}D^{4.87})$ and $n = 1.85$.

2.4.2 Local and minor losses

Local losses are generally caused by increased turbulence as the flow passes through valves or other pipe fittings such as bends, tees or tapers. Figure 2.6 illustrates the head loss generated by a partially shut valve. In this case, the velocity under the gate of the valve is increased significantly by the reduced cross-sectional area as the flow at all pipe sections must be same (conservation of mass). This velocity increase results in some of the pressure energy being converted into kinetic energy as the overall energy must remain constant (conservation of energy).

Figure 2.6 The head loss generated by a partially shut valve

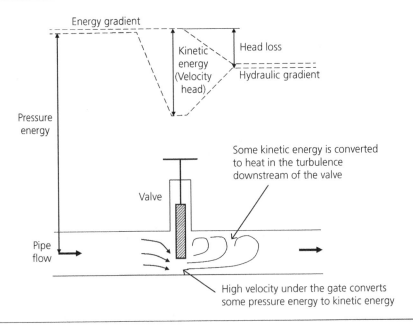

If all the pressure energy is converted to kinetic energy then the pressure at the reduced cross-section under the valve gate may become sub-atmospheric and vapour cavities may form due to water evaporating at the ambient temperature. The violent collapse of these cavities downstream, as the pressure rises, causes cavitation which can damage the valve and the pipeline.

Downstream of the valve, massive eddies are created as the velocity slows. This results in some of the energy being converted to heat, as not all the kinetic energy is reconverted to potential energy downstream of the valve, and an overall head loss occurs across the valve.

As the head loss is related to the kinetic energy lost, this head loss is often expressed as a coefficient, k, of the initial kinetic energy of the pipe flow

$$h_l = k\frac{v^2}{2g} = KQ^2 \qquad (2.24)$$

where h_l = local head loss (L)
k = local head loss coefficient (-)
K = local resistance coefficient (T^2/L^5)
v = characteristic velocity (typically downstream of a local head loss) (L/T)
Q = flow rate (L^3/T)
g = acceleration due to gravity (L/T^2)

Figure 2.7 illustrates that the mechanism of head loss through increased turbulence is the same for other pipeline fittings such as bends and tees. The value of K for various types of pipeline fittings can be obtained from standard hydraulic textbooks such as the 'Manual of British Water Engineering Practice', Table XXXIV (Skeat and Dangerfield, 1969).

Figure 2.7 Head loss due to turbulence through pipe fittings

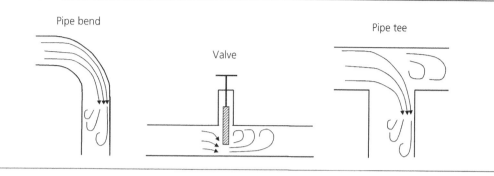

For most water transmission pipelines, the fitting losses are normally small compared with the overall friction losses along the pipe, but they may be significant at points where there are a number of fittings and the diameter of the pipe is constrained – for example, at pumping stations and bridge crossings, and so on.

2.5. Steady flow analysis in networks

A typical water distribution network is a complex system normally consisting of thousands of simpler individual elements (e.g. pipes, valves, pumps and reservoirs). As with many other complex systems, computer simulation is normally the only viable way of analysing the complex hydraulic behaviour in a water distribution network – that is, to determine the flow in every pipe and the pressure at every node in the system. With many network elements linked together to form loops, computer models are used to calculate how much water takes which of the numerous alternative routes and what are the head losses along each route. The two equations used to compute the flows and energy heads are the continuity equation and the energy equation.

When applied to a network node connecting n_p pipes the continuity Equation 2.7 gives

$$\sum_{j=1}^{n_p} Q_{ij} - q_i = 0 \qquad (2.25)$$

where Q_{ij} = the flow in pipe ij – positive for inflow and negative for outflow (L^3/T)
q_i = the nodal discharge at node i (L^3/T).

Note that, although demands are actually distributed along a pipe in reality (by way of property connections), these demands are lumped at nodes for modelling purposes and represented as the nodal discharge. The above equation can then be written for every node in the system.

Similarly, the energy balance equation can be written between any two nodes (e.g. nodes 1 and 2) in the water distribution network

$$H_1 - H_2 = \sum_{path_{1-2}} h_f = \sum_{path_{1-2}} R_{ij} Q_{ij} |Q_{ij}|^{n-1} \qquad (2.26)$$

where $path_{1-2}$ = includes the set of pipes on the path from node 1 to node 2

H_1, H_2 = the heads at path's upstream (1) and downstream node (2)

R_{ij} = the resistance coefficient for each pipe ij of $path_{1-2}$

n = the exponent from the headloss equation (e.g. 2 for Darcy-Weisbach and 1.85 for Hazen-Williams)

Network nodes can be classified into the following three groups

- fixed-head nodes
- variable-head nodes
- ordinary nodes or junctions.

The main feature of a fixed-head node is that its water level does not change during the analysis – for example, a fixed-head node represents external sources or sinks of water to the network, such as lakes, rivers or groundwater aquifers. Variable-head nodes have associated storage capacity where the volume of stored water can vary with time during the analysis. Junctions represent all other points in the network where pipes branch out (with or without demand associated with them) or where there is a net export (or import) of water from (to) the system – that is, water enters or leaves the network (Figure 2.8).

Figure 2.8 shows a branched (a) and a looped system (b), where a loop is defined as a closed circuit that has no interior crossing pipes – also called a simple, independent or 'natural' loop (Epp and Fowler, 1970). It is important to note that in a branched network there are no loops. Furthermore, for a closed loop (e.g. pipes $p_2-p_4-p_6$) – that is, one beginning and ending at the same node (N_2) – the energy loss is zero.

Steady flow analyses in branched networks are performed directly, without iterations, and fall into two main problem categories: *(a)* for known demand at nodes and known pipe characteristics, find pressure at nodes, or *(b)* for known headloss along pipes and known pipe characteristics, find flows in the network. For example, a steady state analysis for a branched system (Figure 2.8) consisting of three pipes, p_1 (D=250 mm), p_2 (D=100 mm) and p_5 (D=200 mm), each 1000 m long and having Hazen-Williams C=100, can be performed using the continuity equation (Equation 2.25) and the energy balance equation (Equation 2.26). If the demand at nodes is known to be q_2=2 l s, q_3=3 l/s and q_5=12 l/s and the known fixed head at N_1 is 20 m, the computations are made in the following manner

Figure 2.8 An example of (a) a branched system and (b) a looped system

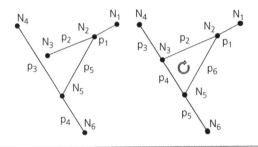

From Equation 2.25

$$Q_{12} - Q_{23} - Q_{25} - q_2 = Q_{12} - 3 - 12 - 2 = 0 \Rightarrow Q_{12} = 17 \, l/s$$

By using Equation 2.26

$$H_1 - H_2 = h_f = R_{12}Q_{12}|Q_{12}|^{0.85}$$

$$= 10.67 \frac{1000}{100^{1.85}0.25^{4.87}} 0.017|0.017|^{0.85} = 0.97 \, m$$

$$H_1 - H_2 = h_f = 0.97 \Rightarrow H_2 = 20 - 0.97 = 19.03 m$$

Similarly

$$H_3 = H_2 - 10.67 \frac{1000}{100^{1.85}0.10^{4.87}} 0.003|0.003|^{0.85} = 19.03 - 3.39 = 15.64 m$$

$$H_5 = H_2 - 10.67 \frac{1000}{100^{1.85}0.20^{4.87}} 0.012|0.012|^{0.85} = 19.03 - 1.51 = 17.52 m$$

Once all pipes, nodes and independent loops are identified prior to performing a steady-state flow analysis for a looped network, a system of non-redundant continuity and energy equations is developed (for the entire network). The number of these equations is related directly to fundamental relations between the number of pipes, number of nodes (including all types of nodes, i.e. fixed-head, variable-head and ordinary nodes/junctions) and number of closed simple loops in the network system (Larock *et al.*, 2000).

Various formulations of the governing equations exist for the solution of network analysis problems (Larock *et al.*, 2000). These are named
(a) **Q-equations** – flow equations in which pipe discharges are treated as unknowns
(b) **H-equations** – node equations in which nodal heads are treated as unknowns
(c) **Q-H-equations** – combined flow and node equations (e.g. Todini and Pilati algorithm)
(d) **ΔQ-equations** – in which a loop flow correction of ΔQ is computed for each of the independent loops and each of the paths between fixed-head nodes in the network.

Irrespective of the formulation of the governing equations chosen for implementation, an approach based on the Newton-Raphson method can be applied to solve the system of equations (Press *et al.*, 2007).

2.5.1   Hardy Cross method
Prior to the introduction of computers, engineers had to simplify the network into a few strategic pipe loops and use an iterative method developed by Hardy Cross (1936). The so-called Hardy Cross method is the first systematic approach for solving a system of equations resulting from writing a set of continuity equations for each node and energy equations for each independent loop in the network.

The method starts with an initial guess of the direction and magnitude of pipe flows, such as the principle of continuity being satisfied at each node. It then calculates the head losses around each independent loop, which normally results in the need to rebalance the network (i.e. change assumed flows) until the sum of the head losses around each loop is close to zero. A correction of the assumed flows, ΔQ, is computed for each independent loop in the network. Iterations are performed such as that

$$\Delta Q^{(i)} = \frac{-\sum h_f^{(i)}}{n \sum \dfrac{h_f^{(i)}}{Q^{(i)}}} \tag{2.27}$$

where $Q^{(i)} =$ flow in a pipe at iterations i (M³/T)
$\qquad \Delta Q =$ corrective flow passing through a pipe in a loop (M³/T)
$\qquad \Sigma =$ the summation operator is applied to all pipes in a loop
$\qquad n =$ the exponent from the headloss equation

This gives the new flow estimate in a pipe for the next iteration

$$Q^{(i+1)} = Q^{(i)} + \Delta Q^{(i)} \tag{2.28}$$

It is important to note that the corrective flow is equal for each pipe in a loop and that, for pipes belonging to multiple loops, the corrective flow is a sum of individual loop corrective flows. The value of n in the Hardy Cross method is assumed to be constant.

After the flows in each pipe are found – that is, the corrective flow gets lower than a predefined error threshold and the headloss in each pipe is also known – the nodal heads are simply calculated by starting at known (i.e. fixed) heads.

The Hardy Cross method depends on the initial estimate of the flows and could suffer from slow convergence. Instead of adjusting flows around each individual loop in the network, the Newton-Raphson method adjusts the flows (Q-equations), or heads (H-equations), or flows and heads (Q-H-equations) along all loops simultaneously. These simultaneous corrections have improved convergence over the Hardy Cross method and the current modelling software relies almost exclusively on the Newton-Raphson approach. Therefore, the Hardy Cross method is of historical interest rather than an approach that can be used for the large water distribution models preferred by today's practitioners.

Flow analysis in large water distribution networks has been made possible by the advances in computer technology, particularly in the 1960s and 1970s when the first research-based simulation tools were developed and then in the 1980s when personal computers were first introduced. A number of methods implemented in software packages, such as GINAS (Coulbeck et al., 1991), PICOLLO (Jarrige, 1993), WATNET (developed by WRc), STONER (now SinerGEE) and so on, have advanced beyond the Hardy Cross method and allowed sophisticated steady-state and dynamic analyses to be performed. The development of the global gradient algorithm (Todini and Pilati, 1988) and its implementation within the widely available freeware EPANET (Rossman, 2000) has led to the widespread use of the method by software developers and vendors.

2.5.2 Todini-Pilati method

At a conference in 1987, Todini and Pilati (1988) presented an elegant formulation (Q-H-equations) and an efficient gradient solution methodology for the system of nonlinear equations. The global gradient solution method has excellent convergence characteristics and is incorporated into a number of software packages currently available as commercial or public-domain software (e.g. EPANET).

Following from Equations 2.24 to 2.26, the flow-headloss and the continuity relations can be derived for any water distribution system as

$$H_i - H_j = RQ_{ij}^n + KQ_{ij}^2 \text{ for all pipes}, n_p$$

$$\sum_{j=1}^{n_p} Q_{ij} = q_i \text{ for all junction nodes}, n_n \tag{2.29}$$

where q_i is the demand at node i.

The continuity equation is developed for each junction node connecting n_p pipes, while the head-loss equation is developed for each pipe between nodes i and j. Note that for pumps, the non-linear headloss equation can be represented by a power law expression (Todini, 2003).

For the gradient solution method, the system of Equations 2.29 can be written in a matrix form. Thus, the simulation of a network of n_p pipes with unknown discharges/flows, n_n nodes with unknown heads (internal nodes) and n_0 nodes with known heads (e.g. tank levels) can be formulated as

$$\begin{bmatrix} \mathbf{A}_{pp} & \mathbf{A}_{pn} \\ \mathbf{A}_{np} & 0 \end{bmatrix} \begin{bmatrix} \mathbf{Q}_p \\ \mathbf{H}_n \end{bmatrix} = \begin{bmatrix} -\mathbf{A}_{p0}\mathbf{H}_0 \\ \mathbf{q}_n \end{bmatrix} \tag{2.30}$$

where
    $\mathbf{Q}_p$ is the $[n_p,1]$ column vector of unknown pipe flows (M³/T)
    $\mathbf{H}_n$ is the $[n_n,1]$ column vector of unknown nodal heads (M)
    $\mathbf{H}_0$ is the $[n_0,1]$ column vector of known nodal heads (M)
    $\mathbf{q}_n$ is the $[n_n,1]$ column vector of demands lumped at nodes
    $\mathbf{A}_{pn}=\mathbf{A}_{np}^T$ and $\mathbf{A}_{p0}$ are topological incidence sub-matrices of size $[n_p,n_n]$ and $[n_p,n_0]$, respectively, derived from the general topological matrix $\bar{\mathbf{A}}_{pn}=[\mathbf{A}_{pn} \mid \mathbf{A}_{p0}]$ of size $[n_p,n_n+n_0]$ as defined by Todini and Pilati (1988)
    $\mathbf{A}_{pp}$ is a diagonal matrix whose elements are given by the entry-wise product $\mathbf{R}_p|\mathbf{Q}_p|$, with $\mathbf{R}_p$ being the vector of the pipe hydraulic resistances

The above matrix equation can then be written as the following non-linear and linear system of equations, respectively

$$\mathbf{A}_{pp}\mathbf{Q}_p + \mathbf{A}_{pn}\mathbf{H}_n = -\mathbf{A}_{p0}\mathbf{H}_0$$
$$\mathbf{A}_{np}\mathbf{Q}_p = \mathbf{q}_n \tag{2.31}$$

Thus, the non-linear mathematical problem has (n_p+n_n) unknowns $(\mathbf{Q}_p; \mathbf{H}_n)$ and its boundary conditions are $(\mathbf{R}_p; \mathbf{q}_n; \mathbf{H}_0)$ – that is, the total number is equal to $(n_p+n_n+n_0)$.

The topological incidence matrix $\overline{\mathbf{A}}_{pn}$ is defined as

$$\overline{\mathbf{A}}_{pn}(i,j) = \begin{cases} -1 & \text{if the flow in pipe } j \text{ leaves node } i \\ 0 & \text{if pipe } j \text{ is not connected to node } i \\ +1 & \text{if the flow in pipe } j \text{ enters node } i \end{cases} \tag{2.32}$$

The global gradient algorithm (GGA) for solving the system of equations (2.31) was given by Todini and Pilati (1988) as

$$\mathbf{B}_{pp}^{iter} = \left(\mathbf{D}_{pp}^{iter}\right)^{-1}\mathbf{A}_{pp}^{iter}$$

$$\mathbf{F}_n^{iter} = -\mathbf{A}_{np}\left(\mathbf{Q}_p^{iter} - \mathbf{B}_{pp}^{iter}\mathbf{Q}_p^{iter}\right) + \mathbf{A}_{np}\left(\mathbf{D}_{pp}^{iter}\right)^{-1}\left(\mathbf{A}_{p0}\mathbf{H}_0\right) + \mathbf{q}_n$$

$$\mathbf{H}_n^{iter+1} = -\left[\mathbf{A}_{np}\left(\mathbf{D}_{pp}^{iter}\right)^{-1}\mathbf{A}_{pn}\right]^{-1}\mathbf{F}_n^{iter} \tag{2.33}$$

$$\mathbf{Q}_p^{iter+1} = \left(\mathbf{Q}_p^{iter} - \mathbf{B}_{pp}^{iter}\mathbf{Q}_p^{iter}\right) - \left(\mathbf{D}_{pp}^{iter}\right)^{-1}\left(\mathbf{A}_{p0}\mathbf{H}_0 + \mathbf{A}_{pn}\mathbf{H}_n^{iter+1}\right)$$

where *iter* is a counter of the iterative solving algorithm and $\mathbf{D}_{pp}$ is a diagonal matrix whose elements are the derivatives of the head loss function with respect to $\mathbf{Q}_p$.

The gradient solution method begins with an initial estimate of flows in each pipe that normally does not satisfy flow continuity. In the next step, the GGA methodology is applied, thus tackling simultaneously unknown flows and heads. The iterative process continues until a convergence criterion is reached.

Further details on the numerical methodology employed by the gradient solution method can be found in Todini and Pilati (1988) and Todini (2003).

2.5.3 Demand-driven or head-driven analysis

Most of the commercial water distribution analysis software packages, including EPANET (Rossman, 2000), are built on the assumption that demand at a node is fixed (predetermined) and fully satisfied, regardless of the pressure at the node – that is, outflow at the node is equal to its demand. This is the so-called *demand-driven analysis* (DDA), which determines the nodal pressures and pipe flow rates that correspond to the specified nodal demands (regardless whether or not they can be satisfied). This assumption is reasonable in well-designed systems and when they operate under normal conditions. In these cases, residual heads at demand nodes are sufficient to supply the demand. However, DDA gives unrealistic results in cases where networks operate under abnormal conditions – for example, due to pipe burst or excessive demands.

Under abnormal conditions the distribution of flow in pipes and the deliverable discharge at the nodes cannot be calculated correctly using DDA. For example, nodal heads or pressure may be insufficient to deliver the required water demand such that the real outflow at the nodes is only a fraction of the original demand. *Head-driven* (or pressure-driven) *analysis*, (HDA), is needed to correctly account for the actual outflow at nodes where the pressure head falls below the minimum service level required to supply the full demand.

Bhave (1981) was the first to investigate the pressure dependency of flow in water distribution systems. Since then, a number of pressure head-flow relationships have been suggested and incorporated into network analysis methodologies. For example, Wagner et al. (1988) suggested the following parabolic relationship

$$q = \begin{cases} q^{(d)} & \text{for } H \geq H_{ser} \\ q^{(d)} \left(\dfrac{H - H_{min}}{H_{ser} - H_{min}} \right)^{0.5} & \text{for } H_{min} \leq H \leq H_{ser} \\ 0 & \text{for } H \leq H_{min} \end{cases} \tag{2.34}$$

where q = the actual outflow at a node (M^3/T)
$\quad q^{(d)}$ = the demand at a node (M^3/T)
$\quad H$ = the actual head at node (M)
$\quad H_{ser}$ = the minimum service head required to supply the full demand $q^{(d)}$
$\quad H_{min}$ = the minimum head below which no outflow is possible

In the UK, normally, the minimum service pressure head H_{ser} = 15 m is assumed to be sufficient to supply demand in full (WRc, 1994). At any pressure head below this threshold the node would supply only a fraction of the demand based on Equation 2.34.

Todini (2003) showed how the global gradient solution method can be extended to perform HDA, by either adopting a head-outflow relationship or by performing an iterative analysis to determine the possible outflow at demand nodes as a function of the prevailing system pressure. The work of Todini was followed by the development of a number of pressure-driven methods. These incude methods based on different analytical formulations of pressure head–flow relationships with multiple authors introducing artificial components such as emitters, pipes, reservoirs and other elements to represent this behavior (see review in Suribabu et al. 2019).

One such method was developed by Mahmoud et al. (2017) who proposed a new approach based on using a specific set of elements added to deficient pressure demand nodes only. This set involves a short pipe with a check valve, a flow control valve and an emitter. This, in turn, enables modelling the minimum pressure head below which no flow occurs, the required pressure above which the full demand required is delivered, and the partial flow conditions for the pressure heads in between these two characteristic values. The method is effective and efficient in simulating the hydraulics of large networks. However, this and other methods that make use of artificially added elements to the network nodes became largely redundant with the introduction of pressure-driven modelling capability in Epanet 2.2, as this enabled the implementation of pressure head-flow relationship directly in Epanet's source code.

Giustolisi et al. (2008) demonstrated how leakage from pipes can be incorporated in a HAD approach by using the model proposed by Germanopoulos (1985). Furthermore, Giustolisi and Todini (2009) and Giustolisi (2010) showed how to deal with actual demand patterns along mains which resulted in preserved model parsimony and energy balance.

2.6. Unsteady flow analysis in networks

2.6.1 Extended period simulation

When applied to a service reservoir for dynamic (unsteady) conditions the continuity equation states that the difference between the flow in and out of the reservoir is equal to the change in storage, or

$$\sum Q_{in} - Q_{out} = \frac{dS}{dt} = A\frac{dH}{dt}$$ (2.35)

where Q_{in} = sum of the inflows into the reservoir (L³/T)
Q_{out} = flow out of the reservoir (L³/T)
S = storage (L³)
H = water level in reservoir (L)
A = reservoir cross-sectional area (L²)
t = time (T)

The above equation is used when it is necessary to take into consideration the slow variation of flow conditions over time. This may result in changes of water level in a reservoir, changes in nodal demand or change in the state of valves and pumps. This type of network analysis is generally known as extended period simulation (EPS) or dynamic analysis.

EPS is normally computed as a sequence of steady-state (snapshot) analyses performed at fixed times, each followed by a mass balance computation based on Equation 2.35. In solving the set of hydraulic equations for the unknown heads and flows at each snapshot – that is, Equations 2.24 to 2.26 are built taking into consideration the changes that occurred at the end of the previous time step. These changes include the dynamic state of reservoirs and the pre-defined changes at the current simulation time, such as nodal demands and the status of the pipes, pumps and valves. Further details of the methodology can be found in Walski *et al.* (2003).

2.6.2 Transient flow analysis

Pressure transients occur when there is a rapid change in the fluid velocity within a pipeline. These pressure changes are often generated by the rapid opening/closure of valves or switching/failure of pumps within the pipe system. An initial estimate of the pressure change generated by a sudden or instantaneous change in velocity can be calculated using the Joukowsky equation (Thorley, 2004)

$$\Delta h = c\frac{\Delta v}{g}$$ (2.36)

where Δh = change in pressure head (L)
Δv = sudden change in velocity (L/T)
c = wave speed at which the pressure transient travels along the pipeline (L/T)

The speed of the transient wave is partly determined by the fluid carried by the pipeline and partly by the pipeline material, diameter and thickness. However, the wave speed is significantly reduced by the presence of free gas in the fluid.

Figure 2.9 illustrates the propagation of a typical transient wave in a gravity pipeline from a reservoir, following the instantaneous closure of a downstream valve. As the velocity of the water in the pipeline

Figure 2.9 Propagation of a transient pressure wave following valve closure

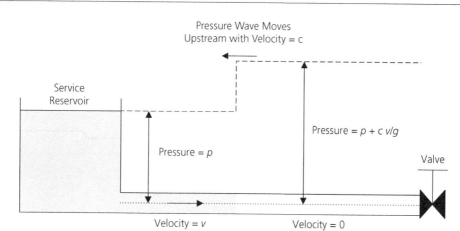

is reduced to zero by the valve, the pressure increases in accordance with the Joukowsky equation and a positive pressure wave is sent upstream to the reservoir. On reaching the free surface of the reservoir, the increased pressure energy is then released creating a return wave at the original pressure but with a velocity in the reverse direction towards the valve. When this negative velocity hits the valve, a negative pressure wave is generated that travels upstream to the reservoir as the velocity of flow is again reduced to zero. On reaching the free surface of the reservoir the pressure again returns to the initial steady state value and a positive velocity is created which travels back to the valve again.

To help visualise this cycle of operation, imagine a goods locomotive pulling a train of wagons. When the locomotive stops rapidly, all the couplings between the wagons compress, as each wagon in turn is stopped, until the last wagon is reached. The last wagon then bounces back pulling the couplings between each of the wagons taut with a velocity in the opposite direction.

In a frictionless pipeline system the pressure wave would continue back and forth but in a real system the energy lost through friction gradually reduces the amplitude of the wave, creating the classic pressure surge wave often recorded by transient pressure loggers (Figure 2.10).

The Joukowsky equation is often used to initially calculate the maximum or worst-case pressure change due to an instantaneous velocity change to assess the potential magnitude of the surge problem. However, where the velocity change is not instantaneous, the pressure change will be less severe. In this case, the behaviour of a fluid in a slightly deformable pipeline can be described by momentum and continuity equations. Details of these equations can be found in standard textbooks on pressure surge (Wylie and Streeter, 1978).

In the past, graphical methods of analysis, such as those developed by Schnyder and Bergeron (Frey and Althammer, 1961), have been used to provide a solution to the momentum and continuity equations and an estimate of surge pressures in most practical situations. More recently, solution by the method of characteristics has become the most popular method of analysis, as it can be applied to most practical systems and can be easily adapted for analysis by computer (Thorley and Enever, 1979). The method of characteristics can also include an allowance for non-linear terms such as friction, column separation and air bubbles.

Figure 2.10 Typical transient pressure wave recorded by a data logger

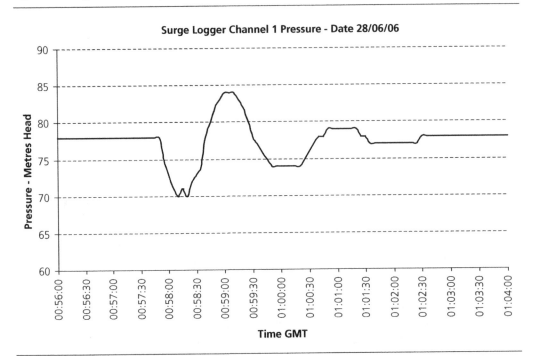

2.7. Water quality analysis in networks

Drinking water quality is of great importance due to its impact on public health. It is not surprising then that considerable attention has been given to the development of algorithms for water quality analysis in water distribution systems. Water quality modelling applications in water distribution systems include simulation of chlorine decay, simulation of blending of multiple sources, estimation of water age, tracing of contaminant propagation, prediction of formation of disinfection by-products and total trihalomethanes in water, and so on. Because of the improvement in public health due to the use of chlorine as a water disinfectant, simulation of chlorine decay in water distribution systems is one of the most often encountered applications of water quality analysis.

Similarly to hydraulic network models, water quality analysis can be formulated as either steady-state or dynamic models. Steady-state models are based on the principles of mass conservation, which is used to determine the ultimate distribution of dissolved substances with constant hydraulic conditions. Thus, conservation of constituent mass for node j is given as

$$\sum_j \overbrace{(QC)}^{\text{incoming}} - \sum_j \overbrace{(QC)}^{\text{outgoing}} = 0 \tag{2.37}$$

where Q = the volumetric flow rate (L^3/T)

C = the volumetric constituent concentration (M/L^3)

The nodal concentrations can be obtained for conservative and zero-order reacting constituents by an explicit analytic approach (Boulos and Altman, 1993). These models assume that either the constituent concentrations do not change over time (conservative) or the time rate of change of a constituent is constant (zero-order). For a general n-th order reaction kinetics, the rate of decay of a constituent is given as

$$\frac{dC}{dt} = -kC^n \tag{2.38}$$

where $t =$ time (T)

$k =$ the reaction rate constant (1/T)

The analytic approach by Boulos and Altman (1993) for solving the steady-state water quality modelling problem was later extended by Chung *et al.* (2007) to include first- and second-order reactions. The rate of the first-order reaction is proportional to the concentration of the constituent and depletes a constant percentage of a constituent with each period of time that elapses (exponential decay), while the rate of the second-order reaction is proportional to the concentration of the constituent squared ($n = 2$ in Equation 2.38). The first-order model for chlorine decay is most commonly employed to simulate the disappearance of residual chlorine due to reactions with materials in the aqueous phase at different residence times, t, in the network. Thus, integrating Equation 2.38 for $n = 1$ gives an exponential decay model

$$C(t) = C_0 e^{-kt} \tag{2.39}$$

where $C_0 =$ the initial chlorine concentration (M/L^3).

With particular reference to chlorine, several studies concluded that chlorine undergoes reactions with particulates in the bulk of the water as well as at and near the pipe wall at separate reaction rates (Rossman *et al.*, 1994). Therefore, much better agreement between observed and simulated chlorine concentration are obtained when the simulated decay of chlorine is separated into a bulk and wall decay model

$$\frac{dC}{dt} = \overbrace{-kC}^{\text{bulk decay}} - \overbrace{R}^{\text{wall decay}} \tag{2.40}$$

where $R =$ the pipe wall demand – for example, due to corrosion rate (M/L^3/T).

The most commonly encountered dynamic water quality analysis is based on a one-dimensional modelling approach, single or extended period simulation hydraulics, instantaneous and complete mixing at nodes, negligible longitudinal dispersion and single constituent (e.g. chlorine), with one or more feed sources (Vasconselos *et al.*, 1996). These models simulate three processes: *(a)* advection in a pipe; *(b)* kinetic reaction mechanism; and *(c)* mixing at nodes. The advection process of a conservative constituent is given (in one-dimensional form) as the following mass-conservation differential equation

$$\frac{\partial C}{\partial t} = \overbrace{-v\frac{\partial C}{\partial x}}^{\text{spatial dispersion}} \overbrace{-r}^{\text{temporal dispersion}} \tag{2.41}$$

where $v =$ the velocity of flow (L/T)

$r =$ the rate of decay of chlorine in the pipe (M/L^3/T).

For a first-order kinetic reaction equation, the rate of decay r becomes k, as seen in Equation 2.39. Finally, mixing at nodes is given by the nodal mass balance principle in Equation 2.37.

The analytical solution to the problem described in the above equations becomes intractable, and numerical solution techniques are employed to simulate the advection process. Within the EPANET toolkit (Rossman, 2000), a modified version of the Lagrangian time-based approach developed by Liou and Kroon (1987) is employed to simulate the variation of chlorine throughout a water distribution system. The method starts with the known direction of flow and velocities in the network provided by the hydraulic model simulation. It then stores the chlorine concentration information in discrete (water) volume elements that are progressed through the network at short time intervals, typically 5 minutes or less. For each time step, the water segments travel through the pipe with the flow velocity. This approach avoids the numerical diffusion commonly associated with numerical solutions (e.g. finite-difference or finite-element schemes) for the advection Equation 2.41.

Shang et al. (2008) developed a general framework for modelling the reaction and transport of multiple, interacting chemical species in drinking water distribution systems. The framework accommodates reactions between constituents in both the bulk flow (through pipes and storage reservoirs) and those attached to pipe walls. The software implementation is provided as an extension to the EPANET programmer's toolkit (a library of functions that simulates hydraulic behaviour and water quality transport in pipe networks).

Historically, EPANET was limited to tracking the dynamics of a single chemical transported through a network of pipes and storage tanks. The USA EPA (Shang et al., 2011) released EPANET-MSX (a Multi-Species eXtension of EPANET) which can be used for the simultaneous modelling of multiple species by way of interacting chemical and biological reactions in water distribution systems.

REFERENCES

Bhave PR (1981) Node flow analysis of water distribution systems. *Journal of Transportation Engineering, ASCE* **107(4)**: 457–467.

Boulos PF and Altman T (1993) Explicit calculation of water quality parameters in pipe distribution systems. *Civil Engineering Systems* **10**: 187–206.

Chung G, Lansey KE and Boulos PF (2007) Steady-state water quality analysis for pipe network systems. *Journal of Environmental Engineering, ASCE* **133(7)**: 777–782.

Colebrook CF and White CM (1937) Experiments with fluid friction in roughened pipes. *Proceedings of the Royal Society of London, Series A* **161 (906)**: 367–381.

Coulbeck B, Orr CH and Cunningham AE (1991) *GINAS 5 reference manual*. Research Report No. 56, Water Software Systems. De Montfort University, Leicester, UK.

Epp R and Fowler AG (1970) Efficient code for steady-state flows in networks. *Journal of the Hydraulics Division, ASCE* **96**: 43–56.

Frey and Althammer (1961) The calculation of electromagnetic transient on lines by means of a digital computer. *The Brown Boveri Review* **48**: 344.

Germanopoulos G (1985) A technical note on the inclusion of pressure dependent demand and leakage terms in water supply network models. *Civil Engineering Systems*, **2(September)**: 171–179.

Giustolisi O (2010) Considering actual pipe connections in water distribution network analysis. *Journal of Hydraulic Engineering, ASCE* **136(11)**: 889–900.

Giustolisi O, Savic DA and Kapelan Z (2008) Pressure-driven demand and leakage simulation for water distribution networks. *Journal of Hydraulic Engineering, ASCE* **134(5)**: 626–635.

Giustolisi O and Todini E (2009) Pipe hydraulic resistance correction in WDN analysis. *Urban Water Journal* **6(1)**: 39–52.

Jarrige PA (1993) *PICCOLO – Users Manual. Potable Water Distribution Network Modeling Software*. SAFEGE Consulting Engineers, Cedex, France.

Larock BE, Jeppson RW and Watters GZ (2000) *Hydraulics of Pipeline Systems*. CRC Press, Boca Raton, USA.

Liou CP (1998) Limitations and proper use of the Hazen-Williams equation. *Journal of the Hydraulic Engineering, ASCE* **124(9)**: 951–954.

Liou CP and Kroon JR (1987) Modeling the propagation of waterborne substances in distribution networks. *Journal of the AWWA (American Water Works Association)* **79(11)**: 54–65.

Mahmoud HA, Savić D and Kapelan Z (2017) New pressure-driven approach for modeling water distribution networks. *Journal of Water Resources Planning and Management* **143(8)**: page range required.

Moody LF (1944) Friction factors for pipe flow. *Transactions of the ASME* **66**: 671–684.

Nikuradse J (1933) *Laws of flow in rough pipes*. Forschungsheft Ver. Dtsch. Ing., No. 361. (In German.)

Press WH, Flannery BP, Teukolsky SA and Vetterling WT (2007) *Numerical Recipes: The Art of Scientific Computing*. Cambridge University Press, Cambridge, UK.

Reynolds O (1883) An experimental investigation of the circumstances which determine whether the motion of water shall be direct or sinuous, and of the law of resistance in parallel channels. *Philosophical Transactions of the Royal Society* **174**: 935–982.

Rossman LA (2000) *EPANET 2, Users Manual*. US Environmental Protection Agency, Cincinnati, OH, USA.

Rossman LA, Clark RM and Grayman WM (1994) Modeling chlorine residuals in drinking water distribution systems. *Journal of Environmental Engineering* **120 (4)**: 803–820.

Suribabu CR, Renganathan NT, Perumal S and Paez D (2019) Analysis of water distribution network under pressure-deficient conditions through emitter setting. *Drinking Water Engineering and Science* **12 (1)**: 1–13.

Shang F, Uber JG and Rossman LA (2008) Modeling reaction and transport of multiple species in water distribution systems. *Environmental Science & Technology*, **42(3)**: 808–814.

Shang F, Uber JG and Rossman L (2011) *EPANET Multi-Species Extension Software and Users Manual*. US Environmental Protection Agency, DC, USA. https://cfpub.epa.gov/si/si_public_record_report.cfm?Lab=NHSRC&subject=Homeland%20Security%20Research&dirEntryId=218488 (accessed 03/01/2023).

Skeat WO and Dangerfield BJ (eds) (1969) *Manual of British Water Engineering Practice*. Heffer Publishers, Cambridge, UK.

Swamee PK and Jain AK (1976) Explicit equations for pipe-flow problems. *Journal of the Hydraulics Division, ASCE* **102(5)**: 657–664.

Thorley ADR (2004) *Fluid Transients in Pipelines*, 2nd edn. Professional Engineering Publishing, MI, USA.

Thorley ARD and Enever KJ (1979) *Control and suppression of pressure surges in pipelines and tunnels*, CIRIA Report 84. CIRIA, London, UK.

Todini E and Pilati S (1988) A gradient algorithm for the analysis of pipe networks. In *Computer Applications in Water Supply*. Research Studies Press, Letchworth, Hertfordshire, UK.

Todini E (2003) A more realistic approach to the 'extended period simulation' of water distribution networks. In *Advances in water supply management*, Maksimovic C, Butler D and Memon FA (eds). Balkema, the Netherlands, pp. 173–184.

Vasconcelos JJ, Boulos PF, Grayman WM *et al.* (1996) *Characterization and modeling of chlorine decay in distribution systems.* AWWA Research Foundation, CO, USA.

Wagner JM, Shamir U and Marks DH (1988) Water distribution reliability: Simulation methods. *Journal of Water Resources Planning and Management, ASCE* **114(3)**: 276–294.

Walski T, Chase DV, Savic DA *et al.* (2003) *Advanced Water Distribution Modeling and Management.* Haestad Methods Press, Waterbury, CT, USA, p.751.

WRc/WSA/WCA Engineering and Operations Committee (1994) *Managing Leakage: UK Water Industry Managing Leakage Reports A–J.* London, UK.

Wylie EB and Streeter VL (1978) *Fluid Transients.* McGraw Hill, NY, USA.

Dragan A. Savić and John K. Banyard
ISBN 978-1-83549-847-7
https://doi.org/10.1108/978-1-83549-846-020242004

Chapter 3
Water demand: estimation, forecasting and management

Adrian McDonald
Professor Emeritus, University of Leeds, UK

Peter Boden
Edge Analytics Ltd, Leeds, UK

Chris Lambert
Thames Water, Reading, UK

3.1.   Introduction and context

In the first edition of this book, we concluded that the '...demand planner is therefore catering for the water needs of age cohorts not yet born, in a climate not yet experienced, and within government policies not yet thought of.' As if these constraints were not demanding enough, we would now wish to add the impacts of pandemics influencing birth rates and death rates, and dramatic changes in energy prices influencing industry planning and priorities, reinforcing the industry message to consumers that reduced water demand saves costly energy.

Countries differ in their approaches to water demand management. The differences arise from variations in local circumstances, such as the nature of the water resource, climatic challenges and predicted demand. We cannot seek to cover the global range of circumstances and our focus will be on the UK, primarily, with some comment on challenges and advances elsewhere.

Water remains fundamental to life on earth. We cannot address the food security, health needs, economic well-being, education provision and governance of people unless they first have water services. The world is rapidly urbanising and so part of the water issue is to provide resources for large, dense communities who do not (perhaps cannot) have a sufficient local supply. Thus, an estimate of demand is required so that supply can be sourced and provision made to transfer and distribute the water resource. But let us be clear, there are some rapidly expanding cities with large areas of informal housing that will not be served with water at house level. Even a local tap, free of exploitation, may be the realistic goal over the medium term.

We need to balance supply and demand. This is a reasonably straightforward task for short periods and modest areas. However, for large conurbations, the scale of works needed and the needs for inquiry and local governance result in a very long lead time. So, demand is always estimated for 25 years ahead, this being the statutory minimum in the UK (see Water Resources Planning Guidelines updated April 2023) but for major resource schemes, a forecast as much as 100 years ahead is necessary.

As the challenge of balancing supply and demand increases, and as different companies might be abstracting from different sectors of a water resource system (be that different river sections or groundwater linked to rivers) the need arises to be, at least, aware of the future plans of neighbouring water companies. France has a long history of regional water planning and economic planning based on large water regions, the Bassin Financiere. The USA and Canada tend to have discrete, fragmented, independent water supply companies operating either from a Harmon doctrine or a riparian owners' rights basis in many regions. Water supply management needs to be overlaid on a complex regional management of water for power and irrigation (examples include the Columbia River, Kooteney River, Skagit, Pend d'Orielle, Colorado, as well as the resurrection of interest in the transfer of the waters of the Mackenzie). Such combative, even anarchic operations are in marked contrast to the UK where the state, through regulatory authorities such as the Environment Agency (EA), grants a licence to abstract.

In the UK, water resource planning has evolved rapidly and significantly since 2019, with the advent of the National Framework for Water Resources (EA, 2020) and the requirement for regional water resources plans, which, although not yet statutory, are a requirement for water companies to follow in producing their statutory plans. The regional plans are expected to be cascaded into the companies' statutory plans, so now water planners are producing demand forecasts at a regional level, and these are subsequently disaggregated to individual water companies' areas. The regional plan is both driven and constrained by the Water Resources Planning Guideline (2023) and so is central to the process both at regional and company level.

It is possible that, in arriving at future water balances, we will be required to recognise further water demand components. For example, every community and every activity has a water footprint – that is, the direct use of water plus an amount of water required to provide other input goods and services. In the UK, we each use about 140.9 litres of water per day, but it is claimed that the goods and services we require have a hidden water requirement of 3400 litres per person per day making the UK the 6th largest water 'importer' in the world (Orr and Chapagain, 2009). There may be pressure to recognise and account for these virtual flows in future (Chapagain and Hoekstra, 2008). In addition, the natural environment has a right to water to maintain landscapes, habitats and species. The rights of the river environment are effectively enshrined in the European Union Water Framework Directive (WFD) which requires the achievement and maintenance of good ecological status (or its *potential* for heavily modified water bodies) for most river reaches. Attaining this, especially in the timeframes currently envisaged, is ambitious and many EU countries will fall back on the derogations and cost exemptions in the directive. The UK government has asserted that the WFD will be retained post Brexit but that it might be greatly constrained in the changed economic picture of debt from pandemic expenditure, increased defence expenditure, the failure to realise the 'tidal wave of inward investment' promised in Brexit, inflation and soaring energy prices.

So, the domestic water supply competes for raw water resources with agriculture, the river and industry. Of course, the UK abstracts only 1% of rainfall and 2% of runoff, so, nationally, it appears that there is an ample water balance, but for some seasons and years, and in some high demand regions such as the south-east of England, this is a very misleading picture! In the future, with climate change probably increasing supply variability (UKCP, 2018) and encouraging an upward shift in demand, it may not be possible to balance all the demands with the supply and the exercise might become one of prioritisation of the allocation of scarce resources. Under current demand structures, compliance, at some sites, under some conditions, may not be achievable. This is an issue only recently being recognised in the UK and Europe as evidenced by the growing requirement in the UK to consider company mutual support in asset plans; a process now formalised, as outlined earlier and in Europe, to consider, for example, water transfers between Occetanie and Barcelona.

Water is fundamental to life, as we have seen, but it is also a fundamental characteristic of the business system for a water service provider. The majority of water service providers source their supplies from within the company boundary or from historically defined external sources. The demand for water in a region and adjacent regions will partially govern the import/export potential, the investment needed to meet demand, and the actions and initiatives needed to control demand and meet efficiency targets, to ensure compliance with standards of service and to meet customer expectations, even in the highest demand periods of a prolonged drought.

Assessing and forecasting how supply and demand will be maintained in balance, given future uncertainties, is a core element in the UK of the economic regulator's (Ofwat) price reviews, particularly the asset management planning process. These are plans drawn up every 5 years by every water service provider. This process has as its foundations a strategic direction statement and a water resource management plan – in essence, the vision and the practical proposals. These are subject to consultation, revision and, possibly, inquiry before a final accepted plan for the next five years is agreed. Since the plan is one influence on the water price for each company and contains firm commitments, it is one of the financial foundations for the water service provider for the ensuing period; the others being the cost of servicing debt and the regulator's assessment of the operational efficiency and delivery of the previous 5 year plan.

Demand is not a simple measure, nor is it static. Figure 3.1 indicates the scale of the components that require to be measured, derived or estimated if a thorough and robust understanding of the water components is to be attained. This is the analysis process map as required by OFWAT for the asset management plan reporting. Starting at the catchments, be they ground or surface waters, the company determines the collected raw waters as the balance of abstracted raw water +/− imports and exports. Only a proportion of the raw water collected is transferred to the water treatment works as some is lost or used as raw water. Some of the water is used or lost at the treatment plant but the majority (with any imported potable water) provides the available potable water stock. This amount less any exported water is the distribution input. Some is lost in distribution (the leakage usually cited in discussions, but note there are several other losses in the overall process) and the remainder is delivered to 'customers' and to distribution system operational use.

Not all customers pay for water so the final distribution is characterised as unbilled or billed to different customers (all of whom also leak some of the water that they have received). Even this detailed water accounting is, however, a summary as we shall see later in this chapter. The IWA Water Loss Task Force (Lambert, 2003) examines the water balance post distribution input only but does so at a slightly finer scale by accounting for meter inaccuracies.

Demand can be measured as aggregate total volume of demand over the long term, as discussed, or it can be measured as peak demand. This peak demand is not so instantaneously expressed as is the case in, say, the electricity industry as there is a 'storage' in the distribution network that does not exist to quite the same extent in some other utilities. The long-term demand for water, whether measured per capita or per household, is a significant component of the overall supply required. So this 'aggregate' demand defines the adequacy of the river flow resource, the storage available, the abstraction licensing agreements and so on. Peak demand on the other hand is more strongly related to short term peaks in the drivers be that temperature, perceived drought, school holidays

Figure 3.1 Water balances and components from raw sources to delivered water (EA 2007)

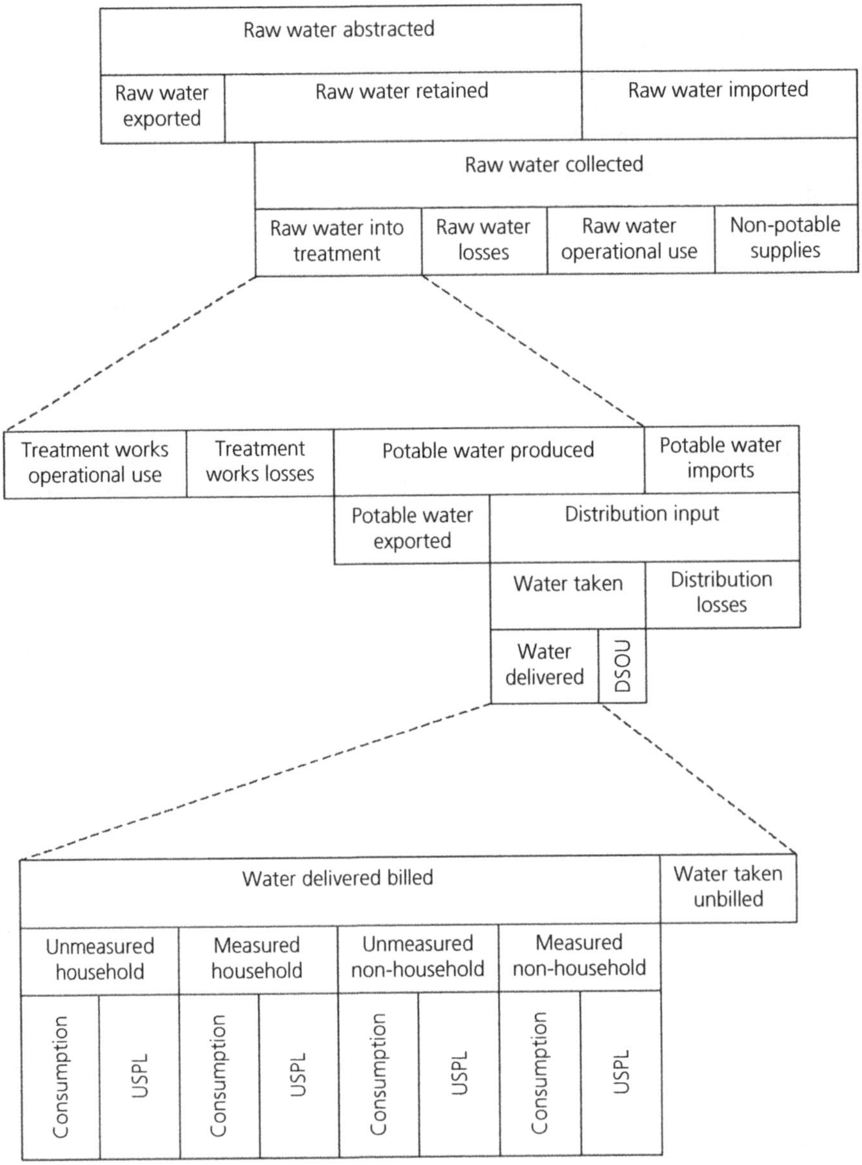

DSOU – distribution system operational use
USPL – underground supply pipe losses

as well as the capacity of the infrastructure to treat, pump and transmit the peak flows demanded. Thus, systems (and so companies or resource zones and, in some cases, whole countries) can either be resource or asset constrained.

Peak demand is determined by every company as part of the planning cycle discussed earlier. Most companies will refer to the methodologies discussed in the UKWIR (2006) report. That report reveals a startling variety of measures and methods used by water companies. The report also observes that the lack of normality in the data is challenging (a point to be recalled in the discussion of regression-based forecasting later in this chapter), and that the data is not all ordinal and seems to equate explained variance with the correlation. In the changing public expectation and policy environment surrounding water supply, an update to the 2006 report is probably necessary. That report noted that some demand component peaks do not automatically coincide. Peak in environmental demand, non-potable demand and embedded demand are not all part of the peak demand determination, although some components are indirectly related to potable peak – for example, environmental requirements as a response to peak low flow may cause a prohibition in abstraction impacting on the amelioration of peak potable demand.

3.2. Variations in water demand

Customer demand for water varies over time. There is a marked diurnal variation with demand falling off significantly overnight. There is both a morning (sharp) and an evening (broader) peak in domestic demand, which is more obvious during the working week and less so at weekends, public holidays and vacation periods. Over the year, domestic demand increases in the growing season and in warmer weather (although only very modestly), but demand on resources can peak in winter as the reduced domestic demand is masked by the increases in leaks (and in the difficulty of addressing leaks when ground is frozen.)

Almost half of the water abstracted from England and Wales is used for public water supply: 52% is for household use, 23% is for non-household use and 23% is due to leakage. The remaining 2% is for other uses such as firefighting, operational and illegal water use (Defra, 2008). Defra (2019) statistics show unchanged abstraction for potable supply but the previous decade shows a major increase (26%) in abstraction overall largely due significant increases in electricity generation.

Demand for potable water varies between countries. The Organisation for Economic Co-operation and Development (OECD) countries, for example, consume three times more water per person than persons in Latin America, East Asia, Africa or India. Households in the most developed countries use only a small proportion of the water abstracted (around 8%) and, in continental Europe, the dominant raw water users are industry and agriculture. There is limited useful UK data on raw consumption as there are big differences between licensed and actual abstractions. Households are the largest consumers of public potable water supplies but, as we noted above, the water consumed by households is a small component of overall river flow. Thus, *generalised* arguments that domestic water saving is required to address lowered river flows arising from climate change are misleading. Instead, we need to concentrate on the water balance interventions needed in specific areas at times of water stress. Saving water is much less appropriate in (wetter) Northumberland than in (drier) Kent, for example.

Per capita consumption varies, even in the developed world, from 100 to 300 litres per person per day. The US and Canada lead this table while the Netherlands and Germany are lower users. Of this water use, only 5% is for cooking and drinking (European Environment Agency, 2001). Making comparisons between the apparent water demand in 'similar' countries can be very misleading. As an example, German water demand at 110 litres per person per day has been held up as evidence that the UK should be able to do the same. German housing stock is c.70% flats, 30% other. The UK is the reverse with c.30% flats and c.70% other. Flat dwellers are unlikely to be washing cars and watering non-existent gardens. There are also differences in housing stock ages, building standards and effective communal facilities which also impact on water demand and explain the error in the initial simplistic conclusions.

3.3. Components of demand

The water balance attempts to balance supply to a number of demand components, the majority of which are unquantified independently. Each is estimated separately and there are few independent measures of error and uncertainty. The main demand components are

(a) household
(b) commercial
(c) industrial
(d) operational
(e) unbilled
(f) illegal
(g) leakage.

Domestic water use by households can, in the UK, be either measured (metered) or unmeasured. In most developed countries all potable water is metered. One might assume high levels of accuracy and utility for measured consumption. However, the water service provider often does not know immediately how much water is used because data is typically collected for billing purposes and many providers still have to take a *sample* from their billing computers to gain an estimate of measured demand. This is changing as meters become smart. The measurement of demand for unmetered users is more complex yet and is discussed in the next section. Unmetered users are no longer the majority in the UK. The EA reports, in March 2017, that about half of all customers in England were now metered but this varies significantly between companies, driven by business philosophy and resources stress. Some companies have permission to compulsorily meter (in water stressed areas such as South East England) all their customers but, in the main, the growth in metering was through customers opting to have a meter or moving into a new house with an existing meter. New houses are now always metered. While metering was optional, metered customers had a lower demand, but ascribing this to the impact of meters modifying behaviour is a false logic, as opting for meters has created a strong bias towards lower users joining and higher users avoiding metering. Smart metering however may well change this picture as the immediate, regularly updated information on water use may influence behaviour. Certainly the information from a growing number of smart meters contains extremely valuable information, giving a more solid foundation on which to build the water balance and informing water managers of the characteristics of the different water users – that is, who and where (and perhaps why).

Domestic demand, measured over the longer term, has been rising but in recent years there is a suggestion of a demand decline. This small signal currently towards a decline in demand may well be a managed decline driven by the water service providers who are, themselves, being encouraged by regulators to promote demand reductions by advice and subsidised devices such as shower timers and water butts. Clearly there is a growing number of lower water use appliances particularly in new homes and as a result, even without behavioural change, there will be a trend to a lowered demand.

Commercial demand tends to be stable, showing only modest short-term fluctuations. Over the longer term it is declining, but the real reduction in water demand has been in raw water and not potable water commercial demand (it can be argued, of course, that this is not a true commercial demand reduction but a commercial water demand transfer to, for example, China). In practical terms, all commercial demand is measured. Commercial demand itself can be subdivided into service and non-service components and for the needs of specific companies into the Standard

Industrial Classification classes. If refined to this level, the discussion starts to focus on industrial demand. In some industries, water demand is for raw waters and, as we saw, has increased significantly, but even in these cases there remains a residual of potable water needed for the workforce. In effect there is a 'domestic' demand component within the commercial and industrial water use. In some food processing industries, say a bottled drinks manufacturer, the demand will be entirely for potable waters. So despite the significant decline in manufacturing output, the demand for potable water in this sector has not reduced to the same extent because the heavier manufacturing industry used raw waters (sometimes partially treated at their own plants). For some water service providers, the industrial water demand is dominated by a small number of very large companies (for example, the Royal Navy dockyards with a small number of large bases, food processing and freezing plants, and major car plants). Here the water service providers tend to account for these as separate lines in the water accounting and to determine the impact of their closure on the water and financial balances. Some very large water users exercise extremely variable water demands – for example, processing the pea harvest for which the water provider will be alerted to the onset of this processing.

Some water is used by the water service providers themselves as part of the operation of the water treatment and distribution system. Typically the water is used *(a)* to flush mains when they are being replaced or repaired, *(b)* as a pump lubricant in water treatment works and *(c)* as a normal potable supply in the company offices. There appear to be differences in the practices between companies as to where that third component should appear in the water balance accounting. Principally, these relate to whether it should be billed internally or remain unbilled, and this choice is governed in part by whether the water is taken pre the works output meter or is taken post the output meter and has entered the distribution system. If not accounted for, this latter water use might well be interpreted as leakage, a value that every company would not wish to see increase.

There are two main components of the unbilled water, that which is legally unbilled and that which is illegally taken unbilled. Legally unbilled is, for example, the use of water for fire-fighting, while illegally taken water is stolen by standpipe in the case of some building sites or by tapping into a neighbours supply pipe (Lambert, 2003). There is a grey area between these two extremes that relates to houses that use water but for which the water service provider has no occupant details. Known in the industry as void properties, a significant minority are not actually void of people but void of the details of the occupants. These voids tend to be concentrated in particular house types, ages and postcodes and are typically new builds, rented flats and student accommodation. Vacant and void properties are widely distributed through the water regions of England and Wales (Table 3.1). It is possible to take the view that it is up to the company to find its customers and until they do so the water is taken unbilled but legally. However it is clear that many customers deliberately withhold details and act to avoid payment by moving without forwarding details. If the owner rather than the tenant could be billed this may ease this situation. Water companies tend to have much less information on their customers than do the other utilities simply because they have to supply water as a right and have no contract, whereas an energy company has a specific contract and will seek considerable customer information before agreeing a tariff and contract. It follows, then, that the drive to expand the proportion of customers that are metered has the added benefit of establishing a contract and thus the details of the customer.

Finally, there is the question of that demand component called leakage (discussed in detail in Chapter 9 of this book). Determining how much leakage occurs in the system is usually approached by water balance accounting, by local departures from expected flows or by public reporting.

Table 3.1 Void residential properties as a proportion of billed residential customers for selected water companies. Derived from water industry database for last complete year 2020–2021

Region	Billed	Void	%
Affinity	532426	34528	6.5
Northumbrian Water	929310	71413	7.7
Severn Trent	1716849	177184	10.3
Thames Water	1872741	135871	7.3
United Utilities	1628602	191641	11.8
Welsh Water	700866	48175	6.9
Yorkshire Water	890310	103920	11.7

Not all leaks will be detected, particularly those leaks occurring within the curtilage of a property. Even if a leak is determined, say, from an unexpectedly moist area of garden, a householder might choose not to investigate further. Such a laissez-faire approach might have been acceptable in the past but not today. If water resources are strongly and consistently in surplus, a private company might choose to determine which leaks are to be repaired using so-called 'economic level of leakage', which is the point at which it costs more to repair leaks than to source new water. This was an approach required by the UK water regulator in the first decade of the twenty-first century (OFWAT, 2002). That is no longer the case. For reasons of resource scarcity, climate change adaptation and public concern, new policies, which embrace new targets, now apply. These new policies and targets are presented in the NFW that water companies are expected to follow. These policy requirements will inevitably strongly influence the production of the overall demand forecast – for example, reducing leakage by 50% by 2050, household per capita consumption (PCC) of 110 litres per head per day and so on. The PCC expectation is included as a scenario in the NFW and, as far as water companies are concerned, would require government support (effectively the provision of tools) if it were to be achieved – for example, mandatory labelling of water efficient white goods, changes in building regulations, removal from the market of all water inefficient devices (compare with incandescent lamps in energy conservation) and so on.

Achieving such ambitious targets will involve significant, additional costs and risks. Ultimately it is the companies who bear the risk when it comes to planning for future water requirements, and this risk is an important aspect that will inevitably influence an individual company's forecasts. For companies who operate in a benign supply situation, these targets can be striven for in the knowledge that resources are available should the targets prove more problematic to attain. There are few companies in the UK that could be considered to operate in such a benign situation. Most companies that built these large-scale 'savings' of leakage and demand into the water balance in the near term – that is, before realising any savings – are taking a very significant risk for both the public and the company.

What is required, then, is a strong, informed debate. Essentially, and at a minimum, for each company the critical bottlenecks in the supply enhancement pathway will need to be determined, their development time identified, and this safety, time-based margin built into the planning framework. Internal grid networks, where they exist, need to be evaluated for flexibility (or otherwise) to be adapted to meet supply needs across the networks. Costs are likely to be substantial given that the

NFW requires security of supply, such that severe water restrictions and standpipes would only occur once in 500 years on average. Treasury subsequently endorsed the need to achieve 1 in 500 year drought protection by 2040 and the requirement for 50% reduction in leakage.

Since the UK has few very-long-term records of water use and water availability, and the variables that drive these characteristics and, further, that we are seeking to 'evaluate' across known climate periodicities that may be modified by anthropogenic influences, we have to accept greatly increased uncertainties and increased planned and contingency costs. The UK faces an affordability crisis, partly home grown (e.g. housing crisis, loss of markets) and partly global (e.g. supply chain failures, energy pricing, seeking net zero, renewable and non-renewable resource scarcity) and costly water ambitions cannot exist in isolation. It might be anticipated then, that water debt, already very large and expanding, will mushroom in the face of other demands on the customers' budget. Further analysis of water debt and company performance can be found in Clarke, McDonald and Boden (2012).

We called for improvement, in the first edition of this book, in the management of leakage and demand, arguing that further 'improvements (from economic levels of leakage) are surely needed... (and) methods...need to be dynamic reflecting changes in resource availability or competition or prioritisation of abstraction or change in abstraction licensing'. The economic level of leakage concept no longer applies. Water resource policy guidelines now require leakage targets to be policy based and expect a 50% reduction in leakage from a 2017/18 baseline and informed by stakeholder and customer expectations. It will be no surprise then if stakeholder expectations include achieving leakage reduction targets at little additional cost. This is an unrealistic expectation because achieving such an ambitious target will entail significant expenditure linked to mains replacement and inevitably involve additional costs for financially challenged stakeholders.

Water balances are vital but problematic, as many of their elements are estimated. Leakage is one such element. The advance of smart metering, particularly by Thames Water which is rolling out smart metering (the progressive metering programme) in London and its Thames Valley area, has revealed the increasingly important role of smart meters in terms of understanding and influencing water use. The company now has detailed daily data from thousands of properties, analysis of which has revealed that wastage (water loss in the home) is much higher than previously thought and, hence, leakage (water loss in the network) is lowered, but further that the households are reducing consumption as an information and behavioural response rather than a punitive tariff response (as no punitive tariff has been applied). Demand reductions, following comprehensive, supported, smart metering, of 13% are credible and the data on which these figures are based have been thoroughly vetted by the regulating agencies. The smart meter data has also shown that there is a very long tail of high water users with a household consumption significantly above the median value. Understanding what is driving the usage in such households will be key in developing targeted water efficiency campaigns. A further benefit of the smart meter data has been to facilitate much more effective targeting of leakage control activity linked to the enhanced understanding of household water consumption.

3.4.   Drivers of demand

What drives water demand? The key influencing factor affecting domestic demand is population change (Defra, 2008; Environment Agency, 2009; Sim *et al.*, 2007). Westcott (2004) has identified several further drivers, namely: water policy (e.g. metering and water regulations), technology (white goods and other water saving devices), behaviour (attitudes to water consumption) and economics (e.g. personal affluence). Studies from around the world have identified similar factors (Jeffrey and Geary, 2006).

Domestic demand breaks down into two key components – that associated with the house and that associated with the occupants; the former is near constant while the latter changes in proportion to the number of occupants. Thus a five person household uses much more water than, say, a two person household but, measured as per capita consumption, the five person household will have much lower individual demand (McDonald *et al*, 2003; Memon and Butler, 2006; Sim *et al*. 2007). These studies note, in addition, the role of the ages of the household residents and the time of year. Therefore, given the demographer's forecast of significantly smaller household size in the UK by 2025, even if total population does not change, water demand will increase. Given that there is expected to be a significant increase in population as well (DCLG, 2007), there are then two separate population influences likely to drive up demand. Against these upwards pressures must now be laid the influence of the Covid-19 pandemic which has lowered birth rates in most countries, early headlines to the contrary notwithstanding. Some countries, the Netherlands for example, have not followed this trend. The question for the demographer is how long will this trend exist.

Herrington (1996), Downing *et al*. (2003), Roaf (2006) and others have also noted and attempted to quantify the potential effect of climate change, with climate forecast to become more variable with drier summers now forecast as somewhat wetter and warmer, and wetter but warmer winters with the extent of this disparity related to regions. Thus we can expect more extremes in water demand with lower winter demand (less exacerbated by frost related leakage increases) and summer demand increases. To achieve a continued resource balance in the face of this volatility may well require additional storage.

A key question remains, namely the reliability of the prediction of increased winter rainfalls. We have, in the last 30 years, seen a greater frequency of the sequential winter droughts. These place considerable stress on the water resource system. We have not yet determined whether a three-dry-winter drought is a realistic and quantifiable prospect and, if so, whether the system can cope with this. We have not fully defined the spatial extent of prolonged, sequential droughts and so our regional support resilience is not yet secure.

3.5. Estimating current demand

While half of UK households remain unmetered, the need to make estimates of demand continues. Two approaches to the estimation of water demand in unmeasured households are employed – either the use of domestic consumption monitors (DCM) or the use of area meters (AM). The choice of approach appears unlinked to location and customer attributes, and appears to relate more to the historic approach developed by the particular water company, with some companies employing both approaches.

A DCM is a sample of 2000 households (in practice, smaller than this) selected at random, to be offered a place in the DCM, which, should they opt to join the survey, are then metered but continue to pay on a fixed tariff related to the rateable value of the property. DCM populations are therefore self-selected and subject to a range of biases (financial advantage, staff inclusion, self-selection, Hawthorne – see McDonald *et al*., 2003). Most of these biases result in under-estimation of demand. In addition, DCMs decay through births, deaths, migrations and opt-out, and so DCM maintenance is a complex and seldom wholly successful operation, particularly if the companies are required (by the regulator) to report outcomes in per person consumptions rather than per household consumption values. The population of households (billing addresses) is known with more certainty than the populations within households and resurveying repeatedly to determine household numbers in the DCM adds to some aspects of bias. Detailed information about occupants is also increasingly difficult to collect as people become more wary of disclosing personal information.

Calculation of demand from DCM data can offer misleading alternatives, as illustrated in the very simplified example in Table 3.2 where data for 10 people in 4 households is presented. The two demand estimates shown in the lower right box are significantly different. The first is determined from the column totals of water use and population, and the second derived from the average demand of each household size. The former is, of course, the correct approach, although mathematicians might argue both are correct but require careful definition. Care will need to be exercised to ensure that the coding of the analysis of the smart meter data does not embed any flaws in the calculation.

This absolute value of 2000 for the DCM sample was the outcome of an industry-wide study. Normally, to arrive at a sample size, you need to know the variability in the population and the accuracy required of the estimate. Since the former will certainly vary between regions, the size requirement must vary by region and thus between companies.

So all DCMs are inherently biased. A good DCM will recognise the possibility of bias and address it in three ways.

(a) Removing a source of bias (if that is an option) – thus, a DCM that has had water company staff encouraged into the membership to get the process started will remove members (but not all members, as there should be equality of representation of the water employees).
(b) Correcting a bias by adapting recruitment – for example, if a local authority has refused to allow its tenants to take part, it is necessary to recruit only from social housing to replace these potential members. If the managers simply return to the list of customers and draw further potential members at random then council rented groups will be under represented.
(c) Analytical correction may be required if a specific group is underrepresented and particularly if they appear to be higher water users – for example, Asian households. Here you would need to correct for the under-representation. In practice, many corrections will be required to deal with, for example, under-representation of specific house types (e.g. flats) or social groups (e.g. ethnic minority communities). It is both a complex and sensitive task.

Table 3.2 Simplified output from a 4-household DCM

Household Consumption (litre/days)	Number of Residents	Per Capita Consumption (litres/person/day)
400	4	100
150	1	150
280	2	140
360	3	110
1190	**10**	**119/127.5**

Area meter based determination of the household demand is more complex and less direct. An area meter gives the water use in a relatively large area. It is, therefore, the sum of many demands, few of which are known. Initially, the nightline is determined. The nightline is the demand in the early hours of the morning and assumes that household use is at a minimum. From the nightline, an allowance for legitimate night use is removed, as is night metered usage and night commercial and industrial use (also metered); the residual is leakage. From the average demand on the area meter, commercial and metered demand is subtracted, as is the now 'known' leakage. Further allowances for operational use, theft and the lowered leakage under the lower pressure, daytime regime may also be entered, but these are all estimates. The result is average demand exercised by the unmetered population.

There is obviously considerable potential for error in this process. Meters typically under-register but the under-registration varies by meter type, size and age, none of which tend to be corrected in an area specific manner. What constitutes legitimate night-time use is a generalised figure and one that changed following a water industry investigation which included the beneficial effect of reducing leakage. Household numbers in an area will be derived from the census but these numbers will be several years out of date. As a result of such issues, there is an understandable tendency to seek simple uncomplicated sites, but these tend to be 'cul de sacs' and urban edge sites that host a different and probably unrepresentative population, and so, as with DCMs, the challenge is to understand how the results relate to the overall customer base. A fundamental limitation of area-derived demand is that the sample size is the number of meters not the number of properties/ customers. Further, variability in demand is not identified, as the measure is of an aggregate sample. It is, however, possible to derive a demand by property type by equating the unknown property demand for each type of residence with the number of such properties within the meter zone and solving that set of simultaneous equations. This may have some utility in supporting regression approaches to forecasting demand.

The use of the term 'unmetered population' perhaps implies a precision that should now be questioned. Populations are derived either *(a)* from census material released through the UK National Statistics Office (NSO) or through commercial firms that pass on this information, somewhat modified, or *(b)* from company billing records for the households in the area under scrutiny. In applying both DCM and AM derived water demands there are issues related to the population figures for the supply area as a whole – indeed, the NSO recognises that there are inaccuracies in their summary figures; figures with which water service providers often try to reconcile their own population components. More fundamentally, however, is the lack of recognition and accounting for hidden and transient populations, which exist in a region, use water, but which are unlikely to be recognised in the water balance and population accounting. Hidden populations are those that do not appear in the census records. They comprise a range of categories from visa over-stayers through asylum seekers who are rejected but not returned to true clandestines. They are a category of population that are typically poorly quantified by governments.

Transient populations are those that live and are recorded in one region but who work in another but are not in a metered establishment and who are not captured by normal accounting (McDonald *et al.*, 2008b). Table 3.3 below provides an anonymised example of part of the categorisations used. In a full analysis there are 14 categories of hidden and transient populations to be considered. Here we simply give some examples, limited by commercial confidentiality and by the sensitivity of the material in some regions. Each category requires analysis. Take, for example, the student population. The total number in higher education in a region is a matter of record. Assume all students

Table 3.3 Examples of the tabulations of hidden and transient populations in a water company area. The totals given here are real but the components are incomplete. Figures are populations in '000

Category	ONS Capture	Legal	Low	Medium	High
Transients					
Students	Yes	Yes	0.8	5.1	9.3
Second Home Householders	Partly	Yes	5.1	9.0	13.0
Short-term Migrants	No	Yes	1.1	2.8	4.5
Hidden					
Clandestine	No	No	8.5	11.8	15.7
Human traffic	No	No	0.1	0.8	1.5
Totals			**50.5**	**64.9**	**80.5**

with a home post code within 50 km of the university will commute and reduce the student number accordingly. Assume all halls of residence and university flats are metered and so their water use is captured. A figure can then be determined of the number of students who do not live at home or in student identifiable halls/flats. To identify the water use of these groups – who will be part-year occupants but in higher household numbers than expected for the size of the property – requires further assumptions (e.g. that they will not opt to be DCM members and that they will have a lower water use). There is a danger of double counting in this process as students are also a major component of the void property problem discussed earlier.

3.6.   Forecasting demand
3.6.1   Commercial
There are two approaches to forecasting commercial demand: *(a)* the empirical statistical and *(b)* the process deterministic. The statistical approach uses the empirical data of the commercial water demand history and plots a trend through the data. Although a simple and easily explained approach, it suffers from the assumption that the drivers over the period of the data will be continued into the future. The process models look either at the inputs (employment) or outputs (gross product volume or value) and establish a relationship between numbers employed, for example, and water demand. Such models tend to work better if there is some degree of disaggregation. Mitchell (1999) developed an alternative approach to forecasting; a more rigorously tested and refined process model which explored a range of possible drivers of commercial demand (including employment and price) and soil moisture deficit (as an index of drought). The model incorporated future water saving activity designed to minimise waste. This component of the model differentiated between large and small firms, differences in rate of uptake and differences in potential saving between industrial sectors.

All of the industrial forecast models have to estimate the significance of business structural changes. This incorporates the changes in the predominance of some industrial types and the changes in the processes involved that use water. Businesses, more than households, are sensitive to the costs of the water used. A particular difficulty arises when demand changes arise from factory closures driven by international and, sometimes, national policies. Thus a water service provider with a large water demand associated with a military establishment in their area will need to pay close attention to defence reviews and possible rationalisations as these could affect its planning. Closures at, for example, car factories in Swindon, Derby or Teeside would similarly create a sudden and largely unforecastable change in water demand.

3.6.2 Domestic

There appear to be many different approaches to forecasting domestic water demand based on demographics, behaviour, micro-components, microsimulation of demand, multivariate analysis and amalgamations of these approaches (see, for example, Memon and Butler, 2006; Clarke *et al*, 1997, Williamson *et al.*, 1996, 2002; Bellfield, 2001; and Jin, 2009). In reality, these are all accounting procedures. In this section we focus primarily on the accounting procedures and on regression-based forecasting.

The basic account for a long forecast takes two key measures: population and water use. To this are added less significant drivers (household size, house type, religion, affluence, conservation commitment and so on). The drivers can be forecast from past trends as in regression, as a panel/ expert judgement or as the policy ambitions of the regulator. Micro-components – for example, frequency and duration of showering – give a mechanistic interpretation of water-using activities but forecasting the technology and behavioural environment some decades hence is challenging. Microsimulation can provide synthetic small area water-using characteristics but the complex mathematical procedures required makes validation challenging. All these approaches may or may not choose to incorporate the effects of tariffs, smart meters and policy information.

The key determinant of water demand in a region is the population. Population forecasts are based on the fecundity of the age cohorts that comprise the population, modified by the balance of inward and outward migration. More sophisticated population models will incorporate fertility differences between different social groups. Thereafter, for water forecasts, it is necessary to estimate the household size class distribution that the population will form. However, because water service providers are encouraged (no longer required) to report per capita consumption for both measured and unmeasured populations, the companies need a measure of occupancy in both measured and unmeasured properties. This is data that is not available through the census and is highly dynamic, being influenced by breakups, deaths, births, migration and so on at a household level. Arriving at robust figures for these two occupancies is one of the most challenging measures facing demand planners. This unnecessary complexity, PCC forecasts, appears to be becoming less important.

Total water demand is accounted most commonly from either household size or house type stratification. In the example in Table 3.4, for one resource zone of a UK water company we use household size for stratification. The advantage is that the DCM no longer needs to mimic the population as a whole. Instead the DCM is simply a sample of 1, 2, 3 and so on person households and the household demand is calculated for each separately.

Usefully, Twort *et al.* (2000) have and lifestyle of the household. While the census examines use. Typical in-house household uses are personal hygiene, drinking and food preparation, and toilet use, while examples of out-of-house use are garden watering and car washing. It is useful because demand management interventions typically focus 'in' or 'out' of house and these two macro-components are greatly influenced by house type; a driver that can be identified externally and which features in planning scenarios and permissions and can, therefore, contribute to forecasting.

At the heart of effective forecasting is population attributes. Let us assume that the gross population forecasts are accurate. The most important characteristics remaining are the household size and

Table 3.4 Demand forecasting using a stratified household size approach

Household Size	Number in WSP Area	Domestic Consumption Monitor demand l/day/hh	Total Demand m³/day	Total Population	Demand l/ Person/ Day
1	545196	221	120615	545196	221
2	636634	322	205109	1273268	161
3	290886	392	114163	872658	131
4	247324	465	115121	989296	116
5+	126257	538	67886	631285	108
Total	**1846297**		**622895**	**4311703**	**144**

the affluence and lifestyle of the household. While the census examines many of these attributes, confidentiality requires that the data be aggregated into large spatial bundles. Microsimulation, more properly micro-analytic simulation, 'unpicks' these bundles and creates small area units of hypothesised household characteristics – a synthetic micro-population. This allows us to use known water demand relationships to forecast and assess demand at a smaller area than would otherwise be possible but it remains an accounting process, albeit at that smaller area. While this is a complex and computationally intensive procedure, it is well documented and of considerable provenance being used in both the UK and US as a basis for economic estimation by Treasury (Roe and Rendell, 2009). But how much simpler (and richer in confidence) would be a system based on 100% domestic water metering.

Regression appears to have become much more popular in current water demand forecasting. Regression might be viewed as a form of accounting in that it uses most of the drivers discussed above but expresses the final water use through an equation. In general a regression consists of a dependent variable and several independent variables – thus the term multiple regressions. The coefficients for each independent variable are derived from a statistical analysis of a block of existing data but then, just as in accounting, the user needs to make a forecast of the driver values in the future. The user also must strive to incorporate all the necessary drivers in the dataset used to formulate the regression equation. Typically, the dependent variable will be water demand at some scale and granularity, and independent variables will relate to population, household characteristics and population make up.

As with any statistical analysis there are a number of assumptions and data characteristics that need to be correct for the outcome to be valid. Here we consider six assumptions/characteristics already examined in the forecasting of industrial water demand (Mitchell and McDonald, 2000).

(a) The dependent variable should be measured on an interval or ratio scale (there are four data types: nominal and ordinal (called categorical), and interval and ratio (called continuous)). If the measure has been water demand of a household, measured in volume per unit time, this would be valid. It seems unlikely but if the developer has been using, say, high, medium, low water users, or some similar Likert scale, then this is an ordinal measure and requires ordinal regression.
(b) It is assumed that there will be two or more independent variables, otherwise it is not a multiple regression. Most analysis software will expect continuous independent variables. If categorical variables are employed, say for house type, then they should **not** be coded 1, 2, 3 and so on,

following a perceived water use ordering but be declared as nominal. It might be appropriate to use dummy variables but these would be dichotomous and some further analysis is required before applying regression analysis.

(c) The observations that make up the training data set should be independent and should be checked for independence of residuals.

(d) A linear relationship must exist between the dependent variable and each of the independent variables. If a linear relation cannot be demonstrated then the data requires to be transformed – for example, by taking the log of the original values. Non-linear multiple regression also exists if data transformation is insufficient but non-linear regression has its own validity challenges.

(e) Multicollinearity occurs when some of the independent variables are correlated. While this causes minor technical issues with the creation of the regression equation, the more serious issue is the interference with the validly of the quantification of the explained variance. The explained variance is used to identify the relative importance of the independent variables and the order in which they should be added to build the equation. Many weather, social, housing, educational and behavioural characteristics that might be used in making forecasts of water demand will be inter-correlated. Although not strictly multicollinearity, serial autocorrelation (where the next value in a series depends on the current) should be considered problematic.

(f) It is assumed that the residuals are normally distributed. Normality should be checked.

Some analysts would consider a further expectation that outliers, which might unduly influence the outcome/efficiency of the regression, be examined and in some cases removed. The view of the authors is that unless there is clear evidence that some specific data is demonstrably incorrect, it should remain in the analysis.

The reader might find the considerations above laboured. However, without considering all of these points, and employing analysis at each stage to quantify/prove that the assumptions are met or to adapt the analysis as required, the outcomes, while appearing to have favourable statistical support, may in reality be invalid. Perhaps more importantly, it is challenging to explain in a fully comprehensible way to all stakeholders the full process involved. Stakeholders need to be aware that the relationships should only be used within the ranges of the data used to set up the equations. We note below that multi-variate analysis extends well beyond 'simple' multiple regression and do so to extend our recognition of the challenge of communication of such complex analysis to stakeholders.

Regression is just one part of multi-variate analysis. Factor analysis and principal components analysis can reveal how initial variables load onto an underlying component or factor. This might be useful in explaining what needs to be addressed. It is possible to regress these underlying factors rather than the original variables. The factors may be 'independent' (or orthogonal) or interrelated (or oblique). Their employment in forecasting would be challenging to explain.

The benefits of regression are apparent: an advanced scientific approach and an explicit statement of the basic data on which the regressions are based.

3.7.   Managing demand

Current UK water policy is to reverse the trend towards higher water demand apparent for several decades arguably since records began (Defra, 2008). This aspirational ambition needs to be backed by an implementation plan if aspiration is to be turned into reality. For potable water demand, the planning is largely in the domain of the companies required to achieve the targets,

Figure 3.2 Components of abstracted water (Based on data from 1997 to 1998 available from ONS (2010))

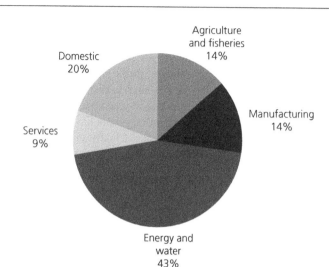

referenced earlier, of a PCC of 110 litres and a leakage reduction of 50%. Figure 3.2 below shows UK abstractions by major industry based on data from 97/98 available from ONS (2010). Worthy of note is the small component of overall water use that is ascribed to the domestic sector.

Methods for demand control fall into two broad categories of technical measures and behavioural adjustment. We will address firstly the technical measures.

3.7.1 Technical methods

We can displace demand by encouraging the use of alternatives to potable waters or by reusing potable water following its first use. Such reuse can be onsite or integrated into the overall supply system of a region. Clearly the former is simpler as it has the implicit consent of the householder.

Rainwater is both a resource and a potential problem. It is a problem when excess rainfall occurs in impervious urban areas and results in flooding or in a problematic combined sewer overflow operation. Harvesting this 'green' water for use in garden watering or toilet flushing displaces the demand for potable water for these functions. It is not a complete solution however. Green water must be stored in relatively large volume to be effective, either in terms of urban drainage amelioration or for demand displacement (approximately 2 m^3 per household if used for water saving purposes only). If stored in the roof space it may require stronger joists to bear the additional weight, and so there is additional embedded carbon as every building material has a carbon cost for extraction, refining and transportation. If stored externally below ground, it uses electricity for pumping which leads to carbon emissions. Although called 'green' with connotations of purity and so on, rainwater will have an avian derived faecal component and so care must be taken to ensure that the green water cannot be inadvertently used, for example, to fill a paddling pool (Ward *et al.*, 2009). Care might also be required if aerosols result from flushing (Fewtrell *et al.*, 2008).

Light grey water (water from a bath or shower) or somewhat more contaminated dark grey water (from kitchen sinks etc.) might also be used for toilet and garden water displacement. Grey water recycling systems require much smaller collection tanks (approximate 0.5 m³ per household) but need more intensive water treatment. Such systems can also be used in combination with rainwater harvesting (Dixon *et al.*, 1999). All recycled water use requires the agreement of the property owner. As the scale of potential contamination increases, so does the available resource but so must the diligence in ensuring that no improper use occurs. Black water (toilet flush water) is currently not widely used as part of direct reuse but has the potential to displace new abstraction if the black water is sufficiently treated and if the product is publicly accepted. Schemes that return treated sewage to the river upstream of water works abstraction points have greater public acceptability, a process whereby the water effectively is treated twice. Clearly in severely water stressed areas this approach has a significant role to play, although it is clearly not without a carbon cost. Thames Water, for example, is examining London reuse schemes which treat water using RO given the potential contaminants in the treated effluent from the urban area before it is subsequently returned to the river upstream of the London intakes on the River Lee. Whether that would count as demand reduction or as resource development is open to question and interpretation.

There is some doubt as to whether rainwater harvesting or grey water recycling are viable options for single households from a purely pragmatic perspective or based on financial or carbon accounting (Butler and Dixon, 2002). Installation and especially retrofit are far from perfected and we do not as yet have a 'fit and forget' technology available.

Water-using appliances grow more efficient each year, as lowered water requirement is perceived as a marketable attribute (but possibly this is simply another side effect of energy efficiency). It has been asserted that the simplest appliance efficiency increase is through the installation of a cistern displacement device which reduces the capacity, and thus water usage, of a toilet cistern by 10 to 15%. The future role of cistern displacement devices is limited however as a six litre flush is now the mandatory standard. Further, data from the Thames Water smart meter initiative indicates that displacement and dual flush systems are not as successful as anticipated and, indeed, may increase water usage because the flush is not effective. Dual flush, six litre drop valve systems appear to have increased leakage vulnerability, with the leaking water running into the toilet bowl and thereby not being easily noticed. Yorkshire Water, in a carefully controlled investigation which encouraged households to self-install simple water saving devices, discovered that many householders had never installed the devices chosen. There are a growing number of alternative water efficiency solutions and Butler and Memon (2006) provide a more comprehensive review of available approaches. A good source of up-to-date information is available on Waterwise's site (www.waterwise.org.uk).

3.7.2   Metering and tariffs

Moving next to issues of behavioural adjustment we consider the impact of metering and tariffs. In the UK, the majority of water supplied is unmeasured and so there is no 'cost' to the customer beyond the annual flat-rate charge. Globally, water is most commonly metered and is charged in relation to the unit price and the volume used but in some cases with a considerable standing charge. The key assumption in advocating metering as a demand management tool has been that by adjusting prices (upward), consumption will be moderated. In fact, water demand has generally been found to be inelastic (i.e. does not change in direct proportion) to changes in price (Young, 2005; Renwick and Archibald, 1998; Nauges and Thomas, 2000; Arbues *et al.*, 2003) and widely varying (Mays, 2001). Peak demand appears to be more elastic than off-peak demand (Lyman, 1992).

In 1967, Howe, based with the organisation Resources for the Future, and Linaweaver, a White House Fellow, examined all the water demand and price data then available in the US (Howe and Linaweaver, 1967) and concluded that 'domestic demands are relatively inelastic with respect to price'. Howe re-examined that data in 1982 showing that elasticity was even lower than first determined and concluded that 'pricing would have essentially no effect on in-house (water) use'. Howe (2005) examined all the data from the US and Canada (a much larger data set) and confirmed that demand was little influenced by price. In Europe, Reynaud (2015) reported to the European Commission that an examination of price/demand across all 28 countries confirmed that the price-demand relationship was very weak. A price rise of 10% reduced demand by 1%, typically, and at most, 5%. Reynaud and Romano (2018) concluded that the consensus is that residential water demand is inelastic to price. Against all this evidence Reynaud and Romano were puzzled (their words) as to why public authorities still view price increases as the most direct tool in shaping water use. Perhaps an unswerving belief in classical economic theory, involving a (yet undiscovered) substitute for water? Or perhaps they embraced, without appreciating the detail, one of the conclusions of earlier studies in the US that the companies could increase price because their income would not be damaged by changing demand?

In the UK, it had been widely asserted that metering of water supplies will reduce demand by approximately 10%, based on a mix of demand suppression and leakage identification. This is based on limited evidence derived, for the most part, from the national water meter trials and unconnected studies conducted by several water service providers. The trials took place on the Isle of Wight and in selected water company areas between 1989 and 1993. Around 50 000 households were compulsorily metered and the ensuing demand profiles were compared against a control. That important study is now aging, certainly the Isle of Wight study site is atypical of the UK and, of course, as with any study of behavioural response, is partly a product of awareness and perception at the time and, perhaps, exclusivity. The response to metering as seen in the Isle of Wight study is not predominantly a price related response but an observational and behavioural response.

In the minor studies conducted subsequently, the subjects are meter optants (i.e. they have chosen to have a meter fitted) and so are a self-selected group. They tend to be smaller households who were already lower water users. Using this group will inherently over-estimate the potential savings from fitting meters. If policy requires a 25% reduction in demand driven by a range of initiatives but underpinned by metering, the reduction would, in turn, cause a significant reduction in water service provider income, while expenditure would not fall by a similar amount as asset investment is driven more by demand peaks than by average consumptions. Recognising that reduction in measured water consumption associated with water efficiencies would penalise companies, the financial regulator, Ofwat, has introduced a revenue correction mechanism to avoid penalising success.

3.7.3 Achievement

To attain the water reductions (misleadingly titled efficiencies) is a difficult task. For example, in England, although there are many existing water bylaws and water related building regulations (CLG/Defra, 2007), and a growing body of new regulation such as the Code for Sustainable Homes (CLG, 2009), there is also a body of evidence that the regulations achieve only modest compliance. Some reports suggest that up to 30% of newly built houses are not regulation compliant, a figure set to increase as public sector employment declines (UKWIR, 2008). Even if they were compliant at the outset, the question of the sustainability of that compliance arises. Regression is the term often used to indicate a slow return to the pre-existing conditions. A cistern displacement device removed by a plumber or the householder during maintenance may not be retained and replaced. A shower delivering a low flow rate of water in a new bathroom may be replaced or re-drilled. Aerating taps may be replaced, as has been observed by the authors in Code for Sustainable Homes compliant dwellings. The energy efficiency drives of the last 40 years are full of well-documented and quantified failures in energy efficiency initiatives, and water efficiency will go the same way unless we learn from energy and, for example, make the water equivalent of an incandescent bulb unavailable on the market. In the energy industry, improved efficiency did not lead to a decline in demand but to a general increase in appliance capacity or house mean temperatures. Efficiency was exchanged for comfort rather than for cost reduction. It will be interesting to see if similar paradoxes occur in the search for more efficient water. The extent of regression to less water efficient but more customer satisfying fitments and appliances is difficult to determine. While the energy initiatives of the late twentieth century were rather fragmented, the more unified approach of Waterwise, the UK NGO focused on decreasing water consumption and building the evidence base for large scale water efficiency, may yet result in effective conservation interventions.

3.8.   Water neutrality

A water neutral development is one in which water consumption post-development is no greater than that before development. This is achieved by minimising water consumption in the development itself and in other areas if required. Clearly, this concept makes a number of assumptions about the scale and sustainability of water efficiency, assumptions that need to be considered carefully. Further, savings made in another water region will have no impact on the water balance of the region issuing the new abstraction licence, unless and until there are potable water transfer agreements between regions covering the volume of the new abstraction, at least in terms of extreme drought situations. Such an approach has been shown to be theoretically possible in the Thames Gateway development to the east of London (EA, 2007), but did illustrate how difficult it would be to achieve in practice (Sims *et al.*, 2006; Butler and Herrington, 2007).

3.9.   Modifying lifestyles

How do we influence the user so that we modify the demand for water? Fundamentally, we must modify behaviour both in the style of their water use and the nature of water-using facilities installed and retained. At the crudest level this might, in theory, be encouraged by price through tariffs but this requires that water is metered and very significant price increases (see earlier discussion in this chapter on price elasticity). However, a strongly escalating tariff raises immediate questions about water justice and affordability. To attain the twin goals of influencing behaviour and delivering social justice requires that the tariff is both dissuasively progressive and socially just. In other words, a tariff that rises rapidly from a low base to deliver protection for the poor while punishing the profligate (we refer to this later as the 4P tariff). This may work in theory, but for a resource that is strongly related to numbers in the household, you are likely to markedly penalise larger families. And we make the final reminder that there is substantial evidence that demand is inelastic.

In the longer term, information on both the cost of water used that might have been saved and on the current state of the water resource will encourage savings. In Victoria, Australia, such information now forms part of the bill. In the UK, Sutton and East Surrey employ similar information on current and past water usage. Information will also justify regulatory interventions prohibiting specific water uses. The public acceptance of this requires that there is a degree of 'ownership' of the state of the resource. Certainly in the US, Canada and Australia, water reserves are published weekly so that the logic of progressive water saving requirements is obvious and unsurprising. Smart meters, already being tested for energy supply, provide immediate cost information for the consumer and may influence immediate reactions – for example, turning off a light or running a kettle for a shorter period. Water has a problem in this respect in that bills are about a third or less than those of energy, so the cost per usage is far less dramatic and so is less likely to promote a beneficial reaction. Although, the progressive meter programme of Thames Water in London is currently proving very valuable at detecting leakage and promoting behaviour change.

In the long term, education is the key to reducing demand but there has to be a case that will withstand scrutiny. If you can influence children then you influence many future generations.

3.10.   Visions for the future
3.10.1   Ultra low use systems
The humble toilet or WC 'consumes' a large fraction of the household potable water demand. However, new approaches are becoming available that use much less water for the same, or better, performance. One such is the toilet that works by locally pressurising air, which is then expelled through the appliance under pressure during the flushing cycle instead of water. Just 1.5 l per flushing cycle is used to cleanse the bowl and refill the water seal trap, and it can be connected to a standard gravity sewer system (Littlewood et al., 2007).

3.10.2   Dual systems
We use between 140 and 150 litres of potable water per person per day in the UK depending on which report one reads. Of this, perhaps 30 litres needs to be of this potable quality, perhaps less. Looking 50 years ahead, should we be planning for a dual system, which in its extreme manifestation, might be bottled water for drinking with somewhat lower quality washing water and recycled water distributed for all other functions? Dual systems are not new and were trialled in North America and Europe. The cost of maintaining separate systems however, appears to be prohibitive with some being discontinued on the grounds of cost, complexity and, at least in the Netherlands, health security.

3.10.3   Advanced tariffs
Some people believe that a **P4** tariff that '**p**rotects the **p**oor and **p**enalises the **p**rofligate' is needed urgently. However, development of such a tariff is not as simple as tariffs for energy, for the water use is strongly driven by individual needs, whereas the energy tariff is much less occupancy dependent. Therefore, the tariff needs to be set either on the occupancy numbers or on the house type as a surrogate for occupancy. For practical efficiency reasons, the tariff needs to be set at an upper expectation for occupancy and an appeal procedure that allows evidence to be presented to initiate a lower tariff needs to be in place. The industry already faces challenging levels of bad debt (Clarke et al., 2012) which will not be helped by raising tariffs, and we note that not all above average users are profligate as some will have valid medical or cultural requirements for additional water use.

3.10.4   Quotas and advanced payment cards
Establishing a quota fairly to an amount of water could be achieved by having prepayment cards that allocated a free 'social' quota of water with every recharge. However, advanced payment cards are not currently approved for water use but may again find favour if a trickle (that is, limited flow rate) supply becomes the permitted default. Water trading has been shown by Mitchell and McDonald (2015) to have potential to reduce demand at a household level. Unpublished data analysed by these authors have shown that the UK is one of the countries with a water stress issue that also has the IT and financial infrastructure to support water trading.

3.10.5 Behavioural change

We have seen major changes in behaviour over the last 30 years – reliance on the web, use of mobile phones and so on. We have also seen a growing environmental awareness and a willingness to engage with the environmental agenda. With information that will stand scrutiny and with a schools-based education on responsible water citizenship, there is every prospect that behavioural changes in our water use will permeate society. However, we must be aware that the prospect of a backlash exists if government policy on water saving is found to be based on dubious arguments, then public reaction may well ignore the many sound arguments for water saving.

3.10.6 Living with environmental and social change

Our climate is changing. The debate on the balance between natural and anthropogenic drivers remains live. But for water managers increased drought prevalence, more variations between 'wet' and 'dry' seasons and greater variations in climate change across the UK must mean that our management of demand becomes more sophisticated and responsive. Add to this the clear need to use less energy in water provision and the result is probably a movement to more local sourcing and more engagement with water saving and an acceptance of more risk. Social change is occurring both nationally and internationally. China, for example, may increase its demand for air conditioning with consequent energy demand rises; the Russian invasion of Ukraine has resulted in rejection of Russian gas as a primary fuel in Europe, further distorting the existing energy market, with an outcome that, potentially, more biofuel crops are planted that may require more water. We may indeed anticipate a move to allocation of water as a scarce resource and the days of total water demand satisfaction at very low price might pass.

3.10.7 Lowered standards of service

The water industry has a defined standards of service – for example, that water supply companies plan to a drought resilience of 1 in 500 years and so standpipes and severe restrictions would occur, **on average**, once per 500 years. Thus, the industry is investing in an infrastructure that will supply the water demanded on effectively all occasions. There is a case for questioning how and why such standards of service arose and how best to communicate this concept to the customers. It may well be interpreted that restrictions will never occur or if they do, they will not be seen again until 2500! Such security will be costly. If policy makers were willing to accept a higher risk – let us say drought resilience of once in 100 years – then levels of investment could be retargeted.

3.10.8 Mutual support vs. local provision

The industry has evolved from a single source-single sink local provision to regional water service providers that, with the exception of some edge rationalisation, still source all their resources internally or from 'traditional' sources external to the region. Within a water service provider, the benefits of water transfer have become apparent and, in most cases, companies have continued to develop resource flexibility and transfer. Each company has a headroom allowance (a reserve capacity) that exists not to meet an unexpected eventuality but to address the situation when all 'uncertainties' work against demand satisfaction; headroom addresses the worst case scenario. If companies agreed, in advance, water transfer agreements (potable or raw as appropriate to local circumstances) with adjacent companies then they would have a larger 'headroom' to draw upon and this is one of several reasons the NFW promotes regional cooperation. This should result in an industry more resilient to external variations or the retention of the same resilience without expanded resource requirements. Of course at the heart of this argument is the impact of climate changes and currently we have no forecasts of the return periods of droughts (longevity and severity) in relation to their spatial extent, a measure that would be key to developing rational, regional water agreements.

3.11. Conclusions

The estimation and forecasting of water demand is fundamental to effective resource planning but depends on good data, which cannot exist without universal metering. Potable water demand cannot be separated from other demands for raw water. Climate and demographic change complicate forecasting and resource planning. We do not start with a blank canvas, and our choices are strongly constrained by existing urban structures and infrastructure. In the longer term, however, movement towards a more informed, flexible, adaptable and resilient water supply infrastructure would be wise.

Acknowledgements

David Butler and Clare Ridgewell contributed material to the first edition. We acknowledge their residual contribution to this edition. We thank the reviewer of this edition for posing useful comment.

REFERENCES

Arbues F, Garcia-Valinas MA and Martinez-Espineira R (2003) Estimation of residential water demand: a state-of-the-art review. *Journal of Socio-Economics* **32(1)**: 81–102.

Bellfield SL (2001) *Short-term domestic water demand: estimation, forecasting and management.* PhD thesis, University of Leeds, UK.

Butler D (1991) A small scale study of wastewater discharges from domestic appliances. *Journal of Institution of Water and Environmental Management* **5(2)**: 178–185.

Butler D and Dixon A (2002) Financial viability of in-building grey water recycling. *International Conference on Wastewater Management and Technologies for Highly Urbanized Cities.* Hong Kong, People's Republic of China.

Butler D and Herrington P (2007) *Towards Neutrality in the Thames Gateway.* Peer reviews, Environment Agency Report SCH01107BNMN-E-P.

Butler D and Memon FA (eds) (2006) *Water demand management.* IWA Publishing, London, UK.

Chapagain AK and Hoekstra AY (2008) The global component of freshwater demand and supply: an assessment of virtual water flows between nations as a result of trade in agricultural and industrial products. *Water International* **33(1)**: 19–32.

Clarke GP, Kashti A, McDonald A and Williamson P (1997) Estimating Small Area Demand for Water: A New Methodology. *Water and Environment Journal* **11(3)**: 186–192.

Clarke M, Boden P and McDonald A (2012) Debtor: debt evaluation, benchmarking and tracking: a water debt management tool to address UK water debt. *Water and Environment Journal* **26**: 292–300

CLG (Community and Local Government) (2009) *Code for Sustainable Homes: Technical guide – May 2009 Version 2.* London, UK.

CLG/Defra (2007) *Water efficiency in new buildings: A joint Defra and Communities and Local Government policy statement.* London, UK.

DCLG (2007) *New projections of households for England and the regions to 2029* (online). http://www.communities.gov.uk/documents/statistics/pdf/1089402.pdf (Accessed 21/03/2024).

Defra (2008) *Future water: The Government's water strategy for England.* Stationery Office, London, UK.

Defra (2019) Water abstraction statistics: England, 2000 to 2017. (Accessed 25/03/2022).

Dixon A, Butler D and Fewkes A (1999) Water saving potential of domestic water re-use systems using grey water and rainwater in combination. *Water Science and Technology* **39(5)**: 25–32.

Downing TE, Butterfield RE, Edmonds B *et al.* (2003) *Climate change and the demand for water.* Stockholm Environment Institute Oxford Office, Oxford, UK.

Environment Agency (2007) *Towards Neutrality in the Thames Gateway*. Environment Agency project number SC060100/SR3.

Fewtrell L, Kay D and McDonald AT (2008) Rainwater harvesting – an HIA of rainwater harvesting in the UK. In Fewtrell L and Kay D (eds. (2008) A *guide to the health impact assessment of sustainable water management*. International Water Association, Amsterdam, the Netherlands, chapter 3.

Gardiner V and Herrington P (1986) The basis and practise of water demand forecasting. In Gardiner V and Herrington P (eds) (1986) *Water demand forecasting*. Geo, Norwich, UK.

Gardiner V and P Herrington (1986b) *Water demand forecasting: proceedings of a workshop*. Geo, Norwich, UK.

Herrington P (1996) *Climate change and demand for water*. HMSO, London.

Herrington P (1998) Analysing and forecasting peak demands on the public water supply. *Journal of the Chartered Institution of Water and Environmental Management* **12(2)**: 139–143.

Howe C and Linaweaver F (1967) The impact of price on residential water demand. *Water Resources Research* **3(1)**: 13–32.

Howe C (2005) The functions, impacts and effectiveness of water pricing: evidence from the United States and Canada. *International Journal of Water Resources Development* **21(1)**: 43–53

Jeffrey P and M Gearey (2006) Consumer reactions to water conservation policy instruments. In Butler D and Memon FA (eds) (2006) *Water demand management*. IWA Publishing, London, UK, pp. 305–330.

Jin J (2009) A *small area microsimulation model for water demand*. Unpublished PhD thesis, University of Leeds, UK.

Lambert A (2003) Assessing non-revenue water and its components: a practical approach. *Water 21*, August, pp.50–51. http://www.iwapublishing.com/pdf/WaterLoss-Aug.pdf

Littlewood K, Memon FA and Butler, D (2007) Downstream implications of ultra-low flush WCs. *Water Practice and Technology*, 2, 2, doi: 10.2166/wpt.2007.037.

Lyman RA (1992) Peak and Off-Peak Residential Water Demand. *Water Resources Research* **28(9)**: 2159–2167.

Mays LW (2001) *Water resources engineering*, first ed. Wiley, NY, USA.

McDonald A, Boden P and Clarke M (2008a) *Void property management*. Unpublished report to Yorkshire Water.

McDonald, Boden P, See and Rees (2008b) *Hidden and Transient Populations in the Severn Trent Region*. Unpublished report to Severn Trent Water. CREH, Leeds, UK.

McDonald A, Bellfield S and Fletcher M (2003) Water Demand: a UK perspective. In Maksimovic, Butler and Memon (eds) *Advances in Water Supply Management*. AA Balkema, London, UK.

Memon FA and D Butler (2006) Water consumption trends and demand forecasting techniques. In Butler D and Memon FA (eds) *Water Demand Management*. IWA Publishing, London, UK, pp. 1–26.

Mitchell G (1999) Demand forecasting as a tool for sustainable water resource management. *International Journal of Sustainable Development and World Ecology* **6(4)**: 231–241.

Mitchell G and McDonald A (2000) An SIC coded strategic planning model of non-household water demand for UK regions. *Journal of the CIWEM (Chartered Institution of Water and Environmental Management)* **14(3)**: 226–233.

Mitchell G and McDonald A (2015) Developing resilience to England's future droughts: time for cap and trade? *Journal of Environmental Management* **149**: 97–107.

Nauges C and Thomas A (2000) Privately Operated Water Utilities, Municipal Price Negotiation, and Estimation of Residential Water Demand: The Case of France. *Land Economics* **76**: 68–85.

OFWAT (2002) Best Practice Principles In The Economic Level Of Leakage Calculation https://www.geocities.ws/kikory2004/25_ELL.pdf (accessed 22/03/2024)

ONS (2010) Water use by industry http://www.statistics.gov.uk/cci/nugget.asp?id=159 (accessed 21/03/2024).

Orr S and Chapagain A (2009) UK Water footprint WWF http://assets.wwf.org.uk/downloads/water_footprint_uk.pdf (accessed 21/03)

Reynaud A (2015) Modelling household water demand in Europe. European Commission. *Joint Research Council*. Report **27310**.

Reynaud A and Romano G (2018) Advances in the economic analysis of residential water use. *Water* **10** p.1162.

Renwick ME and Archibald SO (1998) Demand Side Management Policies for Residential Water Use: Who Bears the Conservation Burden? *Land Economics* **74(3)**: 343–359.

Roaf S (2006) Drivers and barriers for water conservation and reuse in the UK. In Butler D and Memon FA (eds) *Water demand management*. IWA Publishing, London, UK.

Roe D and Rendell D (2009) *Microsimulation at H M Treasury: Methods and Challenges*. ESRC Microsimulation Seminar Series. University of Sussex, UK.

Spey A and Marshallsay D (2023) *High Users Study*. Thames Water, UK.

Sim P, Mcdonald A, Parson J and Rees P (2007) *Revised options for UK domestic water reduction: A review*. Working Paper 07/04. University of Leeds, UK.

Thames Water (2023) *High Users Study*. A report by Spey A and Marshallsay D, Artesia Ltd.

Twort AC, Ratnayaka DD and Brandt MJ (2000) *Water supply* 5th ed. Arnold/IWA Publishing, London, UK.

Ward S, Butler D, Barr S and Memon FA (2009) A framework for supporting rainwater harvesting in the UK. *Water Science and Technology* **60**, **19**: 2629–2636.

UKWIR (2008) *The cost effectiveness of demand measures*. UKWIR Project Report WR25/c.

UKCP (2018) UK panel on climate projections. UK Meteorological Office.

WATERWISE (2008) Evidence Base for Large-scale Water Efficiency in Homes. www.waterwise.org.uk (accessed 21/03/2024).

Environment Agency, Natural Resources Wales and the Water Services Regulation Authority (2023) Water Resources Planning Guideline. https://www.gov.uk/government/publications/water-resources-planning-guideline/water-resources-planning-guideline (accessed 4th April 2024).

Westcott R (2004) A scenario approach to demand forecasting. *Water Science and Technology: Water Supply* **4(3)**: 45–55.

Williamson P, Clarke GP and McDonald AT (1996) Estimating small-area demands for water with the use of microsimulation. In Clarke GP (ed.) *Microsimulation for urban and regional policy analysis*. Pion, London, UK, pp.117–148.

Williamson P, Mitchell G and McDonald AT (2002) Domestic water demand forecasting: A static microsimulation approach. *Journal of the CIWEM* **16(4)**: 243–248.

Young RA (2005) *Determining the economic value of water: concepts and methods*. Resources for the Future, DC, USA.

Dragan A. Savić and John K. Banyard
ISBN 978-1-83549-847-7
https://doi.org/10.1108/978-1-83549-846-020242005

Chapter 4
Water supply assessment, management and planning

Julien J. Harou
University of Manchester, UK

Howard S. Wheater
Imperial College London, UK

Geoff Darch
Anglian Water, Peterborough, UK

4.1. Introduction

Providing a reliable and secure supply of water for human uses, including domestic supply, industry, agriculture and the energy and transport sectors, as well as for the environment, is a multi-faceted challenge. It is helpful to view water resources and their use for water supply as a human-natural system.

Traditionally, the 'natural system' refers to the presence and behaviour of water in the natural environment, before human use, and the human side relates to water use by people and economic production and to managed interventions in water systems. However, the relationships between natural and human parts of the system are complex and intertwined: they act on multiple scales, from the human impact on global climate and hydrology, to local impacts on rivers and groundwater systems, including agricultural intensification and urbanisation as well as efforts that seek to restore terrestrial and water-dependent habitats. Thus, the natural environment is often a managed environment, dependent on human infrastructure and policies.

Water management is addressed by multiple disciplines including science, engineering and the social sciences, including economics and governance (Wheater and Gober, 2015). Given increasing pressures on the freshwater environment due to population growth, economic development, and climate change, water management and planning are increasingly seen as one of the world's major challenges (Rockstrom *et al.*, 2023).

This chapter begins by discussing the different scales of the water resources planning and management problem, and summarising its challenges from the physical and human perspectives. Next, the major sources of water supply are reviewed, then how the discipline and practice of water management and planning have evolved to cope with increasingly complex societal demands and environmental conditions.

4.1.1   Local, regional and national scale water supply decisions

Water managers and planners act within one or more geographical contexts, at the local (service or catchment area), regional (water utility or river basin), national (water ministry or regulator) or international scale (e.g. international transboundary river commission and other international organisations). The chapter begins with outlining what are the main questions and issues at each scale, and what analysis must be done.

People generally live in one place and this means all water management has an essential and foundational local element; everyone is interested about how water works for them in their local area. Is there enough water to supply the household and town with good quality water? Is there enough to water the garden? Is there water in the river where one walks one's dog and how does it look or smell? Is there water in the local lake to swim in and to support fishing, bird watching and other nature-based activities? In other instances, there is too much water, and this can cause damage to property, or make land or roads inaccessible. People also work to earn a living and need local water to support their livelihoods, whether in agriculture, industry or other sectors, and to maintain other infrastructure – for example, through provision of cooling water for power generation.

The preoccupations and preferences of single individuals may seem unimportant, but they cumulate rapidly. When communities and/or industries feel the same way and communicate (which they do increasingly), opinions about local water issues can rapidly snowball into political movements with major consequences. One recent example is the observation of poor quality in multiple rivers and coastal sites in England. This is a complex issue, involving pollution from agriculture, which is regulated but remains a major problem, and releases of sewage from stormwater overflows that, in part, reflect a historic lack of UK water industry priority and partly a water regulation system that has been reluctant to invest in improved infrastructure, and thus raise consumer prices. A similar concern has been raised in France concerning water quality in the Seine ahead of the 2024 Olympics.

More generally, recent attempts by governments in Europe to restrict agricultural pollution from nutrients and farm wastes led to political dissent and farmer protests (e.g. in France, Germany and the Netherlands). Other notable examples in the last decade include the South African water supply crisis where Cape Town's countdown to day zero (no water) was covered by the international press. Similar examples exist in most countries and point to a central facet of water supply, that outcomes of water management are perceived and experienced locally, but are often complex and multi-faceted, with broad impacts across societies and their economies.

Several water management decisions are taken at the local scale – that is, at specific individual locations. For example, how much can a water user abstract (divert) at a single point on a river, how much flow should be left in the river as a minimum for ecological reasons, or what standard of wastewater treatment is appropriate.

Different decisions are needed at the river basin scale. This poses questions such as:

- Where in the basin should new resources like new dams or groundwater well-fields be exploited?
- Should river basin transfers be used to take water from water-rich regions to water scarce areas?
- Where should water treatment plants be installed?
- Where should treated wastewater be released into the river?

It should be noted that river basins can cross administration or national borders, in which case river basin decisions are more complicated.

Water utilities have a municipal or regional area of influence, which is not usually congruent with river basin boundaries; they face similar decisions but with a different spatial unit. For example, transfers might not be within a basin but between basins, where a water company service area spreads over different basins. Companies also face unique questions – for example, to what extent should they interconnect different service zones (with pipelines) or should they opt for more decentralised water supply systems?

At national scale, policies are set that affect multiple river basins and water companies. These may include items such as levels of water supply service, drinking water quality regulations, ecological habitat protection measures and rules, and how flood mitigation and risk management measures may affect water supply management and infrastructure. Conflicting priorities can normally be resolved through national policy decisions. However, conflicting management priorities in shared international waters can lead to tensions between nation states. The United Nations International Court of Justice has thus far heard three water disputes – Hungary v. Slovakia, concerning Danube dams; Argentina v. Uruguay, concerning pulp mill pollution of the river Uruguay; and, most recently, Chile v. Bolivia, concerning the water uses of a small river providing water for the Atacama Desert (all proceedings are available from the ICJ web-site https://icj-cij.org). However, these cases raise wider issues, beyond the scope of this chapter (Wheater *et al.*, 2024).

4.1.2   Physical water challenges

Water issues and human interventions in water systems have both local and larger-scale implications. This is because both the natural manifestations of water resources (rivers, lakes, aquifers, etc.), the engineered components (reservoirs, water transfers) and the uses of water have both local and wider impacts. Some of the central physical global challenges of water supply management and planning are reviewed, and the scales at which they are relevant are described. Below drought, flooding, poor water quality and ecological water needs are tackled, briefly.

Droughts are prolonged periods of abnormally dry weather conditions. Drought can have different implications for different water users. There are four major types, namely meteorological, agricultural, ecological and hydrological drought. Hydrological drought is the most relevant to water supply and will typically include below-average precipitation, low river flows and water levels in natural lakes and reservoirs, decreased groundwater recharge leading to low groundwater levels, and increased water demand associated with higher temperatures. Although not impacting water supply directly, environmental regulators will typically coordinate drought response considering impacts across multiple sectors. This means that agricultural drought, when there is insufficient rainfall and moisture in soils to support crop production or farming practices, and ecological drought, which depletes habitat for aquatic species and exacerbates water quality issues, will also impact water supply. Drought response measures in the UK, for example, may include water use restrictions such as hosepipe bans, public awareness campaigns promoting water conservation, and coordinated efforts by the relevant national regulators to mitigate the impacts on agriculture, ecosystems and industries.

Inland flooding is the inundation of land that is normally dry by water that has escaped or been released from the normal confines of rivers, aquifers, streams, lakes or reservoirs. Floods have a wide range of societal impacts and affect water supply specifically. Floodwaters can have higher than normal concentrations of contaminants such as bacteria, chemicals and debris that can contaminate drinking water supplies. Floods can damage water supply infrastructure such as water treatment plants, distribution pipelines and pumping stations, disrupting deliveries of water or making water supply infrastructure sites inaccessible. Another infrastructure related issue, which manifests in the long-term and can be exacerbated by flooding, is the progressive filling of reservoirs with sediment, thereby decreasing their capacity to store water. Because of the stress it puts on the infrastructure system by way of 'outage' – that is, service interruptions to water supply infrastructure – flooding may temporarily increase demand from water supply assets that are still actively working. These issues can be mitigated through proactive measures such as floodplain management, infrastructure resilience improvements and emergency preparedness.

Poor water quality makes it harder to provide cost effective and safe water supply. The presence of bacteria, viruses and chemical pollutants in raw surface water or groundwater increases human health risks and, therefore, the importance, intensity and short- and long-term cost of water treatment. Like flooding, poor quality events can lead to outage, where certain sources are temporarily deactivated. The presence of corrosive substances can accelerate the deterioration of water supply infrastructure, including pipes, pumps, wells and storage tanks, leading to costly repair or replacement. Recent concerns about health product residuals (such as contraceptive hormones) in wastewater illustrate the wide-ranging nature of some of these challenges.

Finally, environmental water needs that preserve ecosystems by protecting biodiversity and endangered species that rely on freshwater habitats, like wetlands and other natural habitats, have received increasing emphasis but increase the complexity and cost of water supply. These water needs typically constrain water supply diversions because they are mandated and prioritised by environmental regulation. In England, the environmental regulator has begun capping public water supply licenses to prevent deterioration, with more substantial 'sustainability reductions' expected in future in order to facilitate the restoration of water dependent ecosystems (Environment Agency, 2020). Such changes, a reflection of societies' increased valuation of ecosystem services, can lead to further demand for water management infrastructure, including new water storage. Balancing environmental water needs with competing water supply demand is commonly a requirement of water supply planners and managers.

4.1.3   Water management and governance challenges

The challenge of water supply relates to the governing of water resource and supply systems, which includes the management, regulation and allocation of water resources. Some key challenges include water scarcity, legal and regulatory complexity, the need for multi-scale governance, stakeholder involvement, political preferences, skills deficit, and ensuring the economic efficiency and financial sustainability of managed water resource and supply systems. These are briefly reviewed in turn below.

As river basins become 'closed' (Molle, 2003) and aquifers reach the limits of sustainable use, low cost new supplies are increasingly hard to find and competition between water uses becomes more prevalent. Some authors differentiate between physical water scarcity, where there is insufficient water available to meet demand, and economic water scarcity, where the existing infrastructure is insufficient to meet demand.

Physical scarcity leads to increasingly complex interactions between water users from different sectors – for example, by way of water sharing regulatory regimes, like water licensing or rights and, in some cases, water trading. These are further discussed in Section 4.3 on water management. As scarcity increases, water supply governance is increasingly governed by complex regulatory and legal frameworks. Inconsistent or overlapping regulatory requirements, bureaucratic hurdles and legal uncertainties can create barriers to investment, innovation and sustainable water management practices.

Water management to make water supply available is typically undertaken by a range of organisations including water utilities (or 'water companies'), river basin agencies, government ministries and regulators. This results in multiple stakeholders, including various levels of government and non-government actors like private companies and non-government organisations, all having responsibilities relating to different aspects of water services. Within government there is also an overlapping mix of local, regional and national authorities, often sharing responsibilities for different aspects of water services. The coordination and cooperation among such diverse stakeholders can be challenging, leading to fragmented decision-making, conflicting priorities, and ineffective implementation of water policies and programs.

In a related point, water supply aims to serve stakeholder communities (water users) which, in many countries, may be called upon in highly variable levels of involvement in water governance. This can range from being a passive observer, being consulted or becoming actively involved in a co-production decision-making process. Including local communities, water users, and marginalised groups is increasingly recognised as best practice for more inclusive and transparent water governance. However, many water resource and water supply institutions struggle to engage stakeholders effectively, resulting in limited public awareness, distrust and, in some cases, resistance to water policies and projects.

In some contexts, water supply decisions may be influenced by political considerations, leading to short-term planning, and patronage and favouritism in resource allocation or selection of water sector interventions and their siting (Macpherson *et al.*, 2024). Political interventions can undermine the effectiveness and impartiality of water governance, erode public trust, and prevent or distract water supply organisations from effectively addressing long-term water challenges like water scarcity and pollution.

Achieving and maintaining the required institutional capacity for high standard water management is difficult. Not all water sector organisations will be able to meet these skills. Also, from a technical perspective, many organisations are reliant on external consultants and experts rather than internal expertise. Over the last decades the role and scope of water supply engineers has evolved. For example, in England and Wales, whereas in the past water planners and asset managers were expected to select future water supply infrastructure by picking the least cost options, increasingly there is an expectation that planners will consider and value an ever-growing set of criteria, including demands for diverse water services from different sectors.

Ensuring the financial sustainability and economic efficiency of water supply services is a persistent challenge in countries at all levels of socio-economic development and water scarcity. Balancing affordability for consumers with the need to generate sufficient revenue to cover operation, maintenance and infrastructure costs is a delicate balancing act for water utilities and governments. This balance is often different across sectors, with some sectors receiving more subsidies or more favourable water licensing conditions than others. The ways in which the water supply sector can achieve satisfactory financial and economic performance depend on the regulatory context of the country. In some countries this question will be entirely reliant on central government planners, in others, where water utilities are privatised to some degree or where water users compete for water and water access rights are tradable, seeking economic efficiency in water allocation becomes relevant.

Addressing these governance challenges benefits from transparency, accountability, and stakeholder participation in water management and supply investment decision-making. Strengthening institutional capacity, enhancing regulatory frameworks, promoting multi-stakeholder partnerships, and seeking an institutional context where water supply services are financially sound helps achieve a sustainable water environment and water supply services.

4.2. Water supply sources

Water circulates in a hydrological cycle, with only about 2.5% of the world's water being naturally fresh water. Precipitation of atmospheric water vapour provides the water that maintains river flows and lakes (called surface waters) and recharges groundwater (water stored underground in 'aquifers'). Evaporation from the land surface and water bodies, including oceans, changes the liquid water to vapour to rejoin the store of atmospheric water vapour and complete the cycle. The main sources of natural freshwater resources are abstractions of surface water, from rivers and lakes, or of groundwater, from major or minor aquifers. Alternative sources of water include saline or brackish waters, including sea water or water reuse. Thus, water resources can be developed from a range of sources, available at a range of scales. The major water sources are reviewed below.

4.2.1 Surface water

The simplest manner of supplying water is direct abstraction from a surface water body, such as a river or lake. In many countries in temperate climates, river flows are typically high in winter, low in summer. Water providers may, therefore, wish to store some of the high winter flows for use during summer. This can be done by the construction of a dam, to create a reservoir. Water from a reservoir can be used directly by way of diversion structures and pipelines. Alternatively, the reservoir can be used to release additional water into the river to augment summer flows; this is frequently used when upstream reservoirs in hilly or mountainous areas are used to supply downstream urban and agricultural regions.

To design a surface water supply system, one needs to consider the seasonal and inter-annual variability of flows. This requires long series of observed or simulated flow data, the latter obtained using statistical or hydrological rainfall-runoff models. Plotting a flow duration curve (the name hydrologists use for an empirical cumulative distribution plot) will indicate to managers what fraction of a typical year they could expect to obtain a certain daily volume of water simply by direct abstraction from the river. If taking from what's available on any given day is insufficient, then water infrastructure will be needed – this is reviewed in Section 4.3.

The other question to pose about surface water is its quality. A range of pollutants can affect surface waters. The source of pollution can be diffuse (from agricultural fields or urban areas) or from a point source (e.g. a pipe emitting pollution directly into a surface water body). Different pollutants are generated in different situations. Urban runoff can contain heavy metals, oils and grease, and cities generate wastewater, which unless sufficiently treated will bring nitrogen and phosphorus, and can lead to eutrophication. Eutrophication, caused by excess nutrients, promotes the growth of algae and aquatic plants, leading to oxygen depletion, fish kills and degraded water quality. Industrial areas can generate heavy metals, chemicals and solvents in wastewater. Agricultural areas can contain runoff that carries fertilisers, pesticides and animal waste into surface waters or in recharge to groundwater aquifers. Any sewage, from human or animal waste, can be contaminated with bacteria, viruses and other pathogens. Pathogens pose risks to human health through waterborne diseases such as gastroenteritis and hepatitis. Health product residuals can affect ecosystems and, potentially, human health.

4.2.2   Groundwater

Groundwater water supply systems exploit water from aquifers that provide natural storage accessed by wells or boreholes (mainly modern tube wells with pumps). However, in most aquifers, complex structure and variation in properties exist, so that careful evaluation is needed of the potential effects of resource abstraction.

Two central issues arise with groundwater supplies – what is the yield that can be delivered by an individual well, which depends on the local aquifer properties and the well construction and development, and what is the natural recharge of the aquifer, which determines the overall resource availability. An important difference between surface water systems and groundwater is that subsurface properties are difficult and expensive to observe.

The important properties of an aquifer are its transmissivity, or ability to transmit water, which in the case of chalk aquifers is largely governed by the degree of fissure flow in the chalk block and its storativity, which represents the water released from the aquifer for a unit change in pressure. Local estimates of these can be obtained by controlled pumping of a well and observation of groundwater response in an adjacent borehole. However, in fractured rock systems, such as chalk, estimates can be misleading (Butler et al., 2009), and, more generally, a numerical model of an aquifer is needed to assimilate the available information on geology and aquifer properties, to simulate recharge (which depends on the balance of rainfall and actual evaporation, subject to the effects of soil moisture and properties of the unsaturated zone) and compare the results with observed river flows and groundwater levels. With such a model, a wellfield can be designed, and environmental effects (such as impacts on river flows and water tables) can be predicted. It is commonly the case that, where a major groundwater development is implemented, the associated drawdown of groundwater levels will affect existing private wells, so that alternative arrangements for supply may need to be made to affected properties.

Because groundwater provides a naturally occurring storage of water, there are many possibilities for the active management of a groundwater system. One possibility is known as aquifer storage and recovery (ASR), where surplus surface water (typically in winter) is recharged to an aquifer, for later abstraction (in the summer). 'Artificial' or managed aquifer recharge can occur by injection of water into a well system or through the creation of infiltration ponds. Critical issues include the protection of groundwater quality and the prevention of clogging or fouling of the facilities into a wellfield (O'Shea et al., 1995).

Groundwater can be polluted by a range of contaminants from a range of sources. The potential for and timescale of pollution depends on the type of aquifer and whether or not it is confined – that is, covered by impermeable geological materials. Pollutants generated from human activities can include industrial spills, infiltration from agricultural land (e.g. pesticides and fertilisers), leaking underground storage tanks and improper disposal of chemicals used at home or in industries. Contaminants from industrial or agricultural activity can include heavy metals, nitrates and pesticides, and volatile organic compounds can pose health risks if concentrations are sufficiently high.

If wastewater recharges groundwater reserves, groundwater can become contaminated by microorganisms such as bacteria, viruses and parasites, which can cause waterborne diseases like cholera, typhoid fever and gastroenteritis. Other substances can be found in groundwater, not because of pollution of human origin, but rather because of naturally geologically occurring substances. Dissolved salts, commonly referred to as salinity, can degrade the quality of groundwater for drinking purposes. Arsenic and fluoride are naturally or geologically occurring contaminants that leach into groundwater from rocks and sediments. Chronic exposure to arsenic can increase the risk of health problems, a notable issue when Bangladesh turned to groundwater wells, which lead to mass poisoning.

4.2.3 Alternative sources

In addition to the development of surface waters and groundwater, several other sources of water supply are available. Here the following are briefly reviewed: rainwater harvesting, water reuse, desalination and demand management.

Precipitation, when used directly, is called rainwater harvesting. In water-stressed areas it can be an important resource. For example, the Greek island of Cephalonia has no surface watercourses and no exploitable groundwater resources. Traditionally, households have collected water from roofs and paved areas to provide their sole source of supply. In Australia, rainfall on the coastal cities is commonly higher than that over the interior catchments that provide the surface water resources; one consequence of recent droughts has been an increased use of rainwater harvesting. Consideration of the design of a rainwater-harvesting system requires the introduction of some basic principles. The characteristics of the supply need to be known – that is, not only how much rain is available but also information about the periods between rainfall, so that appropriate capture systems and storage can be provided.

Water reuse, also known as water recycling or reclaimed water, is when treated wastewater is purified to a quality suitable for a specific purpose, such as for irrigation, industrial use or for drinking. The most usual form of water reuse is for non-potable uses such as for irrigation or landscaping. It is also used, at times, for industrial applications or for toilet flushing. If advanced treatment methods like reverse osmosis, advanced oxidation or ultraviolet light disinfection are used, potable drinking water standards can be reached. Often, barriers to its use for public water supply are from negative public perceptions, although there can also be regulatory challenges. Other, less controversial, uses of reclaimed water include water supply reservoir augmentation, artificial groundwater recharge, environmental enhancement and supporting aquatic ecosystems of rivers, lakes or wetlands.

Desalination refers to the process of removing salt and pollutants from sea water or brackish water to produce drinking water or for other uses. Its use has expanded rapidly in recent decades in water stressed areas, particularly the Middle East, Australia and other regions. Various processes are used – for example, reserve osmosis. From a water quality perspective, it is generally well-accepted as a potable source once the water is re-mineralised, however its high energy consumption and high capital and operating cost are significant barriers. Additionally, desalination plants can have adverse environmental effects if not properly designed; these include their intake of seawater, which may harm marine ecosystems through entrainment of aquatic organisms. Brine, the concentrated salty wastewater generated during desalination, when discharged can affect marine habitats and water quality. For these reasons, public perception and regulatory approval can be difficult, depending on the local context.

A final 'source' of supply is simply to use less water. This is referred to as demand management or water conservation. While not a new source of water like those above, in practice, despite being an indirect supply, it behaves similarly to improve the supply-demand balance. Decreased water demand is achieved by reducing waste, using water efficient appliances and practices, and changing water usage behaviours. Various pricing policies can be used (Nauges and Whittington, 2017), although it is generally accepted that this must be accompanied by social tariffs so that disadvantaged people are not prevented from using sufficient water. Recently, and inspired by similar pricing policies in the electricity sector, dynamic water prices have been proposed by Rougé et al. (2018) – that is, time-changing prices to ensure that water costs more during peak usage times (of the day) or during periods when it creates the most environmental harm (typically during warmer low-flow periods). Water conservation in the agriculture sector involves adopting irrigation technologies that minimise water losses, using soil moisture monitoring systems, practising crop rotation and drought-resistant crop varieties, and optimising irrigation scheduling so that they closely match crop water requirements. Agricultural water conservation is a complicated topic, the effectiveness of which has been contested because it may increase consumptive water use (Ward and Pulido-Velazquez, 2008) and decrease 'losses' that previously provided groundwater recharge or benefited the environment (Grafton et al., 2018; Perry, 2007).

4.2.4   Modelling water supply sources

An important early step of informed and effective water management is to quantify the available supplies. As water scarcity and water's economic value increases, this quantification is increasingly done with computational tools – that is, computer models.

Surface water is generated when precipitation leads to runoff, which accumulates in rivers. That process is simulated with computers by way of hydrological modelling, perhaps the largest area of specialisation of hydrologists. Flooding, although not within the scope of this review, is also a major driver of hydrological modelling progress. Hydrological models simulate the movement of water through the hydrological cycle, quantifying the major components of precipitation, runoff, evaporation and infiltration. There are hundreds of hydrological models built with different benefits and geographic focus; some are more universally applicable and used globally. There are three major types of classical hydrological models, metric (or data-based), conceptual and physics-based (Wheater et al., 1993), and these may be lumped or distributed. Lumped models use simplified spatial aggregations wherever possible, and distributed models typically represent the landscape using a set of topographically based 'hydrological response units' or with a regular grid ('raster') and quantify the elements of the hydrological cycle for each spatial element. Lumped conceptual models are widely used in practice due to their combination of simplicity and accuracy at required assessment points and their ability to quickly run thousands of years of simulations and scenarios.

Recently, quantification of certain elements of the hydrological cycle by way of earth observation (remote sensing) is being increasingly used to inform model calibration (Yassin et al., 2017) and machine learning models, which use methods like neural networks and other AI pattern search methods to approximate relationships and hydrological variables, are promising new extensions of data-based modelling (Kratzert et al., 2019).

For groundwater systems, an important initial question is what the recharge is, which will heavily influence the sustainable yield of an aquifer (in addition to the presence or not of surface supplies which can be 'captured' by wells) (Bredehoeft, 2002). Subsequent questions arise concerning the magnitude and direction of natural groundwater flows, and the extent to which these are modified by current or proposed groundwater abstractions. Groundwater flows through porous media in response to a gradient of water pressure (elevation), analogous to processes that may appear quite different (e.g. temperature or electrical flows), all of which can be represented by the same Laplace equation. Distributed groundwater models aim to solve the governing equations in one, two or three dimensions, using classical continuous medium mathematical methods. A system of partial differential equations is set up and then solved using standard numerical (approximation) methods, typically using finite difference or finite element methods, where the relevant variables (like groundwater levels or velocities) are quantified based on a 2- or 3D network of node locations.

Physics-based models include a complete description of integrated surface water and groundwater systems. However, such models are complex, can have issues of numerical stability and are relatively time-consuming to run. Given that groundwater is 'hidden', alternative interpretations of groundwater systems can lead to disputes, which in some cases means such groundwater models have been used in court cases (Lagos et al., 2024; Wheater et al., 2024). Alternative approaches to groundwater modelling are being increasingly explored – for example, lumped parameter groundwater models (Keating, 1982), analogous to conceptual lumped rainfall-runoff models. These models capture key relationships between infiltration (often being run in conjunction with recharge models), groundwater storage and outflows, thus allowing assessment of the impacts of abstraction on baseflows. However, they are much faster to run than physics-based models, thus being suitable for analysis using long synthetic or climate model-based timeseries. While not suited to local allocation issues, these models are useful to assess water resources and inform policy questions at broader (regional) spatial scales.

A variety of modelling methods and tools have been designed to assess the impact of climate change on hydrological variables. The global climate models used to simulate potential future climate conditions operate at relatively large spatial scales, and hence there is a need to downscale their results to the spatial and temporal scales needed for meaningful hydrological simulations. Statistical methods have been developed to do this, but generally they rely on an assumption that the statistical interrelationships between climate variables remain unchanged under a future climate. Dynamical downscaling is an alternative, whereby higher resolution regional climate models are run with boundary conditions specified by a global climate model. The regionally relevant future climate variables can then be used to generate climate-change perturbed historical or synthetic hydrological scenarios. Synthetic rainfall data are commonly produced using stochastic weather generators, which generate alternative but statistically plausible weather time series (e.g. Serinaldi and Kilsby, 2012; Dawkins et al., 2022). Then the synthetic rainfall data is perturbed using spatially coherent climate projections (Murphy et al., 2018) and routed through a hydrological model to produce climate change-impacted plausible hydrological time-series (Wade et al., 2015) such as future river flows and groundwater recharge levels. These are the hydrological scenarios used by analysts to study whether water management policies and plans will be resilient in the face of climate change (see Sections 4.3 and 4.4). Recently, hydrological modelling has been informed by large climate model ensembles, including single model initial condition large ensembles, which also provide physically plausible time-series of meteorological data including droughts (e.g. Chan et al., 2023).

Alternative supply sources like rainfall harvesting, desalination and reuse are generally easier to model, and spreadsheet style accounting tools typically suffice to draw conclusions about their viability. The efficacy of demand management campaigns is difficult to predict and depends on human behaviour, hence water conservation measures are typically viewed by water supply planners as less reliable than supply augmentation schemes.

4.3. Water management

Water management is a practice rather than a science. For a country or region to implement advanced water management, multiple conditions must be in place, including multi-disciplinary expertise, specialised institutions, governance frameworks and financial resources.

Below some key needs of good water management are reviewed from the perspective of a water supply manager or decision-maker.

4.3.1 Water demands

The previous section reviewed water supplies, but supply is only part of the picture. At a high level, a water supply planner must balance water supply and demand. This section provides a summary description of water demands.

The first thing to note is that agricultural water demand (for crops and livestock) globally is estimated to be over 70% of total human water demands, or potentially more if you consider consumptive use. This large number underlines the importance of food security and rural livelihoods, but it also points to the relevance of water allocation as a tool of water management; as global hydrological supply, although changing, is outside human control (there is a planetary boundary) and the development of water infrastructure will be increasingly expensive and to be avoided for environmental, social and economic reasons. Global reallocation of water from the agricultural sector to other water demands will likely occur in the coming decades (Garrick *et al.*, 2019), with important consequences for food production and security.

Other water supply demands include water for domestic use (drinking, cooking, bathing, sanitation etc.), industrial water (manufacturing, cooling, cleaning, as a raw material etc.), commercial demands (office, hotels, restaurants, landscaping etc.), environmental (water to maintain ecosystems, wildlife habitats and enhance aquatic biodiversity) and recreational uses (pools, water parks, boating, fishing etc.).

With metering and, more recently, smart metering (automated remote measurement of water consumption, potentially with high temporal resolution) (Cominola *et al.*, 2015; Harou *et al.*, 2014) increasingly detailed data on domestic and industrial water consumption are available. Estimating future water demands, however, has not improved as much; forecasting future water demands remains a major challenge. The simplest method is to assume a future per capita consumption and population growth rate, then extrapolate future water volumes. Both of these input numbers are uncertain (in fact deeply uncertain, see Section 4.4), hence the difficulty of the task. More sophisticated forecasting models will consider other variables, like economic and technological development. Another approach is to tackle the issue as a statistical problem and extrapolate future demand based on past data; methods like regression or time-series analysis are used. Finally, one should mention again here that predicting the effectiveness of demand management schemes is difficult (House-Peters and Chang, 2011).

Other sectoral demands are harder to estimate, in particular water for agriculture, which generally is more difficult to meter and predict (crop choices depend on global commodity prices) and, globally, is more likely to be impacted by political initiatives, such as subsidies and other schemes targeting rural development. Recently, earth observation techniques like remote sensing have allowed water planners to obtain increasingly accurate inferred (indirectly observed) data on agricultural land use and water application on fields. However, these methods are still being refined and cannot replace actual measured water volumes from a water metering device (Puy *et al.*, 2022). Of the water applied to a field, some will be consumed (transpired) by plants, some will percolate deeply (typically recharging local aquifers) and the rest returned to the river ('return flows') by agricultural drainage infrastructure. In representing this, hydrological models partition the irrigation water applied to fields: part is consumed (evaporated directly or stored and transpired by way of plants), but some abstraction is non-consumptive and returns to the environment.

Finally, some sectors have mostly non-consumptive demands, but they still depend on water being available at certain times and locations; this is true for hydropower and for environmental flows and their associated ecosystem services.

4.3.2   Environmental impact of water abstraction

Human water use extracts water from the natural environment, consumes part of it and may return the rest as poorer quality 'waste' water, and/or sometimes at a higher temperature. From the point of view of the environment, and the aquatic species that live within it or the terrestrial species that depend on it for drinking and habitat, the effects of water abstraction and discharges into rivers are generally negative, although some managed water schemes are designed to enhance the environment and create new habitats. In addition to low river flows, wetlands, for example, are sensitive to changes in water availability, and water abstraction can lead to the loss or degradation of these valuable ecosystems. Habitat loss implies a fragmentation and contraction of the aquatic habitat, which stresses fish, amphibians and aquatic invertebrates; a shrinking habitat will reduce the number of species (diversity) and the abundance of individuals. Overall there has been a major loss of habitat globally as more areas are desiccated and fewer are available. Altered or low flow regimes can degrade riverine habitats through erosion, sedimentation, and channel instability and loss of instream habitat diversity, and riparian vegetation change can exacerbate habitat degradation.

Artificially low flows during dry periods will harm aquatic ecosystems, for example, through temperature stress, dissolved oxygen depletion, increased pollution concentrations, and lower availability of organic matter, nutrients and food for aquatic organisms. Water abstraction can disrupt the natural migration patterns of fish and other aquatic species by altering flow regimes and reducing access to spawning and feeding grounds. This can have cascading effects on entire ecosystems and fisheries. These factors act cumulatively and exacerbate negative impacts on aquatic ecosystems leading to aquatic species population declines and biodiversity loss. Furthermore, not only aquatic species are affected, most terrestrial species depend on rivers for food and drinking, from large land mammals to birds.

Excessive abstraction from groundwater sources can lead to capture of surface water bodies (Bredehoeft, 2002) like lakes, wetlands or rivers, or result in saltwater intrusion in coastal areas, all of which will also impact aquatic and terrestrial species.

Protection of environmental needs is typically determined on model-based investigations, often using standardised assumptions about hydro-ecological relationships, i.e. the quantity of water ecosystems require. This is often focused on low flows – that is, reducing the drought impact on aquatic species – but increasingly is examining the ecological flow requirements across the full flow duration curve. Typically, management policies aim to restrict deviation from a naturalised flow, based on the ecological sensitivity of a water body to abstraction. However, this can prevent significant challenges for abstractors and there is significant uncertainty about future requirements as ecosystems attempt to adapt to a changing climate.

4.3.3 Water allocation

Water allocation refers to the apportionment of water between users, whether they are of the same use type or representing different water-using sectors. In some cases, all water using sectors are treated equally; this is the case for example in the western United States, where water rights recognise historical users (by way of the doctrine of prior appropriation, or 'first in time, first in right', rather than assigning priorities by use type). Other countries, like England, Wales or Australia, use water licensing or entitlements, which are government-sanctioned permissions to extract and consume water granted to individual properties or organisations. These specify conditions such as withdrawal limits, usage purposes and environmental safeguards.

In terms of water management, it is not the policy name that matters but the extent to which authorities can limit or change allocations. The more flexibility, the greater the ability of governments to prioritise between water user types and enforce changes in water management. However, lack of water supply security constitutes a barrier to businesses and investment, so even in licensing systems in which jurisdictions have, in theory, a large freedom to enact changes, there is a reticence to make changes to the status quo and, in practice, water licences often act more like water rights. In England, government proposals for abstraction reform would see all licences eventually become time limited and with no compensation payable for curtailment (Defra, 2021), but this has not yet occured.

Whichever system is enacted, it can face a wide set of challenges. The first is overallocation; if more water is consumed than that which replenishes surface water and groundwater or other sources, over the short-, medium- or long-term, there will be depleted water bodies which cannot accommodate abstraction, leading to a degraded environment. Another problem that can occur is that over time, as human or environmental conditions change, previous levels of abstractions can no longer be supported, or they are no longer viewed as acceptable because of societal changes. This is a notable current issue in the western USA's Colorado river allocations. Problems could be that previous allocations are viewed as inequitable, the cost on the environment is deemed unacceptable and the value of environmental services increases, or that the infrastructure or institutions that were needed to implement a previous allocation arrangement are no longer able to do so.

4.3.4 Managing levels of service and water scarcity

Water supply is an essential societal resource and levels of service are an important descriptor of its provision – for example, to describe its reliability (how frequently the service is unacceptable?), its resilience (when it fails, how long does it fail for?) and vulnerability (how bad is the service failure and by how much do users need to reduce their usual water use?) (Hashimoto *et al.*, 1982). Water supply levels of service are directly related to the supply-demand balance; in the short- medium- and long-term (days, weeks and months) service can be provided when supply is greater than demand. For this reason, as detailed in Section 4.4, many water supply engineers track and model the water-supply balance per service zone of their service area (rather than, or in addition to, trying to understand the hydrological water balance of the watershed or catchment they are in).

Water scarcity is a concept related to the supply-demand balance, but a more general description is when the available water resources cannot meet demands. It has multi-disciplinary and multi-sector elements. The supply-demand balance describes the perspective of a public water supply manager with financial and service level objectives, whereas the water scarcity concept is relevant to policymakers of a country or region (state, province, basin etc.) with broader economic, social and environmental objectives.

Firstly, a brief survey of the role of infrastructure in water supply management is presented, with an emphasis on how it is used to maintain water supply levels of service and manage water scarcity. The purpose of water infrastructure is to capture, store, release, convey, treat and distribute water; in this section the focus is on the first four of those.

For both surface and groundwater, an essential infrastructure element is to capture and extract water; in the first case, intake structures and diversion channels or pipes are used, for groundwater, wells and boreholes are used. Next is storage; for groundwater, this is included by default because groundwater is a storage source, but with surface waters in their natural state, storage is included only in the case of natural lakes. When lakes aren't available, storage is fashioned by building dams which impound water in a reservoir. These can be on rivers, so that the dam blocks the river, but also off-river, where the river water is captured and conveyed to a storage facility away from the river. The latter case is used increasingly, particularly in industrialised countries where natural free-flowing rivers are seen to have high amenity, recreation and eco-system service value (Grill et al., 2019) (see Section 4.3.5 on water economics). As described in the surface water source sub-section, storage is needed if the water demand can't be met reliably by normal river flows; this can be assessed rapidly with a flow duration curve of the measured daily river flows (answering the question: 'given the recorded flows of recent years, what demand could be met 95% or 99% of the time'). If natural river flows are insufficient in volume or reliability, then surface water will need to be stored in reservoirs. Not all storage sources are equal. Groundwater, typically, delivers clean water, which generally only requires simple treatment before entering a water supply system. Whereas surface waters generally require multiple stages of potentially sophisticated treatment. While dams have much higher capital costs than wells, their operating costs are typically lower (reservoir water is released by gravity and does not normally require pumping, unlike groundwater where pumping costs will generally increase as the aquifer is exploited). However, there can be exceptions – for example, pumped storage reservoirs (which incur pumping costs), artesian aquifers (where no or less energy is needed to pump groundwater), and in some countries (like the UK) reservoir operational costs are increased due to water treatment requirements.

Conveyance refers to canals and pipelines to transport water. Water is heavy and expensive to pump, so gravity-based canals and pipes are preferred but often not possible, and so pumping or lift stations are required. The larger water management issue brought up by water conveyance is the ability to transfer water into other river basins or administrative regions or even countries. Water transfers bring a range of environmental, economic and political issues, in addition to the water engineering aspects of transferring water over large distances (Lund and Israel, 1995). These considerations, as well as infrastructure option choice and system design, will be further discussed in the Section 4.4.1 on water planning.

Once water infrastructure is in place, the question arises, how can it be used effectively to best manage water. This is a difficult question, because what 'best' means will depend on individual preferences, and perhaps which region you live in and which sector you work in.

Water infrastructure operations are a central part of water management, and closely linked to allocation, as water allocation is often implemented by way of engineered infrastructure operations. Allocation is more about raw water volumes, whereas when water deliveries are made, and at what level of reliability to which sector, is decided by the operation of infrastructure and is a sophisticated risk management problem. The key decisions are what to store, release and convey, and when and how much to keep in storage to prepare for future needs. For a multi-use surface water reservoir or system of reservoirs, this is complex. The major uses of reservoir storage, apart from public water supply, include flood control, hydropower and, possibly, navigation and recreation (flood control is only for in-river dams). Any reservoir operation (release of water x, at time of day y, during season z) will be more or less beneficial to each of these purposes or sectors.

Water supply will prefer greater stored volumes to secure future supply reliability but saving water for later (called 'hedging') means you cannot use it now, so there's a trade-off. Releasing for agricultural production will mean a peak of releases timed to coincide with the irrigation season(s) (which may conflict with other uses). Flood control will prefer the opposite to water supply – that is, for large dams to have lots of available capacity, so that they can store some of the flood peak when it arrives. For hydropower production, it is preferable to keep reservoir water levels high and make releases to meet the needs of the regional or national electrical grid. These needs may be quite specific, ranging from base load, where a constant release generates a constant amount of power, to hydro-peaking, where hydropower is used ('dispatched') at short notice to make up for a lack of energy from other sources (increasingly intermittent 'variable' renewables like solar and wind energy). Managing reservoir releases is, therefore, a decision problem that must meet the basic needs and objectives of many different sectors, each with substantial economic and political interests.

Stored water is like money in a bank account; if you use it today, it is not available tomorrow, which means there is an inter-temporal risk management (multi-scenario) problem embedded in the multi-sector, multi-criteria context. The three uses of the word 'multi' in the description hint at the complex nature of this technical problem (see Section 4.3.6 on water management modelling). Examples of strategies that are used to inform water storage management include hedging (keeping releases at a lower level than ideal to mitigate the likelihood of worse water scarcity if dry conditions persist), hydrological forecasting (to anticipate future water availability based on precipitation forecasts, snowpack observations and hydrological modelling), and reservoir release rule-curve optimisation (this can be done with short- and long-term forecasts and can consider the needs of multiple sectors).

Operating rules may vary with conditions at the reservoir site, typically with reservoir level – that is, the amount of stored water available or the season. This can also be true for allocation. Allocation rules may change under extraordinary circumstances, like during a drought or an outage (interruption) of a major public supply source. For example, in the past in England, drought plans assumed a reallocation from the agricultural sector to the environment and public water supply. This is another important aspect of water allocation; societal priorities need to be reflected in operating rules for extraordinary circumstances, as well as for normal ones. This is also true for flood control and hydropower. In both cases, special circumstances (floods, power failures of other generation sources) will lead to special rules or spontaneous requests for mitigating the circumstances – for example, with adapted dam release operations. Equally, environmental flow releases, like minimum flows or flows that try to mimic seasonal natural river characteristics, can be coded into to the standard operating procedures of a reservoir, but they can also respond to extraordinary circumstances in the natural system, like a fish migration event.

Much of the text above was discussing the specific case of storage infrastructure, but water transfers will also be regulated with rules that vary seasonally or with regional differences in water availability.

It should be noted that rarely is water supply dependent on just one source; the norm is for there to exist a portfolio of infrastructure assets which can and should be used conjunctively to cost-effectively secure good levels of water supply service. The term 'conjunctive use' is typically used to describe the effective joint operation of surface water and groundwater. These have complementary characteristics and can work together well – for example, by using surface water to artificially recharge water in aquifers during wet periods such that aquifers can be more intensively exploited during dry periods (Buras, 1963; Todd and Priestaf, 1997; Sprenger et al., 2017). In other countries where groundwater use is less prevalent, or less complementary with surface water, the term conjunctive use can refer to the joint strategic use of infrastructure in general (Fowler et al., 2007).

To finish this high-level overview of water supply management for supply services and water scarcity the topic of governance needs consideration. The multiple needs and processes described above call for specialised institutions and policies. The word 'institution' can describe a variety of arrangements – for example, the water rights or licences discussed above are institutions. The one further example to discuss briefly here is water trading, also referred to with the term water markets (Streeter, 1997; Chong and Sunding, 2006; Brown, 2006). Water trading takes different forms in different jurisdictions, but the overall concept is that one party with the ability to extract water from the environment will transfer that ability to another party. Two different examples, on each end of the spectrum, are the permanent sale of a water entitlement, or short-term 'spot market' transactions, where the right to extraction is exchanged for a short period of time like a week or month.

Water management institutions can also refer to the organisation tasked with setting policies or regulations of water allocation or management. For the former, government ministries or departments exist, typically at national but sometimes at provincial or state level. Regulators or regulating agencies, also with national or more local jurisdiction, in theory with some level of independence from government, have the role of protecting the public and maintaining basic standards – for example, of water quality or environmental protection. These may be charged with regulating water allocation arrangements. Other very different organisations can play major roles. In some countries, the regulator or water ministry mostly acts by way of river basin agencies; this is the case in Spain, for example, where 'hydrographic confederations' have major investment and decision-making powers.

Finally, water utilities are the providers of public water supplies for people and economic operations; they and their regulation, financing and decision-making arrangements have a large impact on how water services are provided, perceived and paid for. In the agricultural sector, irrigation districts (entities that manage irrigation water allocation and delivery) are the relevant organisation. In transboundary basins, water boards or international river commissions may exist; these are often independent bodies or governing councils appointed to oversee water management and allocation decisions within designated regions or jurisdictions. For some sectors, like hydropower, public-private partnerships may be relevant if dams are managed by private sector actors (e.g. a power company), alternatively these will be overseen by a specific hydropower regulator.

4.3.5 The role of economics in managing water

It is important to differentiate between financial and economic considerations and objectives; both relate to monetary sums and values and so the terms are often used interchangeably despite being different. Financial considerations are central to water management, such as the operating and capital costs of infrastructure. These are typically considered by the economic assessments that engineers undertake when considering infrastructure options.

When assessing a groundwater exploitation, for instance, the engineer will quantify the financial (capital) costs required to dig and equip the well and estimate the (operational) pumping costs. The resource economist will ask different questions, such as: what is the long-term profile of groundwater exploitation that would maximise the region's economic benefits, what institution and pricing structure would most likely preserve equity between different well-owners, or what is the economic value to the region of maintaining groundwater levels such that streams are perennial (flow all year long) and contribute to ecosystem services (which are themselves valuated)?

Economics is relevant for both planning and management of water resources and infrastructure; in this section the focus is on the management. A first economic aspect of water management is water pricing – that is, setting the tariffs of water supply for different sectors, either for piped water supply at a home or business, or the cost of exercising a water license (water rights generally don't imply a cost by default as they are a property). Appropriate water pricing is important to secure the financial sustainability of the organisations responsible for water supply and its regulation. Typically, a water utility will receive fees related to piped potable water supplies, whereas an environmental regulator or river basin agency will receive fees related to water licensing.

Water prices are a demand management strategy, as changing the price of water will affect how much is consumed (Marzano *et al.*, 2018). This is in fact one of the major contributions of economics to water management – clarifying that water demands are not a fixed static number, rather, users are willing to pay different amounts to secure different volumes of water and, also, are willing to pay differently for water with different levels of assurance (likelihood that water deliveries will be made). Water supply tariffs need also to be informed by political and social norms and value – for example, water pricing will typically consider social tariffs so that disadvantaged people are not priced out of water services.

One water management institution that exploits the reality of differing willingness-to-pay for water is water markets. Water trading isn't yet widespread globally but has been used effectively in certain regions of the United States, Australia and Chile. Short-term spot markets, in particular, allow water users to temporarily reshuffle water volumes amongst themselves such that those (sectors, business, individuals) with the highest willingness to pay are able to secure the most resource, while financially compensating those giving up water. If done in an open and regulated market, this can work well and maximise the economic efficiency of water use (i.e. ensure that sufficient water is provided to those producers and services in the region which generate the most economic benefits) (Howitt, 1994). Any environment or social benefit can be embedded within a water market by regulating it and imposing constraints on water trades. In countries were markets are poorly perceived or where there would be no way to regulate the market, they are not an appropriate solution. For industrialised nations facing water scarcity, they are generally well perceived by water managers and decision-makers, although careful enforcement of social and environmental goals is essential for public acceptance.

Water for the environment is a challenge for water managers. How much should be reserved to benefit the environment? There is no single accepted answer or approach to this, but economics provides a practical tool by way of water valuation – that is, the estimation of the economic value of water for certain uses, including ecosystem services like water purification, flood control, habitat provision and recreational opportunities. Economists will use different techniques, like, for example, estimating the recreation value of a water body based on what visitors spend to visit it, or other 'nonmarket valuation' methods, such as hedonics and contingent valuation, to arrive at an estimate of the economic value of water. Another valuation approach is to impose an environmental flow constraint (like a minimum environmental river flow) in a hydro-economic model (Harou *et al.*, 2009; Ortiz-Partida *et al.*, 2023), then estimate water's value by assessing its opportunity cost to other economic water uses. This value of environmental water to society estimated by applied resource economists can be compared to other economic values to inform water allocation or other policies (like minimum environmental flows) and ensure that the environment is appropriately considered amongst competing water demands.

Other direct economic incentives, such as subsidies, grants, tax incentives or rebates, can encourage water consumers to adopt water-saving technologies and practices, invest in water-efficient infrastructure and implement conservation measures.

4.3.6   The use of models to inform water management decisions

Water resources management simulation models (Tomlinson *et al.*, 2020) are computer models that track the presence and use of water throughout the human natural system, such as a water utility service area, a river basin or a national scale resource system. Their data inputs include hydrological inflows, groundwater availability, evaporation rates, allocation priorities, seasonal water demands for all locations of water abstraction, infrastructure characteristics, operating rules and so on, and their outputs are flows, water diversions and storage throughout the system. These models represent the water system as a connected network of nodes (inflow points, infrastructure assets, river junction, locations of water demands) and links that join them (representing conveyances like rivers, canals or pipelines) (Jakeman and Letcher, 2003).

Hydrological inflows, aggregated by river reach and runoff entering the stream over an area, will be 'injected' at a single node directly downstream of the reach. Each demand node will be able to abstract river water following its priority and will consume a fraction of its allocated water volume, then return the rest to the water system by way of return flows. Reservoirs are modelled with storage nodes where the volume stored is tracked and must stay between some minimum threshold and the reservoir's maximum capacity. Reservoir rule curves, for example, that relate release to reservoir level and season, determine how much is released. Aquifers can be included as simplified underground storage volumes with an associated pumping cost or, in a more sophisticated way, by linking to groundwater models of varying levels of spatial aggregation which model the feedback between abstraction (groundwater pumping) and aquifer levels (and therefore baseflow to rivers). In a simulation, the model marches through time, time-step by time-step (typically a day, week or month), allocating water throughout the network and executing infrastructure operations until the model is ready to pass to the next time-step. At the end of a simulation, the user can pick any location in the network – for example, a demand node, a storage location and so on – and view a time-series plot showing how much water was available and used at that location during the time-step of the simulation (i.e. over the whole simulated 'time-horizon').

There are several reasons why computer-based simulation models of managed water resource systems have been widely used by water managers globally over the last 50 years and are being increasingly used.

A first reason is complexity; water systems are often spatially large and they change rapidly over time, so there is no real alternative to using a computerised accounting system to track where and when water enters the system, is managed (by infrastructure, e.g. to store or convey water), is used (and potentially consumed) and is returned to surface water bodies or percolated to groundwater and so on. All of these details build a detailed quantified description of how the water system behaves over time and space.

Secondly, water management models allow characterising and summarising the water system's performance (i.e. how well or poorly the system performs for humans or nature). Performance is tracked by way of 'metrics', which quantify various measures of system performance – for example, reliability, vulnerability and resilience (introduced in previous section) at specific nodes of over wider aggregated areas in the water resource system. These key performance indices can quantify performance from a holistic (societal) point of view, or from the viewpoint of a specific water user or organisation such as a water utility. These metrics describe the performance achieved by making one or more specific changes in the system (e.g. like a new infrastructure asset or operating procedure). This is the main purpose of system simulation – assessing an intervention by simulating the system with and without it. Depending on the sophistication of the simulator, a wide range of interventions could be evaluated under many possible conditions, such as scenarios that characterise future uncertainties or severe and historically unexperienced droughts. A classical water system simulator, for example, could be used to quantify the system scale impacts of a new water transfer or reservoir. More specialised water management models, like a hydro-economic model (Harou *et al.*, 2009), could be asked to evaluate the change in benefits resulting from introducing a water market. Or a multi-agent model (Yoon *et al.*, 2021) could quantify how a policy might affect some sectoral or social group. These different types of water resource models typically all share a node-link structure, also some might be 'fed' by hydrological variables generated over a regular grid (a mesh that digitally overlays the landscape).

A third, relatively recent, use of water resource allocation models is to connect them to an external optimising algorithm (referred to as 'heuristic optimisation') (Maier *et al.*, 2019). This optimiser is led by a 'formulation' – that is, a statement of the objectives and constraints of the decision or design problem; an example would be to optimise reservoir operating rules for cost and reliability subject to (i.e. constrained by) minimum environmental flows. This process allows seeking the most efficient packages of interventions – that is, those water management policies that maximise benefits (for one benefit) or find the most efficient trade-offs between two or more benefits to identify robust portfolios of interventions. An example of the latter is provided by Matrosov *et al.* (2015) for London's water supply system, and these techniques have been extensively used in the east of England to assess trade-offs between public water supply, environmental needs and the agricultural and energy sectors (Water Resources East, 2023).

4.4. Water supply planning

Water supply planning is a central component of water resources planning. The planner's task is ensuring that future water supplies can meet future multi-sector water demands at appropriate levels of service, thus guaranteeing society's resilience from a water and environment perspective. Further objectives will typically insert themselves into the planning problem depending on context, such as meeting the future supply-demand balance at low- or least-cost, maintaining and enhancing environmental quality and ecosystem services, dealing effectively and robustly with future uncertainty in both supply and demand, ensuring conflicts between economic sectors and/ or regions are not exacerbated by water scarcity, and balancing water supply needs with other societal objectives (e.g. social equity, carbon reduction, increasing biodiversity and other social or political objectives).

In addition to the positive goals, there are some more defensive ones. Planners want to avoid investing in and implementing interventions too early or too late. Unused capacity (i.e. extra water supply available too soon), given the time-value of money, is inefficient – that is, it unnecessarily increases societal costs. Conversely, insufficient capacity will lead to human-induced shortfall or scarcity events, whose economic, social and environment impacts could have been avoided.

Insufficient supplies carry a high social cost (due to inability to maintain economic production processes where water is an input – for example, production is temporarily halted in a drought or in the longer term, companies that might have been created but weren't, or companies were located elsewhere because of their inability to secure water supplies).

What is the remit of water planners? They may be involved in a range of tasks, from long-term strategic infrastructure decision-making (10–50 years) to mid-term time-scales (2–10 years) or tactical (seasonal to multi-annual) forecasting. In the mid-term, their work dovetails with that of asset management and maintenance departments, which in turn may interface with short-term management or 'operations' units – that is, those who manage the day to day operations and assets of a water company, including dealing with outage and short-term service quality.

Water planners aim to select and deliver interventions to ensure water supply and water resource systems function well for humans and the environment. By interventions, reference is primarily made to water infrastructure, although the ability to change management policies (reviewed in the previous section) must also be considered by planners. Policies can be changed within months or a few years, and their analysis is shared with other departments of water supply organisations (like regulatory compliance, asset management or even operations). The fact that water supply system interventions can be both policies (management changes) or new infrastructure (changes to the typology of composition of the supply-demand network) is a driver of complexity for water planning as is seen below. The complexity stems from the fact that policies and infrastructure interact, so the problem becomes non-linear (i.e. when different interventions interact amongst themselves, then interventions don't simply lead to a volume of 'new water' that can be cumulated, indeed every portfolio of interventions is unique and will need to be tested to ensure they are compatible and work well together).

Like advanced water management, effective water planning requires specialised skills and institutions. It has appropriately been described as a wicked problem (Rittel and Webber, 1973; Reed and Kasprzyk, 2009) – that is, a problem which even the most sophisticated operations research and risk management methods, from mathematics and statistics to systems engineering, struggle with. This contrasts with the budgets and human resources of water planning departments which are typically limited. Still, a combination of water planning practitioners from utilities, river basin agencies and government ministries, development specialists from multi-lateral donor agencies, and academics have made great strides in solving this problem in an effective and intellectually satisfying way which improves water planning outcomes.

4.4.1 Water planning complexity: multiple uncertainties, sectors and institutions

A central difficulty of water planning is uncertainty. It takes many forms, and the easier ones to consider are discussed first.

The first is hydrological variability. This describes the natural fluctuations in weather and, therefore, in hydrological fluxes and states (river flows, groundwater levels etc.). These can be characterised statistically, but a first complication is climate change and the fact that historical data may be increasingly inaccurate to characterise those sources in the future. Another term for this is non-stationarity – that is, the statistical descriptors (mean, variance etc.) of water availability data (river flows, aquifer levels etc.) change over time (Milly *et al.*, 2008). In some cases, climate-change linked uncertainty can be even more problematic, and the very direction of change is unknown (e.g. will it get wetter or drier?). In that case the situation is one of deep uncertainty – that is, where statistical prediction methods become difficult or impossible to use because analysts do not know or cannot agree on the probabilities of future events or their statistical distributions. This is true for many important variables of water supply, beginning with water demands. For example, public water supply demands per capita increased for decades but have begun to drop in many countries (as environmental consciousness grows, supplies become metered, and water conservation regulations on building and appliances tighten). For a planner facing a river basin anywhere in the world, will water demand for public water supply, irrigation and energy generation plant cooling increase or decrease in the next 20–50 years? Unfortunately, it is often difficult to say. The technologies or societies of the future are unknown, how and where will they grow food, how they will generate electricity, what appliances will they have in the home, what social norms they will be modulated by, and what national and global economic conditions will prevail? These factors, decisive in setting future water demands, are deeply uncertain.

Further consideration of institutional uncertainty is explored. Future water management may differ from today's, and this will determine how fit for purpose infrastructure investments ultimately are. An example of this is the city of Santa Barbara, in California, investing in a desalination plant which ultimately wasn't used because, at the time of its commissioning, a drought led to the legalisation of water trading in California – purchasing water during droughts from farmers was cheaper for Santa Barbara then operating a desalination plant, which was eventually decommissioned. Water system uncertainties, and in particular deep uncertainties, can lead to severe social, economic and political costs.

To conclude on water planning uncertainties, it should be mentioned that uncertainties will manifest together, and this will compound risks (a term used when probabilities are relatively known) and deep uncertainties (where probabilities are relatively unknown). Simultaneous presence and interaction of multiple sources of risk and deep uncertainty amplifies the complexity of planning water systems.

Next, consideration is given to the multiple sectors using water. It would be much easier if planners responsible for public water supply could work amongst themselves and consider just drinking water needs – this can occur in locations without water scarcity. However, these situations are increasingly rare, and water supply planning is increasingly impossible to do properly without increasing its scope to water resources planning, where, in addition to public water supply for individuals, towns, cities and businesses, the planner must consider supplies for the agricultural, energy and environment sectors. Since the 1980s the term 'integrated water resources management' has been used to describe the challenge of balancing competing water demands and achieve sustainable and equitable water allocation, but the challenge has grown in scope. Today, scientific interest has moved to new terminologies; researchers and practitioners now speak of human-natural systems (Moallemi et al., 2020) and multisector dynamics (Reed et al., 2022). These new terms stem from a realisation that a water-centric approach (where planners consider a sector's numerical demand, rather than what is actually generating water demand in that sector) will be insufficient to generate the inter-sectoral collaboration and synergies needed to adapt to increasingly scarce water resources and high demands.

Also, shared challenges have increased the scope of water resource planning – for example, green energy transformation calls for the water sector to contribute to the effort of lowering greenhouse gas emissions, not only from making water pumping and heating more efficient, but also, for example, from enabling electrical grids to take on more intermittent renewables. Hydropower is a dispatchable energy source which can act as a battery, allowing an electrical grid to use variable sources like solar and wind energy; but doing so means deviating reservoirs from water supply objectives towards energy transformation objectives – see Gonzalez *et al.* (2023) for a recent example at national scale. Benefits can flow both ways – for example, it has been shown that energy trades can decrease water conflicts between countries or regions (Etichia *et al.*, 2024).

Finally, institutionally, politically and legally, water supply and water resources planning problems can be complex for many reasons. One is the multiple cultural, economic and social facets of water. Water is at the same time an input into economic production, but also an environment sustaining good which, in many cases, has cultural and even religious significance. This multifaceted water value means that, in addition to engineering and economic considerations, water plans may have a social and political element, which will manifest to a greater or lesser degree depending on the context. Public and private water planning institutions must recognise and adapt to these different facets, which means that water planning bodies, and their government and regulatory overseeing organisations, must adapt uniquely to their country. In summary this means that each country's water planning institutions are different and, therefore, the process of water planning needs to be fashioned (adapted) to those institutions, in addition to the geographical context (hydrology, demands etc.).

It should be mentioned that, because hydrological boundaries (alternatively referred to as catchments, watersheds or river basins) often don't correspond to those of water supply organisations (e.g. utilities), of local political jurisdictions (states, provinces) or even national boundaries (many river basins are transboundary), water plans can fall under multiple political and legal jurisdictions. Whenever underground water (aquifers) or basins cut across political boundaries, the transboundary institutional context can become considerably more complicated. Approximately 50% of the world's landmass (excluding Antarctica) is in transboundary basins, and 40% of the world's population lives within them.

4.4.2   Water planning approaches, methods and tools

The approaches, processes and tools of water planning continuously evolve to adapt to water supply challenges, including those described above. In the last 20 years, the pace of progress (i.e. sophistication of water planning approaches and methods) has quickened with heightened water scarcity and a greater appreciation of the importance of water security and the need to transform the world's resource systems to lower greenhouse gas emissions.

This section begins by reviewing traditional methods and finish with the state of the art. Water supply planning can be (and was frequently) viewed as an infrastructure capacity investment problem. In engineering this is referred to as the capacity expansion problem, and can be posed similarly across the water, energy and transport sectors. These sectors can represent their system as a network of locations of either supply or demand of a resource – their objective is to route the resources through the network, from supply points to demand 'nodes', at least cost. This approach can be used to solve two problems: minimising the cost of operating the system (moving the resource from supply nodes to demand nodes) and minimising the cost of investment in new supply nodes or connections (e.g. new dams, wellfields, pipelines or canals). Such a framework has been used since the 1970s in power system engineering. In the early 2000s it was adopted as part of the national regulation system for water companies in England and Wales (UKWIR, 2002). Utilities must use optimisation models that implement this approach to justify their selection and timing of investments in new infrastructure and demand management programmes. The model 'sees' net water demands changing each year (typically increasing) and activates new supply or conveyance options throughout the network in order to minimise the sum of operating (e.g. treatment, pumping) and capital costs (investing in new infrastructure) (Padula et al., 2013). Each new option would be located somewhere in the network, with its associated supply volume, operating cost and capital cost (just three numbers), and wait for its turn to be 'chosen' as part of the least-cost investment mix.

The supplies available from each source (potential infrastructure option) are determined by water utility (or company) hydrologists or their consultants. To do this they characterise the productivity of aquifers and measure river flows throughout the basin; analysing such data statistically and potentially simulating the hydrological and infrastructure system gives water planners an idea of what minimum daily water volume supply they can count on from an individual source, even in dry years. Water supply planners refer to this as the yield or safe yield ('deployable output' in the UK) of a source and the concept can be extended to evaluate the conjunctive yield of a water supply system.

Based on the discussion above on uncertainty, the reader will see a first problem: if past data can no longer be counted on to plan future water supply (because of a changing climate), then reliable estimates of future yields cannot be made, as they become deeply uncertain. This was an existential crisis for water planners, who had depended on planning water supplies using historical data for decades (Milly et al., 2008). But other 'cracks' in this least-cost supply-demand planning began to appear and were discussed in more detail by Hall et al. (2012), Padula et al. (2013) and Matrosov et al. (2013a). In summary, demand and new supply options became recognised as non-stationary and potentially uncertain; stakeholders expressed the view that they would prefer a method to maximise various benefits rather than just minimising monetary costs, and it became clear that, as infrastructure systems became more complex and interconnected (i.e. non-linear), assets were interacting amongst themselves and, therefore, a planning method that considers each supply option as individual, linearly adding an independent cumulative 'new volume' of reliable water to the supply-demand balance, was increasingly inaccurate.

The 'predict and provide approach' outlined above was shared across water, energy and transport engineers – it was going to be hard to dislodge. It was challenged by two ideas widely shared by planners but, until then, mostly ignored by academics and consultants.

(a) Planners are risk averse. They are less concerned with being optimal (e.g. least cost) and more concerned with avoiding spectacular failures. Indeed, planners have always been aware of deep uncertainties (in short, things that could happen) and, therefore, are typically interested in investment options that work robustly (i.e. that work well across a wide range of plausible futures) (Herman et al., 2015). This was the idea behind 'robust decision-making' (Lempert et al., 2006), an approach which became influential with water planners and experts in different contexts around the world (Matrosov et al., 2013b; Dessai and Darch, 2014; Hurford et al., 2020).

(b) Planners have long recognised the value of flexibility. An investment theory called 'real options' conceptualised this as early as 1977. The idea is that there is great value in selecting investments which allow later adaptation, if predictions of the future turn out to be inaccurate. In water supply planning, this encourages adoption of adaptive planning approaches (Haasnoot et al., 2013) and even advanced optimisation models, when early interventions (e.g. new assets) are selected by also considering how economically they can adapt to later surprises (unexpected levels of supply or demand).

Changing views on the importance of uncertainty and deep uncertainty impacted water planning analysts, but other ideas motivated new approaches as well. Least-cost water supply planning was largely unidimensional, or other objectives had to be monetised (which is difficult, time-consuming and open to criticism). Being able to optimise across multiple objectives would be very handy, and, in fact, had been possible (for simple cases) since the 1970s using methods of multi-criteria analysis and multi-objective optimisation. The problem was that water supply systems are anything but simple: supply options are often mutually dependent which creates non-linearities, and infrastructure assets are typically managed with rule books ('if this happens, then do that') rather than smooth mathematical functions. Both of these points greatly complicate the use of mathematical optimisation. This led to the appearance from the 2010s of heuristic search methods (i.e. a new crop of search algorithms which used AI tools rather than maths) which work reliably with real-world systems, and can optimise in up to 10 dimensions, creating new possibilities for water planners. The most influential is the AI-assisted design approach using multi-objective evolutionary algorithms (Maier *et al.*, 2019; Reed *et al.*, 2013) already mentioned at the end of Section 4.3. Because the search algorithm is independent of the simulation, this frees the simulation to represent real-world detail, including the existence and function of institutional governance processes and multi-sector linkages (Yoon *et al.*, 2021; Gonzalez *et al.*, 2023).

In summary, water planning methods have advanced considerably in the past two or three decades. Progress has been driven by the recognition that, given the large costs of infrastructure investments and their potential to disrupt environmental and human systems, their selection, sizing and timing of implementation should be justified using rigorous methods. These methods couple reliable water resource simulation models (discussed at the end of Section 4.3 on water management) and link them to multi-objective design under uncertainty approaches, which can optimise considering multiple dimensions and allow stakeholders to assess the performance trade-offs implied by the most efficient packages of interventions (water supply investment options and policies). They can handle the existence of multiple sources of probabilistic and deep uncertainty by outputting either robust plans (plans that work acceptably over a range of scenarios) or adaptive pathways (i.e. adaptive investment rules which select assets progressively as time goes on and information on future supply and demand is progressively revealed). Recent examples of applications using approaches on real case-studies are available – for example, in UK water systems at regional scale (Water Resources East, 2023) and for individual water companies (Murgatroyd and Hall, 2021; Pachos *et al.*, 2022) or in multi-sector resource systems in Africa (Basheer *et al.*, 2023; Gonzalez *et al.*, 2023).

4.4.3 Economics, finance and governance in water supply planning

Water planning and economics have a special relationship: major concepts of modern microeconomics (Ekelund and Hebert, 1999) and public economics (Banzhaf, 2009) were first conceptualised in water planning exercises. Harou (2023) summarises the use of economic methods and concepts in water planning over the last 100 years or so.

Here it is highlighted that the modern water planning approaches, outlined at the end of the previous subsection, are compatible with economic theory in that they assist with selecting infrastructure investments and policies which efficiently meet the goals of multiple stakeholders. Choosing water interventions that are Pareto optimal (i.e. where improving one dimension of performance can't be achieved without simultaneously diminishing one or more other measures of performance) is a form of economic optimisation. The multi-objective optimised design approaches discussed in Sections 4.3 and 4.4 can optimise for cost-benefit criteria expressed in monetary and other metrics.

In addition to multi-dimensional efficiency (Pareto-efficiency), human time-preferences (people prefer benefits sooner than later) and the value of adaptability can also be considered by way of economic approaches in water planning. Optimised water planning infrastructure selection models consider net present values of future assets (i.e. discounting costs and benefits that occur in the future) and adaptive planning approaches recognise the value inherent in adaptive or 'flexible' interventions (i.e. low regret options which can adapt to a variety of futures with minimal loss). The economic investment theory of real options is conceptually similar to and compatible with adaptive planning approaches that are increasingly popular with researchers (Kwakkel et al., 2015) and regulators (Ofwat, 2022).

In addition to deciding what interventions to fund as part of a future water plan, another pertinent question is how to fund them. Alongside processes and tools of classical infrastructure and utilities finance, the emerging area of climate finance will become increasingly relevant as the water sector aims to adapt to and mitigate climate change. Climate finance refers to the financial resources and instruments designed to support climate change mitigation and adaptation efforts. Increasingly, water planners can access advantageous infrastructure financing funds or conditions if they are able to demonstrate how their plans support climate change adaptation and mitigation goals. Again, the multi-objective design approaches outlined in sections 4.3 and 4.4 allow this by adding such goals (by way of suitable metrics) into the planning problem (i.e. into the list of objectives the plan is being optimised for).

Financial incentives and goals of human-natural systems like water resource and supply systems are set out in policies designed by water governance organisations. Governance arrangements, and the organisations that implement and regulate them, vary by country as discussed above. In Section 4.1 on governance challenges, water utilities, river basins agencies, government water ministries and environmental regulators were mentioned. International organisations, including funders like bilateral development agencies or multi-lateral development banks, may also be relevant if they are providing funding; international or national non-governmental organisations may also get involved, in particular in locations of outstanding natural beauty or ecosystem importance. Civil society organisations like community and indigenous groups will be relevant stakeholders in many countries.

Depending on the national context, typically, several organisations will play a role in setting the rules and policy objectives of water supply and/or water resources planning processes. Often particular agencies will intervene at different spatial scales (Section 4.1) and there may be partially overlapping jurisdictions. In England, for example, national environmental and financial regulators oversee nearly two-dozen privatised water companies. But an increasing amount of planning activity and analysis is now undertaken by multi-sector regional groups (Leonard et al., 2024) and is reviewed and complemented with national scale analysis (Harou, 2019). In other UK nations, a single regulator and a water service provider exist. Evolving institutions which are created to address context-specific national or regional challenges, as part of a societal or social learning process (Pahl-Wostl, 2009), is the norm. As discussed in the first section, human-natural water resource systems will develop governance bodies and technical model-supported planning tools suited to the planning tasks needed to address their geographical, historical, political and economic context-specific issues.

4.5.   Future challenges and opportunities

Water resources management and planning encompasses a wide range of activities, from managing water supplies and their quality to securing environmental services and climate adaptation. Its success requires integrating scientific data and models, governance processes and stakeholder engagement, to ensure multiple simultaneous dimensions of performance can be reached and together achieve wider societal goals like water security and resilience.

With growing water scarcity and increased availability of data and computing, the use of models in water management and planning has become increasingly sophisticated and ubiquitous, allowing for better assessment and design of water resource system interventions. The inclusion of economic drivers, multi-sector interactions, governance processes and institutional detail will allow simulators to more realistically explore and describe the multi-faceted performance of human-natural water systems. The ability to link these simulators to AI-driven multi-dimensional search algorithms means water planners can already derive efficient, robust and adaptive intervention portfolios or pathways; the demand for such capabilities will expand.

The importance and expectations of water governance will grow commensurably with supply-demand imbalances, multi-sector interactions, and increased expectations for stakeholder consultation to evolve towards co-production, all set in a context of increasing pressures on water resources and uncertain futures due to a changing climate. To deal with the complexities of human-water resource systems, institutions and regulatory frameworks will need to be increasingly robust, adaptable and efficient, ideally being deployable at multiple scales. They will need to balance economic efficiency, environmental sustainability and social equity, ensuring that water policies and investment plans are fair and transparent.

In conclusion, water supply and resource planning and management is a multifaceted field that requires a coordinated, interdisciplinary approach to address complex human-natural system challenges. Given these, effective governance will be most advanced by a new crop of technical analysts keen and able to secure and protect water resources for current and future generations.

REFERENCES

Banzhaf HS (2009) Objective or multi-objective? Two historically competing visions for benefit-cost analysis. *Land Economics* **85**: 3–23.

Basheer M, Nechifor V, Calzadilla A *et al.* (2023) Cooperative adaptive management of the Nile River with climate and socio-economic uncertainties. *Nature Climate Change* **13**: 48–57.

Bredehoeft JD (2002) The water budget myth revisited: Why hydrogeologists model. *Ground Water* **40**: 340–345.

Brown TC (2006) Trends in water market activity and price in the western United States. *Water Resources Research* **42**.

Buras N (1963) Conjunctive operation of dams and aquifers. *Journal of the Hydraulics Division*, ASCE (American Society of Civil Engineers) **89**: 111–131.

Butler AP, Mathias SA, Gallagher AJ, Peach DW and Williams AT (2009) Analysis of flow processes in fractured chalk under pumped and ambient conditions (UK). *Hydrogeology Journal* **17**: 1849–1858.

Chan, Wilson CH, Arnell NW *et al.* (2023) Current and future risk of unprecedented hydrological droughts in Great Britain. *Journal of Hydrology* **625**: 130074.

Chong H and Sunding D (2006) Water markets and trading. *Annual Review of Environment and Resources* **31**: 239–264.

Cominola A, Giuliani M, Piga D, Castelletti A and Rizzoli AE (2015) Benefits and challenges of using smart meters for advancing residential water demand modeling and management: A review. *Environmental Modelling & Software* **72**: 198–214.

Dawkins LC, Osborne JM, Economou T *et al.* (2022) The advanced meteorology explorer: a novel stochastic, gridded daily rainfall generator. *Journal of Hydrology* **607**: 127478.

Defra (Department for Environment, Food and Rural Affairs) (2021) Changes to the regulatory framework for abstraction and impounding licensing in England – Moving into the Environmental Permitting Regulations regime – Consultation Document. **In, edited by DEFRA**.

Dessai S and Darch G (2014) Managing uncertainties in adapting water resource systems to a changing climate in England and Wales. In Prutsch A, Grothmann T, McCallum S, Schauser I and Swart R (eds.) *Climate Change Adaptation Manual: Lessons learned from European and other industrialised countries*. Routledge, Abingdon, UK.

Ekelund RB and Hebert RF (1999) *Secret origins of modern microeconomics: Dupuit and the engineers*. University of Chicago Press, Chicago, IL, USA.

Environment Agency (2020) Meeting our future water needs: A national framework for water resources **In, edited by Environment Agency**. Environment Agency.

Etichia M, Basheer M, Bravo R *et al.* (2024) Energy trade tempers Nile water conflict. *Nature Water*.

Fowler HJ, Kilsby CG and Stunell J (2007) Modelling the impacts of projected future climate change on water resources in north-west England. *Hydrology and Earth System Sciences* **11**: 1115–1124.

Garrick D, De Stefano L, Yu W *et al.* (2019) Rural water for thirsty cities: A systematic review of water reallocation from rural to urban regions. *Environmental Research Letters* **14**: 043003.

Gonzalez JM, Tomlinson JE, Martínez Ceseña EA *et al.* (2023) Designing diversified renewable energy systems to balance multisector performance. *Nature Sustainability*, 2023.

Grafton RQ, Williams J, Perry CJ *et al.* (2018) The paradox of irrigation efficiency. *Science* **361**: 748–750.

Grill G, Lehner B, Thieme M *et al.* (2019) Mapping the world's free-flowing rivers. *Nature* **569**: 215–21.

Haasnoot M, Kwakkel JH, Walker WE and ter Maat J (2013) Dynamic adaptive policy pathways: A method for crafting robust decisions for a deeply uncertain world. *Global Environmental Change: Human and Policy Dimensions* **23**: 485–498.

Hall JW, Watts G, Keil M *et al.* (2012) Towards risk-based water resources planning in England and Wales under a changing climate. *Water and Environment Journal* **26**: 118–129.

Harou JJ, Garrone P, Rizzoli AE *et al.* (2014) Smart metering, water pricing and social media to stimulate residential water efficiency: Opportunities for the SmartH2O project. *Procedia Engineering* **89**: 1037–1043.

Harou JJ (2019) Towards a national water resources planning framework in England. *Proceedings of the Institution of Civil Engineers – Water Management* **172**: 271–72.

Harou JJ (2023) Policy note: Artificial intelligence enables multi-dimensional economics of water. *Water Economics and Policy* **09**: 2371003.

Harou JJ, Pulido-Velazquez M, Rosenberg DE *et al.* (2009) Hydro-economic models: Concepts, design, applications and future prospects. *Journal of Hydrology* **375**: 334–50.

Hashimoto T, Stedinger JR and Loucks DP (1982) Reliability, resiliency, and vulnerability criteria for water-resource system performance evaluation. *Water Resources Research* **18**: 14–20.

Herman JD, Reed PM, Zeff HB and Characklis GW (2015) How should robustness be defined for water systems planning under change?'. *Journal of Water Resources Planning and Management* **141**: 04015012.

House-Peters LA and Chang H (2011) Urban water demand modeling: Review of concepts, methods, and organizing principles. *Water Resources Research* **47**, 1–15.

Howitt RE (1994) Empirical-analysis of water market institutions – the 1991 California water market. *Resource and Energy Economics* **16**: 357–371.

Hurford AP, Harou JJ, Bonzanigo L *et al.* (2020) Efficient and robust hydropower system design under uncertainty – A demonstration in Nepal. *Renewable and Sustainable Energy Reviews* **132**: 109910.

Jakeman AJ and Letcher RA (2003) Integrated assessment and modelling: features, principles and examples for catchment management. *Environmental Modelling & Software* **18**: 491–501.

Keating T (1982) A lumped parameter model of a chalk aquifer-stream system in Hampshire, United Kingdom. *Groundwater* **20**: 430–36.

Kratzert F, Klotz D, Shalev G *et al.* (2019) Towards learning universal, regional, and local hydrological behaviors via machine learning applied to large-sample datasets. *Hydrology and Earth System Sciences* **23**: 5089–5110.

Kwakkel JH, Haasnoot M and Walker WE (2015) Developing dynamic adaptive policy pathways: a computer-assisted approach for developing adaptive strategies for a deeply uncertain world. *Climatic Change* **132**: 373–386.

Lagos MS, Muñoz JF, Suárez FI *et al.* (2024) Investigating the effects of channelization in the Silala River: A review of the implementation of a coupled MIKE-11 and MIKE-SHE modeling system. *Wiley Interdisciplinary Reviews – Water* **11**.

Lempert RJ, Groves DG, Popper SW and Bankes SC (2006) A general, analytic method for generating robust strategies and narrative scenarios. *Management Science* **52**: 514–528.

Leonard A, Amezaga J, Blackwell R, Lewis E and Kilsby C (2024) Collaborative mutli-scale water resources planning in England and Wales. In *EGU (European Geophysical Union) Conference*, Vienna.

Lund JR and Israel M (1995) Water transfers in water-resource systems. *Journal of Water Resources Planning and Management*, ASCE **121**: 193–204.

Macpherson E *et al.* in press. Setting a pluralist agenda for water governance: Why power and scale matter. *WiresWater*.

Maier HR, Razavi S, Kapelan Z *et al.* (2019) Introductory overview: Optimization using evolutionary algorithms and other metaheuristics. *Environmental Modelling & Software* **114**: 195–213.

Marzano RR, Rougé C, Garrone P *et al.* (2018) Determinants of the price response to residential water tariffs: Meta-analysis and beyond. *Environmental Modelling & Software* **101**: 236–248.

Matrosov ES, Padula S and Harou JJ (2013a) Selecting portfolios of water supply and demand management strategies under uncertainty-contrasting economic optimisation and 'robust decision making' approaches. *Water Resources Management* **27**: 1123–1148.

Matrosov ES, Woods AM and Harou JJ (2013b) Robust Decision Making and Info-Gap Decision Theory for water resource system planning. *Journal of Hydrology* **494**: 43–58.

Matrosov ES, Huskova I, Kasprzyk JR *et al.* (2015) Many-objective optimization and visual analytics reveal key trade-offs for London's water supply. *Journal of Hydrology* **531(3)**: 1040–1053.

Milly PCD, Betancourt J, Falkenmark M *et al.* (2008) Climate change – Stationarity is dead: Whither water management?. *Science* **319**: 573–574.

Moallemi EA, Kwakkel J, de Haan FJ and Bryan BA (2020) Exploratory modeling for analyzing coupled human-natural systems under uncertainty. *Global Environmental Change – Human and Policy Dimensions* **65**.

Molle F (2003) Development trajectories of river basins: A conceptual framework. **In.: International Water Management Institute**.

Murgatroyd A and Hall JW (2021) Selecting indicators and optimizing decision rules for long-term water resources planning. *Water Resources Research* **57**: e2020WR028117.

Murphy JM, Harris GR, Sexton DMH *et al.* (2018) UKCP18 land projections: science report.

Nauges C and Whittington D (2017) Evaluating the performance of alternative municipal water tariff designs: Quantifying the tradeoffs between equity, economic efficiency, and cost recovery. *World Development* **91**: 125–43.

O'Shea MJ, Baxter KM and Charalambous AN (1995) The hydrogeology of the Enfield-Haringey artificial recharge scheme, north London. *Quarterly Journal of Engineering Geology* **28**: S115–S29.

Ofwat (2022) *Price Review 2024 and beyond: Final guidance on long-term delivery strategies.* Ofwat, Birmingham, UK.

Ortiz-Partida JP, Fernandez-Bou AS, Maskey M *et al.* (2023) Hydro-economic modeling of water resources management challenges: current applications and future directions. *Water Economics and Policy* **0: null**.

Pachos, Huskova I, Matrosov E, Erfani T and Harou JJ (2022) Trade-off informed adaptive and robust real options water resources planning. *Advances in Water Resources* **161**.

Padula S, Harou JJ, Papageorgiou LG *et al.* (2013) Least economic cost regional water supply planning – optimising infrastructure investments and demand management for South East England's 17.6 million people. *Water Resources Management* **27**: 5017–5044.

Pahl-Wostl C (2009) A conceptual framework for analysing adaptive capacity and multi-level learning processes in resource governance regimes. *Global Environmental Change: Human and Policy Dimensions* **19**: 354–65.

Perry C (2007) Efficient irrigation; Inefficient communication; Flawed recommendations. *Irrigation and Drainage* **56**: 367–78.

Puy A, Sheikholeslami R, Gupta HV *et al.* (2022) The delusive accuracy of global irrigation water withdrawal estimates. *Nature Communications* **13**.

Reed PM, Hadka D, Herman JD, Kasprzyk JR and Kollat JB (2013) Evolutionary multiobjective optimization in water resources: The past, present, and future. *Advances in Water Resources* **51**: 438–56.

Reed PM and Kasprzyk J (2009) Water resources management: The myth, the wicked, and the future. *Journal of Water Resources Planning and Management – ASCE* **135**: 411–413.

Reed PM, Hadjimichael A, Moss RH *et al.* (2022) Multisector dynamics: Advancing the science of complex adaptive Human-Earth Systems. *Earth's Future* **10**: e2021EF002621.

Rittel HWJ and Webber MM (1973) Dilemmas in a general theory of planning. *Policy sciences* **4**: 155–169.

Rockstrom J, Mazzucato M, Andersen LS, Fahrlander SF and Gerten D (2023) Why we need a new economics of water as a common good. *Nature* **615**: 794–797.

Rougé C, Harou JJ, Pulido-Velazquez M *et al.* (2018) Assessment of smart-meter-enabled dynamic pricing at utility and river basin scale. *Journal of Water Resources Planning and Management* **144**: 04018019.

Serinaldi F and Kilsby CG (2012) A modular class of multisite monthly rainfall generators for water resource management and impact studies. *Journal of Hydrology* **464**: 528–40.

Sprenger C, Hartog N, Hernández M *et al.* (2017) Inventory of managed aquifer recharge sites in Europe: historical development, current situation and perspectives. *Hydrogeology Journal* **25**: 1909–1922.

Streeter R (1997) Tradeable rights for water abstraction. *Water and Environment Journal* **11**: 277–281.

Todd DK and Priestaf I (1997) Role of conjunctive use in groundwater management. In Kendall DR (ed.) *Proceedings of AWRA Symposium on conjunctive use of water resources: aquifer storage and recovery.* American Water Resources Association, Long Beach, California, USA.

Tomlinson JE, Arnott JH and Harou JJ (2020) A water resource simulator in Python. *Environmental Modelling & Software* **126**: 104635.

UKWIR (2002) *The Economics of Balancing Supply & Demand (EBSD) – Main Report.* **In, edited by UKWIR.** London: NERA.

Wade SD, Lloyd-Hughes B, Sanderson M *et al.* (2015) H++ climate change scenarios for heat waves, cold snaps, droughts, floods, windstorms and wildfires. **In, edited by Adaptation Sub Committee Climate Change Risk Assessment. Met Office, University of Reading, Centre for Ecology and Hydrology.**

Ward FA and Pulido-Velazquez M (2008) Water conservation in irrigation can increase water use. *Proceedings of the National Academy of Sciences of the United States of America* **105**: 18215–18220.

Water Resources East (2023) Regional Water Resources Plan for Eastern England. **In.: Water Resources East (WRE).**

Wheater HS and Gober P (2015) Water security and the science agenda. *Water Resources Research* **51**: 5406–5424.

Wheater HS, Jakeman AJ and Beven KJ (1993) Progress and directions in rainfall-runoff modelling. In Jakeman AJ, Beck MB and McAleer MJ (eds.) *Modelling Change in Environmental Systems.* Wiley, NY, USA.

Wheater HS, Peach DW, Suárez F and Muñoz JF (2024) Understanding the Silala River – Scientific insights from the dispute over the status and use of the waters of the Silala (Chile v. Bolivia). *WIREs Water* 11: e1663.

Yassin F, Razavi S, Wheater H *et al.* (2017) Enhanced identification of a hydrologic model using streamflow and satellite water storage data: A multicriteria sensitivity analysis and optimization approach. *Hydrological Processes* **31**: 3320–3333.

Yoon J, Klassert C, Selby P *et al.* (2021) A coupled human–natural system analysis of freshwater security under climate and population change. *Proceedings of the National Academy of Sciences* **118**: e2020431118.

Dragan A. Savić and John K. Banyard
ISBN 978-1-83549-847-7
https://doi.org/10.1108/978-1-83549-846-020242006
Emerald Publishing Limited: All rights reserved

Chapter 5
Water treatment

Emile Cornelissen
KWR Water Research Institute, Netherlands

Kees Roest
KWR Water Research Institute, Netherlands

Dragan A. Savić
KWR Water Research Institute, Netherlands and University of Exeter, UK

5.1. Introduction to water quality

The primary objective of potable water treatment is to provide water that is of appropriate quality for the intended use at a reasonable cost. Potable water should be microbiologically and chemically safe and aesthetically acceptable to the consumer. In addition, the water quality may be adjusted to avoid or minimise undesirable effects such as plumbosolvency, corrosion, lime-scale deposits or biological growth. The quality of water is defined in terms of individual physical, chemical and biological parameters, and these parameters are used for both operational and regulatory purposes. For the latter, they form the basis of legal standards that define the health risks and palatability of drinking water, and this will be discussed in detail in Section 5.2. These drinking water quality standards are set by national governments to protect human health in their countries. Most water quality parameters are unambiguous in that they refer to specific chemicals (e.g. NO_3, Pb, Mn), particular micro-organisms (e.g. E.coli) or physical properties (e.g. temperature, electrical conductivity), but others are surrogate values which give a general indication of water quality that can be conveniently measured manually on-site or by on-line instrumentation (e.g. turbidity, colour, pH).

Among the very large number of parameters that define water quality (>100), several key parameters are universally applied in the context of drinking water; these include turbidity, conductivity, colour, taste, odour, hardness, residual chlorine (in the case of chlorination) and indicator bacteria. In the context of water distribution systems, additional parameters may be important such as pH, dissolved oxygen, redox potential, conductivity, Langelier index, lead, dissolved organic carbon, chloramines and disinfection byproducts. The presence of particulate contaminants is usually quantified by turbidity, which is an indirect measure determined by nephelometry (relative light scattering). Turbidity is expressed as NTU (nephelometric turbidity unit) by comparing the degree of light scattering with standard solutions. Impurities that cause water to have a visible colour are also determined optically, either by manual comparison with standard colour solutions (units: degrees Hazen), or spectrophotometrically (units: light absorbance at 400nm per metre).

Residual chlorine in drinking water after disinfection will exist either as free chlorine (in the form of hypochlorous acid and hypochlorite ion) or combined chlorine (where the chlorine is bound to organic substances) or as chloramines (e.g. monochloramine, NH_2Cl); total chlorine is the sum of free and combined chlorine. While combined chlorine has less disinfection power than free chlorine, it persists longer and this is believed to be advantageous in large distribution networks. However, in several Western European countries (e.g. the Netherlands, Germany and Belgium) chlorine is not used in drinking water provision and disinfection is achieved by the use of advanced water treatment schemes, including advanced oxidation and membrane filtration.

Routine monitoring of microorganisms involves the enumeration of a limited number of indicator organisms. These comprise total coliform bacteria, faecal coliforms, faecal streptococci and colony/plate counts. The absence of the first three organisms (per 100 ml), and no significant change in colony/plate counts, is assumed to signify the microbiological safety (absence of pathogens) of the water.

Apart from turbidity, colour, free and total chlorine, pH, temperature and conductivity, which are normally measured at water treatment plants by online instrumentation, the large majority of other water quality parameters are determined offline in the laboratory from grab samples. Full descriptions of all water quality parameters and standard methods of determination can be found elsewhere (Rice *et al.*, 2012).

Typical sources of water for drinking water supply have been mentioned in the previous chapter. As in most countries, the proportion of drinking water in the world arising from groundwater or surface water sources varies region by region depending on the local topography and rainfall patterns. In very general terms, some regions in countries which receive above average rainfall depend on surface water as the predominant source of drinking water, usually by way of direct supply reservoirs. Over 50% of the drinking water in countries such as Canada, Australia, the UK, Portugal, Ireland and Norway comes from surface water. In contrast, regions in countries which are much drier or are lowland areas utilise impounded reservoirs and groundwater from greensand or limestone/chalk aquifers. Typically, countries such as Denmark, Germany, France, the Netherlands and Belgium use groundwater as their primary source for drinking water production[1,2]. Consequently, current treatment process streams vary widely depending on the nature of the source water. Upland waters are often rich in organic content, low in dissolved solids and salts, and acidic. In contrast, lowland waters are nutrient rich, high in dissolved solids (e.g. phytoplankton) and salts (highly buffered), and contain substantial quantities of microorganisms, arising from wastewater effluents, urban drainage and runoff. Groundwaters are a preferred source owing to the low level of contaminants as a consequence of natural filtration and microbiological processes. However, specific water quality problems may exist owing to the presence of natural contaminants (e.g. arsenic, iron, manganese) or the accumulation of pollutants from surface sources (e.g. NO_3, pesticides, solvents).

The minimum quality of raw water that can be used for drinking water supplies has been defined by the EU in relation to the degree of treatment that will be required to achieve potable water standards. In addition, water service providers may have the flexibility to blend either raw or treated waters to meet the required standards in the most cost-effective way. While source waters are, and have been, adequately protected by well-established pollution control measures and legislation for many years, there remain a number of challenges to the water service providers in the developed world in terms of fully complying with prevailing water quality regulations. Many of these relate to greater expectations from consumers plus increased scientific knowledge on the long-term health impacts of certain pollutants (e.g. disinfection by-products). Furthermore, recent advances

---

1 https://www.vewin.nl/SiteCollectionDocuments/Publicaties/Cijfers/Vewin-Dutch-Drinking-Water-Statistics-2022-ENG-WEB.pdf
2 https://www.eureau.org/resources/publications/eureau-publications/5824-europe-s-water-in-figures-2021/file

in analytical techniques have resulted in the detection of emerging compounds (e.g. endocrine disrupting agents, pharmaceutical compounds, pesticides, fluorinated compounds) in much lower concentrations.

Deterioration of source waters can also be caused by the effects of changing land use and climate (e.g. organic colour, algal blooms), intensified agricultural practices (e.g. protozoan cysts, fertilizers, pesticides) and urban runoff/wastewater discharges (e.g. NO_3, PO_4, trace organic compounds, pharmaceutical/healthcare products). In the UK and in other countries, specific problems currently include the prevention of cryptosporidium (a protozoan cyst) and pesticide breakthroughs, taste and odour incidents, in addition to reducing the leaching of iron, particulates and micro contaminants from, for example, pipe lining materials (e.g. polynuclear aromatic hydrocarbon (PAH)) within water distribution networks.

5.2.   Water standards, monitoring and regulation

The required quality of drinking water is defined in terms of guidelines and/or legal values for the various physical, chemical and biological parameters that represent no significant risk to health or undesirable aesthetic impact, sometimes described as 'wholesomeness'. For virtually all these parameters the specified values are maximum concentrations, but in the case of water that is softened during treatment, there is a minimum limit for the hardness and alkalinity concentrations. Internationally, the quality of water suitable for public water supplies is based on the World Health Organisation (WHO) Guidelines for Drinking Water which provides extensive reference material for all aspects of drinking water quality; this is summarised in report form (WHO, 2022) or by way of the WHO website[3]. In the UK, water companies must comply with legal standards defined in The Water Supply (Water Quality) Regulations 2018 (England and Wales, 2018), and associated amendments, which incorporated, and added to, those of the 1998 European Union Directive (EU Council, 1998). In the USA, drinking water standards are set by the Environmental Protection Agency[4].

The UK regulations set out the maximum (and minimum) values at particular locations in the water supply system (mostly at the consumer's tap), the designation of monitoring areas ('zones'), the number, frequency and location of samples, and specifications for the collection and analysis of samples. Each year the water company must designate the names and areas within its area of supply that will be the water supply zones for the year, and the population served by these zones must not exceed 100 000 per zone. The annual sampling frequency per zone, and from treatment works/supply points, depends on the population and flow rate, respectively, and the particular parameter. For the majority of water quality parameters, the location of compliance is at the consumers' taps, as indicated in Table 5.1.

Water service providers also monitor many parameters at the water treatment plant and in the water distribution network to detect changes in water quality through the water supply system to the consumer's tap. Some of the regulated parameters that might change as water passes through pipe networks are shown in Table 5.1. The effects of continuing chemical reactions in the bulk water and long contact times with pipe surfaces may lead to reductions in parameter concentrations, such as residual free and combined chlorine and dissolved oxygen, while other parameters may increase simply due to corrosion and biofilm effects, such as iron, manganese, PAHs and turbidity.

---

3 https://www.who.int/publications/i/item/9789240045064
4 https://www.epa.gov/wqs-tech

Table 5.1 Some water quality standards relevant to potential changes in water distribution systems

Parameter	WHO Guideline[a]	UK Regulations[b] Value	Monitoring point
Turbidity	0.1 NTU	4 NTU (1 NTU)	Consumers' taps (WTW[c])
Colour	< 15 TCU	20 mg/l Pt/Co	Consumers' taps
Coliform bacteria		0 /100 ml	Service reservoirs and WTW
Escherichia coli	0 /100 ml	0 /100 ml	Service reservoirs and WTW, and Consumers' taps
Nitrate[d]	50 mgNO$_3$/l	50 mgNO$_3$/l	Consumers' taps
Nitrite[d]	3 mgNO$_2$/l	0.5 mgNO$_2$/l	Consumers' taps
Lead	10 µgPb/l	10 µgPb/l	Consumers' taps
Iron	0.3 mgFe/l	0.2 mgFe/l	Consumers' taps
Manganese	0.1 mgMn/l	50 µgMn/l	Consumers' taps
PAH	0.7 µg/l[e]	0.1 µg/l[f]	Consumers' taps
Total THM[g]	Sum[h]≤1	100 µg/l	Consumers' taps
Chloroform	0.3 mg/l		
Bromoform	0.1 mg/l		
Dibromochloromethane	0.1 mg/l		
Bromodichloromethane	0.06 mg/l		

[a] WHO, 2006
[b] England and Wales, 2000
[c] Water treatment works
[d] [NO$_3$]/50 + [NO$_2$]/3 < 1
[e] benzo(a)pyrene
[f] benzo(b)fluoranthene, benzo(k)fluoranthene, benzo(ghi)perylene, indeno(1,2,3-cd)pyrene
[g] chloroform, bromoform, dibromochloromethane, bromodichloromethane
[h] Sum of the ratio of concentration of each compound to its respective guideline value

In the UK, specifically England and Wales, the regulatory organisation responsible for ensuring drinking water quality complies with the standards is the Drinking Water Inspectorate (DWI)[5]. The DWI is an autonomous government body which is part of the Water Directorate of the Department for Environment, Food and Rural Affairs (Defra). It monitors all aspects of drinking water supply that is carried out by the private water companies in England and Wales, including the collection and reporting of compliance samples, assessment of water quality incidents, technical audits, European standardisation, research on emerging issues and hazards, and approval of the use of products and substances. Each year over 2 million compliance samples are taken and compliance typically exceeds 99%; details of sample compliance and other aspects of water company performance can be obtained from the DWI annual reports[6]. Among the most common reasons for compliance

5 http://www.dwi.gov.uk
6 https://www.dwi.gov.uk/what-we-do/annual-report/

failure are discolouration, microbiological contamination, disinfection/treatment failures, source contamination and loss of supply.

In recent years, the DWI has adopted the WHO recommendation for the creation of a holistic risk management approach to the provision of drinking water in the form of drinking water safety plans (DWSPs)[7,8]. This applies risk management to all stages of the water supply chain, from the raw water catchment, through the treatment and distribution, and finally to the consumer. Water companies are expected to establish DWSPs for each of their supplies and these have three key components: a comprehensive system assessment; operational monitoring; and documentation of management arrangements. At the heart of the DWSP procedure is the identification of hazards and assessment of the risks posed by each hazard, often by way of a risk-scoring matrix. The DWI provides independent verification of DWSPs and expects all significant risks to be appropriately identified and assessed, and clear roles and accountabilities to be established.

5.3.   Introduction to (drinking) water treatment

Public drinking water supply plays an important role in the development of modern societies. The availability of safe and reliable drinking water supply results in a notable decrease in communicable diseases (Craun, 1991; Fewtrell *et al.*, 2005). Furthermore, public drinking water supply is important for economic growth and development. However, with increasing threats regarding the quality and quantity of source water, the cost of water treatment is rising and novel technologies can only be introduced when new standards are required by legislation or the existing plant is nearing its end of life.

Generally, drinking water supply consists of water abstraction, treatment and distribution in which water treatment is the main control system to improve the water quality of the abstracted water. Traditional freshwater sources are, typically, local groundwater and surface water sources (lakes and rivers), which are increasingly being depleted both in terms of quantity and quality due to different drivers, such as salinisation, pollution, drought and climate change. Alternative sources for drinking water supply include seawater and (treated) wastewater, which are increasingly used for potable water reuse (Tang *et al.*, 2018).

The most important goals of drinking water treatment are *(a)* to provide bacteriological safe drinking water, by removing bacteria and viruses to obtain efficient disinfection (Von Gunten, 2003) and *(b)* to remove harmful constituents, such as nitrate, ammonium, iron, manganese, arsenic (Ahmad and Bhattacharaya, 2019), pesticides, endocrine-disruptor compounds, pharmaceuticals and personal care products (Westerhoff *et al.*, 2005). Additional drinking water treatment goals might include softening (Gorenflo *et al.*, 2003) and the removal of taste, colour and odour sometimes caused by the presence of natural organic matter (Matilainen *et al.*, 2010). Specifically, when seawater or brackish water is used as a source for drinking water the removal of salts (desalination) is an additional treatment goal.

5.4.   Typical treatment methods

Different treatment processes are available, which can be grouped into *(a)* conventional processes, *(b)* adsorptive processes, *(c)* oxidative processes and *(d)* membrane processes.

---

7 https://www.who.int/publications/i/item/9789240067691
8 https://www.dwi.gov.uk/water-companies/water-safety-plans/

Conventional processes have been used extensively for many decades and include coagulation, flocculation, sedimentation and rapid sand filtration to remove particulate and natural organic matter from raw water (Matilainen *et al.*, 2010), and sometimes chlorination for disinfection. However, the rising demand for water by society and increasing impacts on water quality by human activities have led to new challenges for the water treatment industry. Furthermore, advances in analytical chemistry methods have allowed the detection of contaminants whose presence in water was previously unknown. These new pressures and improved knowledge base have driven the development of more stringent regulations and treatment goals. These developments have also led to innovation in practice and implementation of advanced treatment technologies and efforts to optimise existing processes to achieve even higher levels of treatment. This chapter summarises some of the current and future water quality challenges and trends and the state of the art in potable water treatment processes.

Adsorptive processes include, for example, activated carbon filtration and ion exchange processes to remove natural organic matter and nonpolar organic micropollutants (Reungoat *et al.*, 2010). Oxidative processes include ozonisation and ultraviolet sometimes followed by hydrogen peroxide (UV-H_2O_2) treatment for disinfection and oxidation (e.g. taste and odour control, decolouration, elimination of micropollutants) (Von Gunten, 2003; Reungoat *et al.*, 2010).

Membrane processes include low- and high-pressure processes, such as micro-, ultra- and nanofiltration and reverse osmosis. Reverse osmosis (RO) is increasingly used in drinking water supply for disinfection, desalination, softening and removing harmful constituents. Although only a 2.0-log reduction credit for bacterial and virus removal is assigned by WHO guidelines (WHO, 2017), a much higher 7.0-log reduction for virus removal was recently reported for intact RO membranes operating on surface water (Hornstra *et al.*, 2019). RO membranes are furthermore reported to remove 99.7-99.8% monovalent salts (Greenlee *et al.*, 2009), > 99% calcium, magnesium and sulphate (Perpov *et al.*, 2000), > 90% nitrate, > 98% fluoride (Sehn, 2008) and > 90% toxic ions (e.g. Cr(VI), As(V) and ClO4- (Yoon *et al.*, 2009). On the other hand, limited 50-90% boron removal (Tu *et al.*, 2010) and 80-95% ammonium removal (Kurniawan *et al.*, 2006) was reported.

In general, the removal of pesticides, endocrine-disruptor compounds, pharmaceuticals and personal care products by RO is high, but can be insufficient depending on the specific molecular properties of these micropollutants. Specifically, low molecular weight and hydrophilic neutral organic compounds are known to pass RO membranes (Albergamo *et al.*, 2019; Bellona *et al.*, 2004; Kimura *et al.*, 2004). Although not designed to remove suspended or particulate matter from raw water, intact RO membranes are known to fully remove these constituents. Apart from the removal of ammonium, boron and small neutral organic micropollutants, all drinking water treatment goals are met using one single treatment step which cannot be matched by any other previously mentioned treatment process. RO is, therefore, the most versatile and robust technique known in drinking water treatment. Today, however, most existing RO membranes are pretreated by other processes, such as coagulation, flocculation, sedimentation, rapid sand filtration, micro- or ultrafiltration, predominantly for the removal of particulate and natural organic matter.

The treatment of raw water to meet drinking water standards involves a number of individual steps, or unit processes, where the type and precise number depend on the nature of the raw water. The processes may be grouped into broad categories as shown in Table 5.2.

Table 5.2 Water treatment processes

Preliminary treatment	Raw water storage, screening, oxidation (aeration and/or chemical addition), pH adjustment
Conventional processes	Coagulation, flocculation, sedimentation, flotation, rapid filtration, slow sand filtration, chlorination
Adsorptive processes	Activated carbon filtration and ion exchange
Oxidative processes	Ozonisation and ultraviolet sometimes followed by hydrogen peroxide (UV-H_2O_2) treatment
Membrane processes	Micro-, ultra- and nanofiltration and reverse osmosis
Final treatment	pH stabilisation, fluoridation, phosphate (lead stabilisation), mineralisation, disinfection (e.g. chlorination or chloramination)

At the simplest level in terms of water treatment, a high quality groundwater may only require the application of a disinfection residual, such as a few milligrams per litre of chlorine. However, it is more usual that groundwaters will contain low levels of undesirable chemicals, such as free and bound metals (iron, manganese and arsenic), anions (fluoride and nitrate) and organic compounds (solvents, fuel additives and pesticides). To achieve compliance with water quality standards, such waters will need additional treatment depending on the contaminant but typically involving one or more processes such as oxidation (aeration, chlorination), membrane filtration (nanofiltration or reverse osmosis), ion exchange or filtration (granular media or membrane), and adsorption (Figure 5.1).

Surface waters are normally abstracted either directly from a river with sufficient flow reliability, a natural lake, or an impounded reservoir constructed to provide adequate capacity for continuous supply. Direct supply from a river is advantageous because of the avoidance of substantial costs in constructing a reservoir, but the variability in flow associated with seasonal changes leads to marked changes in source water quality, which can present difficulties in providing consistent treatment. Lakes and raw water reservoirs provide more stable water quality, in general, but excessive algal activity can arise in nutrient rich eutrophic waters during spring through to autumn in temperate climates. In dealing with high algal waters which can cause high solids loadings on works, taste and odour issues plus the release of algal toxins, various technologies have been applied, including dual-layer filtration (rapid rate granular media or micro-strainers, followed by slow sand filtration), pre-oxidation (chlorine or ozone), membrane filtration (nanofiltration or reverse osmosis), ion exchange and DAF (dissolved air flotation). For moderately contaminated surface waters, such as a lowland urban river source, a typical treatment train would comprise pre-oxidation, chemically assisted clarification (by sedimentation or flotation), rapid rate dual-layer filtration, intermediate ozonation, granular activated carbon and final chlorination.

In the USA and Europe, there is continuing concern about the presence of potentially harmful organic micropollutants in drinking water. Emerging compounds include pesticides, endocrine-disruptor compounds, pharmaceuticals, personal care products and fluorinated compounds (e.g. polyfluoroalkyl substances (PFAS)). Particularly, the last group of compounds form an increased threat to drinking water quality due to their persistent nature, as they are difficult to destroy ('forever chemicals'). Also, compounds of concern include byproducts of disinfection by chlorination or (advanced) oxidative processes. Since disinfection must be carried out rigorously

Figure 5.1 An example of (a) a conventional pre-treatment scheme, (b) a conventional dune water treatment scheme and (c) an advanced dune water treatment scheme

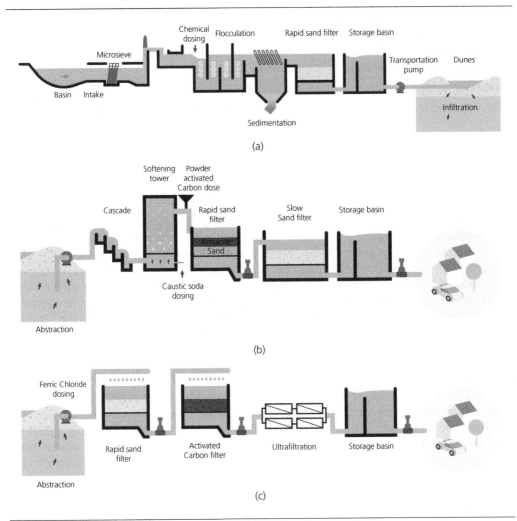

and without compromising microorganism inactivation, emphasis is being placed on improving the removal of byproduct precursor materials prior to disinfection. In many treatment plants chemical coagulation is the principal unit process for removing natural organic substances. Therefore, methods for improving the efficiency of enhanced coagulation continue to be of interest and the subject of research investigation. New types of coagulant chemicals offer greater performance and/or lower cost, and much development work in recent years has concentrated on low basicity, highly charged polymeric aluminium and ferric salts, and the combination of metal salt coagulants with organic polyelectrolytes. The beneficial combination of new inorganic coagulants and organic polyelectrolytes is also being actively considered for particular types of water quality.

At present, coagulation/flocculation followed by floc separation using sedimentation or flotation are widely used in practice. Particularly, for particle removal and the removal of natural organic matter (NOM) (Sillanpää et al., 2018). Examples are the use of sludge blanket clarification or dissolved air flotation (DAF). Both processes have their advantages and disadvantages regarding cost, treatment performance and operational requirements. In recent years various innovations have been applied to improve their performance, such as the addition of fine sand suspensions to increase the density of chemical flocs (e.g. the Veolia 'Actiflo' process) and the combination of DAF and rapid filtration in one unit (DAFF).

A new approach nearing full-scale application at the moment is that of using an electro-coagulation-flotation (ECF) cell to generate the coagulant and bring about floc separation by flotation (Cerisier *et al.*, 1996; Jiang *et al.*, 2002; Bracher *et al.*, 2022; Mao *et al.*, 2023). This process system offers many potential advantages, particularly in providing a more compact process configuration, telescoping into one unit the three conventional stages of coagulant mixing, mechanical flocculation and flotation separation. In addition, ECF avoids the need for storage and dosing of chemical solutions, and it may provide other benefits in terms of lower energy demand, more efficient treatment and some microbiological inactivation. Various aspects of the technology are still under investigation, such as the avoidance of anode passivation and the optimal electrode arrangement (e.g. monopolar or bipolar).

For several major European cities (e.g. London, Paris, Amsterdam, Stockholm and Zurich) slow sand filtration (SSF) was historically the basis of the water treatment process, and this filter type remains in operation today owing to its ability to achieve a high treatment performance for both particulate and microbiological contaminants. The enhanced performance is partly due to the intense biological surface layer ('*schmutzdecke*') that develops with time (Figure 5.2a), which is a unique feature of SSF, assisted usually by an upstream rapid filter stage to lessen the solids loading on the SSF (Figure 5.2b). With SSF either an incorporated granulated active carbon (GAC) layer is used in the filter or a separate GAC adsorption stage is needed to facilitate the removal of micropollutants such as pesticides. Micropollutant removal is also enhanced by biosorption and biodegradation of the biological surface layer on SSF and GAC.

Figure 5.2 Typical water filtration processes: a) biological surface layer in a slow sand filter; b) rapid sand filtration

(a)   (b)

In certain circumstances, pre-treatment of river water may be achieved by drawing the water through sufficiently permeable river bank strata by way of infiltration wells, and the treatment performance of such systems is receiving considerable attention at present (Partinoudi and Collins, 2007).

Finally, both ground and surface water can contain undesirably high concentrations of calcium and magnesium (e.g. $> 400\,mg/l$ as $CaCO_3$), requiring a softening step involving either precipitation (e.g. by pellet softening) or ion exchange. The removed $CaCO_3$ can be utilised as a product as well.[9]

Another treatment process innovation, which is receiving considerable interest, is that of anion exchange (AEX). AEX involves the exchange of anions electrostatically bound on a granule made of a resin that has a fixed cationic charge. These resins are supplied with a particular anion, typically chloride, during regeneration and will absorb other anions from the solution and release an equivalent amount of chloride during use. The exchange is based on the affinity of anions for a specific resin. Molecules with a very high affinity can contaminate the resin by starting to permanently block some of its charges. This is often the case with hydrophobic NOM molecules such as humic acids (Wiercik et al., 2020). AEX is typically applied in packed columns filled with resins, which requires a robust pretreatment to prevent the bed from clogging. Alternative AEX designs are suspended ion exchange (SIX) (Metcalfe et al., 2015), fluidised ion exchange (FIX) (Cornelissen et al., 2009) and magnetic ion exchange (MIEX) (Ixom Watercare, USA). The latter process employs a reusable suspension of fine magnetic particles coated with an anionic resin to remove NOM. As will be described later, NOM is a major concern in water treatment because of its role as a disinfection by-product precursor material and as a nutrient for biological activity in water distribution pipe networks. As mentioned earlier, conventional water treatment is relatively poor at removing NOM and this has led to the need for enhanced coagulation. AEX has been studied as a pretreatment for conventional coagulation processes and generally has been found to achieve the higher organics removal expected (e.g. Shorrock and Drage, 2006). The economic advantages of AEX over other methods of achieving enhanced organics removal appear to be case- and design-specific and this necessitates the need for pilot scale testing to justify its application.

In recent years, there has been a widespread interest in, and application of, oxidant chemicals for a range of water treatment purposes, specifically disinfection and chemical oxidation. Disinfection is aimed at inactivating pathogenic microorganisms (bacteria, viruses and protozoa) and is achieved by common disinfectants including chlorine, chloramine, ozone and ultraviolet (UV) light. Chemical oxidation is aimed at removing or neutralising organic and inorganic contaminants and is achieved by common oxidizing agents including chlorine, ozone, permanganate and hydrogen peroxide. Such chemical oxidants can be applied at different stages of treatment depending on their purpose, as shown in Figure 5.3. Often this has been done to either replace, or reduce, the use of chlorine because of the concern over the formation of halogenated by-product compounds.

While the predominant oxidant applied so far has been ozone, there have been successful applications of chlorine dioxide and potassium permanganate in the treatment of surface waters (Ma et al., 1997). Ozone has been of particular interest because of its ability to degrade pesticide compounds and other organic micropollutants. This normally occurs as an intermediate treatment step, after the conventional processes (clarification and filtration), in combination with granular activated carbon (Figure 5.3). However, typical water treatment conditions limit the effectiveness of ozone treatment

---

9 https://www.kwrwater.nl/en/projecten/high-value-reuse-of-lime-pellets-from-drinking-water-production/

Figure 5.3 Application of oxidants in water treatment (DAF – dissolved air flotation; FBC – flat bottomed clarifier; RG – rapid gravity; GAC – granular activated carbon; BAC – biological activated carbon; UV – ultraviolet irradiation)

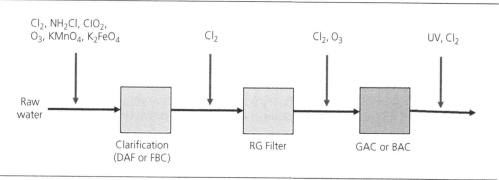

by minimising the generation of highly reactive radical species from ozone. Research interest is currently focused on methods of enhancing radical formation, including combinations of ozone with either hydrogen peroxide (Lambert *et al.*, 1996), UV irradiation, metal catalysts (Ma *et al.*, 1999) or activated carbon (Jans *et al.*, 1998). For example, studies of the ozonation of the herbicide, atrazine, have shown that the presence of a small concentration of manganese (c. 0.5 mg/l) can catalyse the degradation of atrazine (Ma *et al.*, 1999). A possible application of this phenomenon is the ozone treatment of contaminated surface and ground waters that also contain low levels of manganese.

Recently there has been a change in the type of micropollutants being observed in raw waters to more environmentally sustainable and lower human toxicity chemicals, especially regarding pesticides. These new chemicals tend to be more polar in nature and therefore are not as easily removed by the current levels of ozone and GAC treatment. It has been shown that increased concentrations of ozone including hydroxyl radical enhancement are required to break down these compounds. However, increased ozone can produce elevated concentrations of bromate which is a reaction product of ozone (with background bromide) and has a legislated maximum limit of 10 µg/l. Therefore, alternative highly oxidative treatment and adsorptive processes are being explored to achieve the most cost-effective treatment; an example of the former is UV+H_2O_2, which has been successfully used at full scale in the Netherlands (Kruithoff *et al.*, 2007). The combination of UV and H_2O_2 is called advanced oxidation. Advanced oxidation processes (AOPs) are a group of techniques that rely on the use of radicals to destroy complex organic compounds into simpler end products.

Other treatment chemicals that combine oxidation and coagulation-precipitation capabilities are also of interest and undergoing evaluation. While permanganate, as mentioned above, can also assist coagulation through the precipitation of solid phase manganese dioxide (Ma *et al.*,1997), Fenton's reagent (ferrous ions and hydrogen peroxide) and ferrate (FeO_4^{2-}) are also able to produce iron coagulating species as a result of powerful oxidation reactions. While the application of Fenton's reagent is limited to low pH waters, the optimal pH conditions for ferrate are typically above neutral which corresponds to a strong oxidation potential and adequate chemical stability (Jiang *et al.*, 2004). Recently, the potential combination of ferrate and photocatalysis (UV-TiO_2) has been studied (Yuan *et al.*, 2008) since there may be an advantageous synergy arising from the ferrate scavenging of conductance band electrons.

Membrane treatment continues to provide an important alternative or addition to other established technologies for particular raw water qualities and to deal with specific contaminants where it is cost-effective to do so (Taylor and Wiesner, 1999). Membranes can be classified in simple terms according to the nominal size of the pores, molecular weight cut-off or rejection values, the membrane material (ceramic, polymer) and module configuration (spiral wound, plate-and-frame, capillary, hollow fibre); the most common membranes based on pore sizes are microfiltration (MF) (0.1 to $1\,\mu m$), ultrafiltration (UF) (0.01 to $0.1\,\mu m$), nanofiltration (NF) (1 to $10\,nm$) and reverse osmosis (RO) ($< 1\,nm$). Currently, the principal applications of membrane technology to water treatment are versatile ranging from particulate removal (using MF), disinfection (using UF), NOM removal (using UF/NF), softening (using NF), removal of emerging contaminant (pesticides, pharmaceuticals, PFAS) (using NF/RO) (Kim et al., 2018), and desalination of brackish and seawater desalination (using RO).

Two examples demonstrating the versatility of combining membrane processes with conventional treatment have been reported. Firstly, the use of UF after clarification for the treatment of low alkalinity and highly coloured upland water (Hillis, 2006) and, secondly, the conversion of a conventional sludge blanket clarifier to a UF unit to treat heavily laden algal waters (Redhead, 2007). The use of membrane technology is expected to continue to grow in the future as more efficient membrane materials are developed, the causes of membrane fouling are better understood and methods of mitigation applied (Guo et al., 2012). However, some drawbacks of membrane solutions require careful consideration of necessary pre-treatment, lowering of the energy demand and waste stream (concentrate) management, as they are critical to the process, especially for NF and RO applications. The robustness of membrane processes opens up opportunities for potable reuse schemes including advanced membrane technology (Tang et al., 2018).

A significant development for the water industry (particularly UF/MF membrane development) was the crypto outbreak in Milwaukee in 1993 (Fox and Lytle, 1996). This was supported by the wording in the UK Water Regulations, in January 2008, regarding protozoa such as cryptosporidium. The previous legislation required the removal of the oocysts independent of their viability and the main process adopted to achieve compliance at high-risk sites was UF/MF membranes. The change in regulations has allowed other technologies such as UV to be considered to inactivate those organisms and provide a more cost-effective solution. The UV doses required for cryptosporidium are also adequate to inactivate other pathogenic bacteria. This potentially provides an opportunity to simplify the final disinfection stage on some very high-quality waters and good distribution network systems to use UV and marginal or no chlorination, thereby reducing infrastructure and chemical requirements.

While current shortcomings in drinking water quality can be overcome, to a large extent, by new investment in conventional treatment technologies and by renewal of the capital infrastructure, new technologies and methodologies are likely to offer advantages of greater process efficiency and reliability, capital and operational cost savings, more compact plant, and a lower dependency on energy and chemicals. In addition, future developments will be required to provide higher standards of treated water quality and a greater degree of sustainability, principally in terms of the reuse and recycling of chemicals and materials, waste minimisation, valorisation of byproducts and less energy consumption. These developments will be evaluated within the new approach to managing drinking water quality based on the establishment of DWSPs, as described earlier. Thus, conventional and new water treatment processes will be subjected to detailed analysis increasingly in the future within DWSPs to identify whether improvements are required or whether existing processes should be replaced with superior technologies.

Currently, the level of process monitoring and direct computational control at water treatment works is fairly extensive. In contrast, real-time process simulation and optimal operation remain undeveloped due to the complexity and inadequacy of unit process mathematical models. These models are generally based on assumptions that are difficult to apply in practice or contain parameters that rely on real-time measurements. While considerable effort has been invested over the last 30 years, the success of full-scale modelling systems has been site-specific and partial. Advances in computing power and artificial intelligence (AI) provide a new opportunity for improving process simulation and optimal operation. AI techniques can handle complex nonlinear relationships and provide insights into the overall dynamics of water treatment processes (Li et al., 2021). They, however, have their own challenges, including high data requirements, increased uncertainty in model validity, challenges in interpreting model behaviour and decision logic, and increased likelihood of incorporating biases from training data (Aliashrafi et al., 2021).

Among the techniques of computer modelling being applied increasingly to drinking water treatment processes is computational fluid dynamics (CFD) modelling. This approach has matured considerably in the last 10–15 years with the increasing availability of affordable commercial software. CFD is useful for quantifying flow patterns, mixing regimes and contact times. Modelling allows consideration of alternative designs and extremes of conditions that may not always be possible to test onsite using physical tests, such as tracer tests.

5.5. Management of residuals

Associated with the water treatment processes is the production of waste flows ('residuals' or 'sludges') containing solid matter, principally arising from the particle separation processes. These processes include screens, sedimentation and flotation clarifiers, granular media filters and activated carbon adsorbers, membrane processes and softening processes. The large majority (70–80%) of sludges arise from the coagulant solids separated in the clarifiers, and most of the remainder (15–20%) is from softening. Although such waste flows may be small compared to the flow treated (e.g. 1–5% by volume), they cannot be discharged directly to the environment and, therefore, represent a significant cost and operational burden to water service providers in terms of handling and final disposal. A survey of waterworks sludge management in the UK in 1997/98 estimated the annual total production of sludge in the UK to be 131,000 tonnes of dry solids (tDS), and indicated that approximately 50% of sludges (in tDS/per year) were disposed offsite in landfills and a further 25% was discharged to wastewater sewer (Simpson et al., 2002). With the rising cost of disposal to landfill (c between 160 and 310 €/tonne of dry matter for WTW sludge in various EU countries) (Kacprzak et. al., 2017), water service providers are actively investigating novel, more sustainable options for sludge recycling, such as incorporation into building materials, land reclamation and application to agricultural land (Owen, 2002). Since waste flows are initially very low in solids content (<5%), some degree of thickening and dewatering is normally carried out at the treatment plant to substantially reduce the flow volume. Depending on the final disposal method, the sludge may receive thickening by sedimentation (solids content up to 10%) and dewatering by a mechanical plant, such as a filter plate press or centrifuge (solids content up to 30%). Transport costs also have to be taken into account as any benefit from recycling may be negated if a local application cannot be adopted.

Furthermore, the management of residual streams from IEX and NF/RO is a topic of increasing interest. The vast majority of these residual streams are currently discharged untreated into surface water or by way of a sewage treatment plant. Tighter guidelines from regulatory bodies will make this practice less desirable. More research is being dedicated to the *(a)* discharge-proofing of the residual streams (e.g. by removing organic micropollutants) and *(b)* far-reaching treatment of residual streams enabling full or partial water, energy and substance recovery. Examples of substance recovery can be found in the Netherlands. A specific organisation was initially established for all drinking water companies in the country, but they have since also become engaged with a Belgian drinking water company and a number of Dutch water authorities. The organisation seeks destinations for the material streams that are generated in water treatment processes and developed suitable chains, which are then supplied and/or operated in a qualitatively high-value manner. In this way, they ceased seeing the generated residual streams as waste, and certainly not as a problem. The current situation is that functional applications for most of the streams have been developed, which results in sustainable and also financial benefits. This also fits in the concept of a circular economy.

5.6.   Water distribution systems

This section provides a brief introduction to water distribution systems since subsequent chapters will review in much greater detail the key elements of such systems, their design and modelling, their operation, maintenance and performance, and long-term management.

5.6.1   Introduction to water distribution systems

Once water has been treated to the required standards at the water treatment plant, it is transported to the consumer by way of a water distribution system. For the consumer, the water must arrive at an acceptable pressure and in sufficient quantity, and the water quality must retain its wholesomeness and aesthetic acceptability during its passage through the pipe network.

In the UK, the requirements for pressure and quantity are defined by the Office of Water Services, which specifies for domestic consumers an indicative minimum water pressure of 10 metres static head at the boundary of the consumer's property, and a legal minimum of 7 metres static head (Ofwat[10]). To achieve the pressure requirements, the water is conveyed from the water treatment plant through a large network of pipes by means of energy provided by pumping plants and/or elevation head (potential energy). Since water treatment is best achieved under relatively constant flow rate conditions, and the demand for water by the consumer is highly variable, there is a need to balance the two flow patterns by way of storage. Thus, the major components of water distribution systems are typically the following: storage tanks (water towers) or (service) reservoirs, pumping plants and a network of pipes arranged in an optimal layout (Figure 5.4). The layout invariably involves interconnections of pipes in closed loops to provide security of supply in case of failure of a pipe element and to minimise pressure loss.

The overall supply network is divided into operational zones based on service reservoirs, pumping stations, pressure zones or other operational considerations. In the UK, these operational zones are further divided into district metered areas (DMAs) whose boundaries are defined by isolation valves, so that leakage can be monitored and managed by metering the quantities of water entering and leaving the DMA. The subsequent analysis of flow and pressure, especially at night when a high proportion of users are inactive, enables leakage specialists to calculate the level of leaks in the district. DMAs were first introduced to the UK at the start of the 1980s by the then UK Water Authorities Association, and typically cover a part of the supply network corresponding to 500 to 3500 connections. The supply network includes three principal types of pipe made from various materials, such as iron, steel, asbestos cement and plastic (uPVC and MDPE). Trunk mains are the largest in size (c.300 to > 1200 mm dia.) and carry large flows to/from treatment and to the reservoirs and towers in the distribution network area. Distribution mains are smaller (>90 mm dia.)

---

10 https://www.legislation.gov.uk/uksi/2008/594/regulation/10/made

Figure 5.4 Schematic of water distribution systems

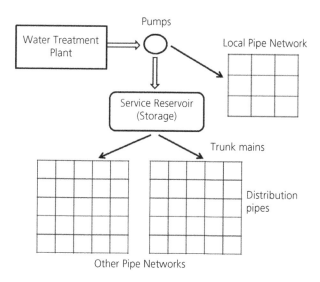

and take the flow from the trunk mains to the streets/roads outside the consumers' properties, from which service pipes (c.20 mm) provide the connection to domestic property. Operation of the distribution network is facilitated by the inclusion of hydraulic devices in appropriate locations, and these include control valves, air valves, washout/flushing chambers, pressure-reducing valves, flow and pressure meters, and fire hydrants; these are described in later chapters. Developments in such equipment in the future will improve operation of the networks and reduce pressure, with direct beneficial impacts on energy, leakage and bursts.

Innovative tools and techniques into the integrity of the existing distribution network are also being explored, to provide greater resilience and minimise disruption. These include novel pipe relining materials and non-intrusive mains assessment methods, which are potential solutions to the current challenge of ageing assets in the network system.

Some water service providers are in the process of developing a more integrated approach between the water treatment works and the distribution network systems by utilising modern data management systems in real time. The advent of digital twins and smart water systems promises to provide a way for the integration and real-time management of drinking water systems. The Smart Water Network (SWAN) Forum[11] defines a digital twin as 'a dynamic digital representation of real-world entity(s), and their behaviours, using models with static and dynamic data that enable insights and interactions to drive actionable and improved outcomes.' If successful, this could provide the most cost-effective water supply for the water service provider and their customers.

5.6.2   Water quality

The quality of potable water is likely to incur minor changes as it passes from the treatment plant to the consumer since the flow may take many hours, and even several days in large networks, to reach the consumer. These changes are inherently undesirable since the quality invariably deteriorates

---

11 https://swan-forum.com/digital-twin-work-group/

with flow time, owing to the reactivity, and loss, of the residual disinfectant and contamination from internal pipe surfaces. The latter has various components which include, principally, sediments that have accumulated during low flow conditions, corrosion products from pipes and fittings, and biofilms that have developed owing to the presence of opportunistic microorganisms and biodegradable organic substances in the bulk flow.

Although the water leaving treatment plants complies with the necessary drinking water standards, low levels of particulates (<0.5 NTU) and dissolved organic substances (<2 mgC/l) are sufficient precursors for the development of sediments and biological activity in pipe systems over long time periods. Under normal conditions, increasing water age leads to a reduction in disinfectant residual (i.e. free and combined chlorine), an increase in trihalomethane compounds and a reduction in dissolved oxygen, but, generally, these are not unacceptable and disinfectant residuals can be maintained by booster chlorination within the network. However, in some networks, intermittent problems with water quality arise, often associated with sudden changes in flow rate caused by bursts, pumping or valve operations that lead to unacceptable discolouration, indicated by high turbidity and colour concentrations. In these situations, the water companies will carry out repairs and flushing of the relevant distribution mains to remove accumulated sediments, corrosion products and biofilms. When flushing occurs the water companies will always try to achieve a higher flow rate than the calculated demand under normal operation to reduce further discolouration events. Sometimes this is difficult to achieve due to the size of some of the larger mains and disposal routes for the flushing water. Therefore, it is important to invest in developing so-called self-cleaning networks.

To assist water companies in addressing customer complaints, particularly regarding occurrence of brown water, the Self-Cleaning Networks tool has been developed[12]. Research into the origins and nature of discolouration problems shows that the major cause is the resuspension of accumulated sediments in pipes. The Self-Cleaning Networks software provides the user with design instructions on how to make such networks self-cleaning. This entails maintaining a certain velocity in the network, which keeps the dynamic process of settlement and resuspension from leading to sediment accumulation. This prevents problems associated with discolouration and long residence times.

Water quality in the distribution network is monitored by manual sampling and analysis either in the field by portable test kits for some parameters (e.g. pH, conductivity, dissolved oxygen, residual chlorine) or in the laboratory, if the parameters are stable or require more advanced instrumentation, such as pesticide concentrations. Currently, a limited number of commercial in-line sensors are available. Those are able to detect tens of basic physico-chemical parameters, including residual chlorine, turbidity, pressure, temperature and conductivity. This type of instrument could provide a valuable audit tool and in the future may be further deployed to assist with the real-time management of the distribution network if incorporated with a suitable data management and control system.

In the future, it is expected that more attention will be given by water companies to instigate a more integrated management system for the supply of water to the customer. DWSPs and the associated monitoring and control of water quality from source to tap will provide a strong foundation. This will assist in the greater use and inter-mixing of alternative water sources to optimise water resources, plus changes in network operational practices to reduce chemical and energy consumption and improve the supply/demand balance.

---

12 https://www.kwrwater.nl/en/tools-producten/self-cleaning-networks/

REFERENCES

Ahmad A and Bhattacharya P (2019) Arsenic in Drinking Water: Is 10 μg/L a Safe Limit? *Current Pollution Reports* **5(1)**.

Albergamo V, Blankert B, Cornelissen ER *et al.* (2019) Removal of polar organic micropollutants by pilot-scale reverse osmosis drinking water treatment. *Water Research* **148**: 535–545.

Aliashrafi A, Zhang Y, Groenewegen H and Peleato NM (2021) A review of data-driven modelling in drinking water treatment. *Reviews in Environmental Science and Bio/Technology*, pp.1–25.

Bellona C, Drewes, JE, Xu P *et al.* (2004) Factors affecting the rejection of organic solutes during NF/RO treatment – a literature review. *Water Research* **38(12)**: 2795–2809.

Bracher GH, Carissimi E, Wolff DB, Glusczak AG and Graepin C (2022) Performance of an electrocoagulation-flotation system in the treatment of domestic wastewater for urban reuse. *Environmental Science and Pollution Research International* **29(32)**: 49439–49456.

Cerisier SDM and Smit JJ (1996) The electrochemical generation of ferric ions in cooling water as an alternative for ferric chloride dosing to effect flocculation. *Water SA* **22(4)**: 327–330.

Cornelissen ER, Beerendonk EF, Nederlof MN, van der Hoek JP and Wessels LP (2009) Fluidized ion exchange (FIX) to control NOM fouling in ultrafiltration. *Desalination*, **236(1-3)**: 334–341.

Craun GF (1991) Causes of waterborne outbreaks in the United States. *Water Science and Technology* **24(2)**: 17–20.

England and Wales (2018) The water supply (water quality) regulations 2018, 2018 No. 647 (W. 121), London, UK.

EU Council (1998) European Council Directive 98/83/EC of 3 November 1998 on the quality of water intended for human consumption. *Official Journal of the European Communities* **L330**.

Fewtrell L, Kaufmann RB, Kay D *et al.* (2005) Water, sanitation, and hygiene interventions to reduce diarrhoea in less developed countries: A systematic review and meta-analysis. *The Lancet Infectious Diseases* **5(1)**: 42–52.

Fox KR and Lytle DA (1996) Milwaukee's crypto outbreak: Investigation and recommendations. *Journal of the American Water Works Association*, **88(9)**: 87–94.

Gorenflo A, Velázquez-Padrón D and Frimmel FH (2009) Nanofiltration of a German groundwater of high hardness and NOM content: Performance and costs. *Desalination* **151(3)**: 253–265.

Greenlee LF, Lawler DF, Freeman BD *et al.* (2009) Reverse osmosis desalination: Water sources, technology, and today's challenges. *Water Research* **43(9)**: 2317–2348.

Guo W, Ngo HH and Li J (2012) A mini-review on membrane fouling. *Bioresource Technology* **122**: 27–34.

Hillis P (2006) Enhanced coagulation, flocculation and immersed ultrafiltration for treatment of low alkalinity and highly coloured upland water. *Journal of Water Supply: Research and Technology – Aqua* **55(7–8)**: 549–558.

Hornstra LM, Da Silva TR, Blankert, B *et al.* (2019) Monitoring the integrity of reverse osmosis membranes using novel indigenous freshwater viruses and bacteriophages. *Environmental Science: Water Research and Technology* **5(9)**: 1535–1544.

Jans U and Hoigné J (1998) Activated carbon and carbon black catalyzed transformation of aqueous ozone into OH-radicals. *Ozone Science and Engineering* **20(1)**: 67–90.

Jiang C, Li X-Z, Graham N, Jiang J-Q and Ma J (2004) The influence of pH on the degradation of phenol and chlorophenols by potassium ferrate. *Chemosphere* **56(10)**: 946–956

Jiang J-Q, Graham N, André C, Kelsall GH and Brandon N (2002) Laboratory study of electro-coagulation-flotation for water treatment. *Water Research* **36**: 4064–4078.

Jiang J and Graham NJD (1998a) Preparation and characterisation of an optimal polyferric sulphate (PFS) as a coagulant for water treatment. *Journal of Chemical Technology and Biotechnology* **73(4)**: 351–358.

Jiang J and Graham NJD (1998b) Evaluation of poly-alumino-iron sulphate (PAFS) as a coagulant for water treatment. *8th International Gothenburg Symposium on Chemical Treatment, Prague. Czech Republic.*

Kacprzak M, Neczaj E, Fijałkowski K *et al.* (2017) Sewage sludge disposal strategies for sustainable development. *Environmental research* **156**: 39–46.

Kim S, Chu KH, Al-Hamadani YAJ *et al.* (2018) Removal of contaminants of emerging concern by membranes in water and wastewater: A review. *Chemical Engineering Journal* **335**: 896–914.

Kimura, K, Toshima, S, Amy, G and Watanabe, Y (2004) Rejection of neutral endocrine disrupting compounds (EDCs) and pharmaceutical active compounds (PhACs) by RO membranes. *Journal of Membrane Science* **245(1)**: 71–78.

Kruithof, JC, Kamp, PC and Martijn, BJ (2007). UV/H_2O_2 treatment: a practical solution for organic contaminant control and primary disinfection. *Ozone Science and Engineering* **29(4)**: 273–280.

Kurniawan TA, Lo W-h and Chan GYS (2006) Physico-chemical treatments for removal of recalcitrant contaminants from landfill leachate. *Journal of Hazardous Materials* **129(1)**: 80–100.

Lambert SD, Graham NJD and Croll BT (1996) Degradation of selected herbicides in a lowland surface water by ozone and ozone-hydrogen peroxide, *Ozone Science and Engineering* **18(3)**: 251–269.

Li L, Rong S, Wang R and Yu S (2021) Recent advances in artificial intelligence and machine learning for nonlinear relationship analysis and process control in drinking water treatment: A review. *Chemical Engineering Journal* **405**: 126673.

Ma J and Graham NJD (1999) Degradation of atrazine by manganese-catalysed ozonation: Influence of humic substances. *Water Research* **33(3)**: 785–793.

Ma J, Graham N and Li G (1997) Effectiveness of permanganate pre-oxidation in enhancing the coagulation of surface waters – laboratory case studies. *Journal of Water Supply: Research and Technology – Aqua* **46(1)**: 1–10.

Mao Y, Zhao Y and Cotterill S (2023) Examining current and future applications of electrocoagulation in wastewater treatment. *Water* **15(8)**: 1455.

Matilainen A, Vepsäläinen M and M Sillanpää (2010) Natural organic matter removal by coagulation during drinking water treatment: A review. *Advances in Colloid and Interface Science* **159(2)**: 189–197.

Metcalfe D, Rockey C, Jefferson B, Judd S and Jarvis P (2015) Removal of disinfection by-product precursors by coagulation and an innovative suspended ion exchange process. *Water Research* **87**: 20–28

Owen PG (2002) Water- treatment works sludge management. *Journal of CIWEM (the Chartered Institution of Water and Environmental Managemen)t* **16**: 282–285.

Partinoudi V and Collins MR (2007) Assessing riverbank filtration removal mechanisms. *Journal of the American Water Works Association* **99(12)**: 61.

Pervov AG, Dudkin EV, Sidorenko OA *et al.* (2000) RO and NF membrane systems for drinking water production and their maintenance techniques. *Desalination* **132(1)**: 315–321.

Redhead A (2007) Membrane technology at St. Saviours water treatment works, Guernsey, Channel Islands. *Water and Environment Journal* **22(2)**: 75–80.

Reungoat J. Macova M, Escher BI *et al.* (2010) Removal of micropollutants and reduction of biological activity in a full scale reclamation plant using ozonation and activated carbon filtration. *Water Research* **44(2)**: 625–637.

Rice EW, Bridgewater L and American Public Health Association (eds) (2012) *Standard methods for the examination of water and wastewater* (vol. 10). American Public Health Association, DC, USA.

Sehn P (2008) Fluoride removal with extra low energy reverse osmosis membranes: three years of large scale field experience in Finland. *Desalination* **223(1)**: 73–84.

Shorrock K and Drage B (2006) A pilot plant evaluation of the Magnetic Ion Exchange process for the removal of dissolved organic carbon at Draycote water treatment works. *Water and Environment Journal* **20**: 65–70.

Sillanpää M, Ncibi MC, Matilainen A and Vepsäläinen M (2018) Removal of natural organic matter in drinking water treatment by coagulation: A comprehensive review. *Chemosphere* **190**: 54–71.

Simpson A, Burgess P and Coleman SJ (2002) The management of potable water treatment sludge: present situation in the UK. *Journal of CIWEM* **16**: 260–263.

Tang CY, Yang Z, Guo H *et al.* (2018) Potable Water Reuse through Advanced Membrane Technology. *Environmental Science and Technology* **52(18)**: 10215–10223.

Taylor JS and Wiesner M (1999) Membranes. In *Water Quality and Treatment: A Handbook of Community Water Supplies*. American Water Works Association, Denver, CO, USA, Chapter 11.

Tu KL, Nghiem LD and Chivas AR (2010) Boron removal by reverse osmosis membranes in seawater desalination applications. *Separation and Purification Technology* **75(2)**: 87–101.

Von Gunten U (2003) Ozonation of drinking water: Part II. Disinfection and by-product formation in presence of bromide, iodide or chlorine. *Water Research* **37(7)**: 1469–1487.

Westerhoff P, Yoon Y, Snyder S *et al.* (2005) Fate of endocrine-disruptor, pharmaceutical, and personal care product chemicals during simulated drinking water treatment processes. *Environmental Science and Technology* **39(17)**: 6649–6663.

WHO (World Health Organization) (2006) *Guidelines for drinking-water quality*, 3rd edn., vol. 1 Recommendations. World Health Organisation, Geneva, Switzerland.

WHO (2022) *Guidelines for drinking-water quality*, 4th edition: Incorporating the first and second addenda. World Health Organization, Geneva, Switzerland.

WHO (2017) *Potable reuse: guidance for producing safe drinking-water*. World Health Organization: Geneva, Switzerland.

Wiercik P, Frączek B and Chrobot P (2020), Fouling of anion exchanger by image and FTIR analyses, *Journal of Environmental Chemical Engineering* **8**.

Yoon J, Amy G, Chung, JinWook CJ *et al.* (2009) Removal of toxic ions (chromate, arsenate, and perchlorate) using reverse osmosis, nanofiltration, and ultrafiltration membranes. *Chemosphere* **77(2)**: 228–235.

Yuan B-L, Li X-Z and Graham N (2008) Aqueous oxidation of dimethyl phthalate in a ferrate-TiO_2-UV reaction system. *Water Research* **42**: 1413–1420.

Dragan A. Savić and John K. Banyard
ISBN 978-1-83549-847-7
https://doi.org/10.1108/978-1-83549-846-020242007

Chapter 6
Distribution network elements

Tiku Tanyimboh
University of the Witwatersrand, Johannesburg

Myles Key
South West Water Ltd, Exeter, UK

6.1.　Introduction

The water distribution network is an essential element of the service provision to water using customers. It is too simplistic to think of this service as simply processes (treatment) and pipelines. Beyond the water treatment works, the distribution network is a complex system that stores, transports and reconditions the water both from a quality and a pressure perspective. The largest asset base of any water service provider (WSP) is the pipeline network. This chapter is concerned with the main elements of water distribution networks, including pipelines, pumps, valves and service reservoirs. The chapter begins with an overview of the history of water transportation and explains how pipelines are designed and maintained. The importance of valves and fittings for managing pressure and controlling water as it is transported around the distribution network is explained. To complement this, some practical aspects of surge control are also covered. Types of pump and pump selection are explained. WSPs must cope with a wide variety of normal and abnormal operating conditions such as very large flows for firefighting, routine repairs and maintenance and the inherent variability in water consumption. This chapter also includes service reservoirs that play a key role in this respect.

6.2.　Pipes
6.2.1　A brief history of water networks and pipe materials

To satisfy our early desires for water it was sufficient for water to be found and used at its source. The Romans were the first civilisation to consider transporting water over distances from the water's source to the populated cities. Ancient Rome built aqueducts to deliver water to the populace but the aqueducts were open channels exposed to the elements and the quality of the water was poor by today's standards.

As civilisations developed and the population boomed, the resourcefulness of the human race, combined with improved understanding of material properties and advancements in technology, saw water transportation systems move from open channels to purpose made pipelines.

Pipe construction moved through various stages, from the development and use of lead piping in the Roman Empire, through cast iron pipes as early as the fourteenth century, to asbestos, steel and plastic products in the twentieth century (discussed in greater detail in Chapter 1: 'Historical development of water distribution practice').

This chapter discusses pipeline design and material selection in modern water distribution networks.

6.2.2   Pipeline design and material selection

Pipeline material selection historically was relatively straightforward: iron or steel for high pressure applications, highways, rivers, and other difficult environmental conditions and plastic pipes for lower pressure applications, rural networks and housing developments. However, advancements in modern plastic pipes with improved pressure ratings and the ability to create continuous plastic pipelines has provided much greater choice and, at the same time, a selection headache for pipeline designers.

The main considerations for a water engineer designing for a new distribution pipeline are

- the environment (ground conditions) in which the pipe will be installed
- the volume of water the new pipeline needs to carry, considering immediate and future demands
- the static pressure and pressure variation that the pipeline will be subjected to.

6.2.3   The interaction of pipelines with the local environment

When selecting the pipe material, consideration must be given to the natural environment in which the pipe will be laid. Pipes are generally laid at a minimum depth of 750 mm of cover although some engineers will insist on 900 mm of cover being provided. This will generally protect the pipeline from damage through frost penetration and from third party activity such as a plough blade.

Pipes laid in unmade ground are subjected to forces exerted by the natural movement of the ground. Typically, shrinkage and expansion of the ground will occur in the UK because of the temperate climate, therefore cold winter weather can lead to ground movement through rain penetration followed by freezing temperatures which causes the water to expand as it turns to ice. On the opposite end of the scale, the warmer summer months cause further ground movement as it dries out and shrinks.

Pipelines need to be able to absorb the stresses created by ground movement otherwise stress fractures will occur. As a consequence, WSPs will incur costs through the detection effort expended in locating the leaks and in dealing with customer contacts and resultant compensation payments due to the supply interruptions caused while the repairs are carried out.

Many pipelines are laid in the highway, which creates different stresses to the pipes due to traffic loadings. Highways, of course, are designed to withstand traffic loadings and are less influenced by weather-related movement as the surface wearing course is designed to carry water away from the sub-structure, however surface cracking and wear and tear on the road surface can lead to localised water penetration which will permit expansion and shrinkage to occur.

Pipes laid at normal depth in narrow trenches are afforded support by the surrounding undisturbed ground so that traffic loadings are effectively dispersed. However, if pipes are laid too close to the surface or if extra-width trenches have been excavated then traffic (surcharge) loading may be significant and could have a detrimental effect on the pipe. The selected pipe bedding material and the effectiveness of the pipe layer's compaction technique are significant factors in affording adequate support to any pipeline but particularly where adverse conditions prevail.

When considering the effects of loadings applied to pipelines, pipes are generally classed as rigid or flexible. With rigid pipes, the weight of the soil and any imposed load is transmitted around the pipe wall from the crown to the invert which, by its nature, resists deformation. With flexible pipes, any imposed load that is not absorbed by the pipe surround can cause the pipe to deform, pushing the pipe wall out horizontally into the surrounding bedding. All pipes have an acceptable degree of deformation, buckling pressure or ovalisation which they are able to withstand before joints leak, or stress fractures or lining cracks occur.

In non-trafficked areas, the weight imposed on a pipeline (dead load) is fairly constant varying only through changes in moisture or air content of the soil and can largely be ignored. In trafficked areas it is important to understand the weight of traffic that can be applied to a buried pipeline and to ensure that the applied traffic weight together with the over burden weight of the soil does not exceed the capacity of the pipeline to resist the resultant wall deformation.

The following formula, adapted by Saint Gobain Pipelines is modified from the Spangler Iowa formula and can be used to calculate the deformation (ovalisation) percentage for ductile pipes in order to ensure the design capacity of the pipe material is not exceeded. The acceptable percentage of ovalisation can be determined for a given pipe material and size by reference to the manufacturers' technical details

$$\Delta = \frac{100K\left(Pe + Pt\right)}{8S + 0.061\ E^{1}/DLF} \tag{6.1}$$

where

Δ = ovalisation (%)
K = bedding coefficient
Pe = earth load (kN/m²)
Pt = traffic load (kN/m²)
S = pipe diametral stiffness (kN/m²)
E^{1} = modulus of soil reaction (kN/m²)
DLF = deflection lag factor

The bedding coefficient K reflects the angle of support at the pipe invert and the quality of the bedding and sidefill material used in the pipe trench. The DLF is influenced by the nature of the native soil, sidefill and the working pressure of the pipeline; it is assumed that for pressure pipelines the DLF = 1.

In addition to stresses through ground movement, material selection also needs to be considered in light of the natural soil type, which will influence the longevity of the pipeline through chemical or electrolytic reaction with the pipe wall. This consideration extends beyond the natural environment as, increasingly, pipelines are being laid to support new housing and industrial developments in land which has been subjected to previous use, typically referred to as brownfield sites. In circum- stances where the previous use has been for heavy industry, fuel stations, and other instances where petroleum products have been used, the ground is often contaminated and pipelines need to be able to provide a barrier against penetration from the ground contaminant.

There is no single product solution for any of the conditions described in the preceding paragraphs. The pipeline designer needs to ensure the selected product is able to provide the expected service life considering the external stresses created by the environment.

6.2.4   Pipe selection (size and pressure rating)

The next consideration is the size (diameter) of the pipe that is selected. Distribution networks, typically, are categorised into trunk mains, distribution mains and supply pipes.

6.2.4.1   Trunk main

A trunk main is classified as a large diameter main (nominally $\geq 300\,$mm internal dia.) which is used for transferring bulk supplies of water, usually between treatment works and service reservoirs but also between one service reservoir and another.

6.2.4.2   Distribution main

Distribution mains, generally, supply water into communities and are supplied predominently from service reservoirs but are sometimes linked directly from trunk mains. Distribution mains typically range in size from 80 mm to 300 mm in diameter, but regional WSPs will have examples of distribution mains that fall outside of this typical size banding.

6.2.4.3   Service pipes

Finally, service pipes connect customers to distribution mains and range in size from 15 mm to 50 mm but, as with distribution mains, there will be examples where supply pipe diameters extend beyond the 50 mm size range.

It is important to remember that demands on networks will often change through the lifetime of the pipeline. This is particularly true on the larger diameter pipes simply because there is greater scope for influence due to the variety of networks that are supported downstream. Pipeline designers must therefore consider potential future growth and allow headroom in their design to enable the downstream network demands to grow before the pipeline needs to be reinforced. The headroom allowance could be based on known future growth which can be established through discussion with local planning authorities. Pipeline design is discussed in greater detail in Chapter 8: 'Design of water distribution systems'.

6.2.5   Protection systems

The ground conditions in which a pipeline is laid influence the longevity of the pipe as the soil might be aggressive and cause external damage to the pipe if it is not adequately protected. Similarly, the water flowing within the pipe will have the potential to cause internal degradation. Knowledge of previous occupancy of brownfield sites is also essential as this can influence a pipeline designer's material selection, as a barrier product might be required to avoid any potential permeation of ground contaminants through the pipe wall.

6.2.5.1   External protection

With increasing awareness of our carbon footprint and with landfill costs rising, WSPs are encouraged to re-use excavated material as backfill wherever possible. In addition to the landfill cost, importing selective backfill material is also expensive and, even then, water logging of ground will still create the potential for corrosion to occur to iron or steel pipelines even though non aggressive material has been utilised to surround the pipe. Designers are, therefore, left with the option to choose the best pipe to suit the prevailing environmental conditions.

Historically, grey iron pipes in either the spun or cast forms were laid with little or no external protection and relied mainly on the wall thickness to provide longevity. In corrosive soils with low resistivity, pipes would often fail due to localised pitting causing pinhole sized breaks to occur which, if not located and repaired quickly, would erode further to create larger holes and consequential leakage losses.

When ductile iron pipes were introduced in the UK water industry in the 1960s, a crude form of loose polythene wrapping was employed to protect the pipeline. While the theory was good, the reality proved that the wrapping was easily damaged and provided minimal protection to the pipeline. Close fitting polythene wrapping was added later to the pipes but they still suffered from damage during transportation, handling and installation activities.

External protection systems on ductile pipes are now highly developed with the addition of zinc, bitumen and epoxy coatings in various combinations to suit the ground conditions in which the pipeline is laid.

Steel pipes, from their inception in the UK water industry, have invariably been protected by way of cathodic protection systems. Metal structures are prone to damage through electrolytic action from the surrounding soil and water which will interact with the steel pipe through imperfections in the external protective coating, commonly referred to as 'holidays'. Although every precaution is taken by the installers to avoid damage to the protective coating, sharp objects in the pipe surround or 'holidays' that occur during the manufacturing process mean that the pipeline requires additional protection in order to achieve its anticipated design life.

Cathodic protection works by creating a sacrificial anode either by using an impressed current (rectifier) or natural anodes such as magnesium or zinc. Current flows from the protection system (anode) through the natural electrolyte (soil) to the steel pipeline. This must be sufficient to reduce the electrical potential of the metal in the pipe below its corrosion potential. Cathodic protection systems need to provide protection to all parts of the pipeline and must be regularly checked and maintained to ensure sufficient potential remains to protect the pipeline.

Theoretically, cathodic protection systems could be used to protect iron pipelines however individual mechanical joints would need to be bonded across to ensure the continuity of the current along the pipeline. This would add significant cost to the pipe laying project and would therefore render this process prohibitive in most cases.

Steel pipe manufacturers, like their ductile iron competitors, have refined and developed their protective coating systems to an extent where some now claim that additional external protection is no longer necessary. However, the potential for damage to occur to the coating during the pipe laying process would persuade most pipeline designers to include the cost of a cathodic protection system in the scheme budget.

6.2.5.2 Joint and fittings wrapping

The majority of pipelines will require that fittings such as tees and valves are added to provide the necessary control and management of the pipeline for the asset manager. Inevitably the addition of a fitting requires that the pipe has to be cut which breaks the continuity of the pipeline and interferes with the protective coating. In order to protect the integrity of the external protection system, joints created through this process need to be wrapped using an oil-based hemp or polyethylene shrink wrap in the case of ductile iron and/or over-bonded in the case of steel. Failure to do this adequately can create corrosion points around the joints which, over time, can result in a reduction in the mechanical strength of the pipeline leading to failure through bursting.

6.2.5.3   Barrier systems

As development land becomes increasingly scarce in the UK, developers inevitably look towards brownfield sites to meet the housing needs for a burgeoning population. In many cases the previous use of the land has left a legacy of 'contaminated' ground, this is particularly true of former garage forecourts and industrial sites.

Developers are presented with the option of 'remediating' the land to remove the contaminant, which is often a prohibitively expensive option, or asking WSPs to install barrier pipe products to eliminate the risk of the contaminant affecting water supplies to the new houses. Invariably housing developments are serviced using smaller diameter polyethylene (PE) or polyvinyl chloride (PVC) water pipes for both mains and services. Chemicals contained in the contaminant may have the potential to permeate through the wall of a plastic pipe, imparting a taste to the water and affecting the life expectancy of the pipe.

Plastic pipe manufacturers have risen to the challenge and many now produce a barrier pipe system. Pipes need to provide a barrier to prevent the potential contaminant from reaching the pipe bore in order to protect the water supply. This is achieved by incorporating an aluminium layer additionally an outer protection layer is required to maintain the long-term strength of the pipe.

In order to maintain the integrity of the barrier system, joints need to be protected either by incorporating barrier fittings or by wrapping with overlapped aluminium tape which is then wrapped again with an oil based waterproof tape.

6.2.5.4   Internal protection

Iron and steel pipelines historically were laid with either crude, at best, or non-existent internal linings. Many WSPs have seen significant elements of their funding programmes expended in relining grey iron mains that have corroded internally causing dramatic reductions in internal bore and discoloured water problems.

Developments in lining technologies have given rehabilitation engineers a range of products to consider for developing a solution to extend the life of pipelines and for providing improved water quality to customers. A spin-off benefit of mains rehabilitation is the recovery of headroom in the network and the resultant hydraulic improvements. There are many examples of water pumping stations being rendered redundant as a consequence of mains rehabilitation projects.

Epoxy resin, PE, including high build PE, and close-fit non-structural and semi-structural linings are some of the alternatives that have been used to rehabilitate water mains. High-build and semi-structural products can be used where the host main has lost some of its mechanical integrity and, therefore, provide an alternative to main replacement in a limited number of cases.

The decision to re-line rather than renew is complex and is covered in greater detail in Chapter 11: 'Finance, regulation and risk in project appraisal'. Re-lining is much less disruptive to customers and to the environment than replacement and is far less expensive.

Modern ductile iron and steel pipes are supplied from the manufacturer with applied internal linings to protect the pipeline from internal corrosion. Cement mortar linings were used from the 1970s with later additions of epoxy coatings to control pH and extend residence times of water in transit.

6.2.6   Pipe restraint

In an earlier subsection we discussed pipe material selection with due consideration to ground movement. In pipelines where socket and spigot joints are utilised it is necessary to consider the need to provide thrust restraint at the points where bends, blank ends, tees and, to a lesser extent, tapers are incorporated into the pipeline. This is required to prevent the pipe from moving due to the pressure exerted on the pipe at these points of deviation by both static and dynamic thrusts acting within the pipeline.

Traditionally, this is achieved by the provision of a thrust or anchor block which is designed to provide additional weight to oppose the direction in which the thrust is acting on the pipe. The size and shape of the block required to overcome the thrust is dependent on the pressure acting on the pipe, the direction in which the pressure is applied and the bearing pressure of the ground in which the block will be constructed.

Designers should also consider the possibility that flow can be reversed in certain pipelines, for example a flow and return main between a pump station and a service reservoir. In such cases the direction of thrust may change and the thrust block must be designed to restrain the thrust forces in both directions. Consideration must also be given to the accessibility of the joint so that repairs can be made, if necessary, without the need to destroy the anchor block.

There are alternative methods of restraining bends such as tie-bars, self-anchoring joints, stub flanges and end restrained fittings but these forms of anchorage often serve to move the potential point of separation to another nearby joint. They should therefore be considered as a complimentary form of restraint rather than a substitute for a well-designed thrust block.

There are two types of thrust exerted on a pipeline: static and dynamic. Both types act in the same direction within a pipeline but dynamic thrust, which occurs under high velocity, can create an additional burden at direction changes. The following formula, adapted by Saint Gobain Pipelines, can be used to calculate combined static and dynamic thrust at bends

$$T = \left(P + 0.01v^2\right)10^2 A_e \; 2 \; Sin\theta/2 \qquad\qquad (6.2)$$

where

T = thrust force acting on bend (kN)
P = internal pressure (bar)
v = velocity of water (metres/second)
A_e = cross sectional area of pipe external diameter (m²)
θ = angle of bend (°)

Once the thrust acting on the bend is known then a thrust block can be designed to restrain the force (Figure 6.1). Thrust block design is complex and expert advice should be sought in order to ensure that pipelines are properly restrained. Disregarding this need can result in catastrophic failures of pipelines, disruption to customers and expensive compensation and repair bills for WSPs.

6.2.7   Pipe jointing

Most pipe materials have a selection of jointing techniques that can be applied. This short section describes the techniques that can be used for joining pipes and pipe fittings on the most commonly used pipe materials.

Figure 6.1 The direction of thrust acting on a selection of bends and fittings and the typical location of thrust blocks installed to resist the thrust. Arrows represent the direction of thrust (Reproduced with permission. ©Saint Gobain PAM UK, www.saint-gobain-pam.co.uk)

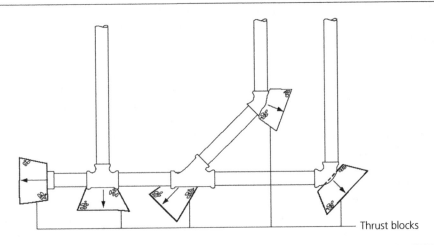

Thrust blocks

6.2.7.1 Ductile iron pipes

Socket and spigot push-fit joint – This is a male (spigot) to female (socket) flexible joint which uses a rubber ring gasket to provide the leak tight seal. The joint provides (typically) 5° angular deflection to accommodate some ground movement and to aid installation where consecutive pipes are not perfectly aligned. It can be joined with an anchor gasket to provide a restrained joint.

Restrained flexible weld bead joint – This is a socket and spigot joint but with the addition of a bead welded to the spigot end and a locking ring which locates adjacent to the bead to stop the joint from pulling apart. Typically used for high pressure applications or where the use of thrust blocks is impractical.

Tied joint – It is a socket and spigot joint with the addition of a flange that is welded to the spigot end of the pipe. A second loose flange is located behind the socket and tie bars between the two flanges tie the joint together to prevent separation (Figure 6.2).

Figure 6.2 Example of a tied socket and spigot joint (Reproduced with permission. ©Electrosteel Castings (UK) Ltd, www.electrosteel.co.uk)

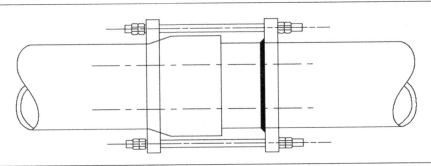

Flanged joint – Generally used for joining pipes to fittings, this is a fully rigid joint connecting flange to flange with nuts, bolts and flange gaskets to provide the watertight seal. The flanges carry a range of pressure ratings, typically 16, 25 or 40 bar.

General – Historically, iron pipes have been jointed using other techniques that are largely redundant on modern distribution networks but will still be present on old spun and cast, grey iron pipelines. These include run lead, bolted gland, victaulic coupling and dresser coupling joints.

6.2.7.2   PE pipes

Butt fused joint – Butt fusion uses an electrically heated plate against which two pipe ends are drawn together simultaneously. When the pipe ends are sufficiently heated, the plate is withdrawn and the pipe ends are brought together at a predetermined pressure until the polyethylene has cooled to form a continuous pipe. The resultant 'joint' will be end load resistant and possess similar properties to the parent pipes.

Electrofused joint – Pipe ends are pushed together within an electrofusion coupler after appropriate preparation of the pipe ends and pipe wall. The coupler and the pipe ends are then fused together by use of an electric current to form a continuous pipe. A correctly fused joint has an equivalent pressure rating to the parent pipes that have been fused together. Electrofusion has an added advantage over butt fusion in that pipes of different diameters can be fused together.

Puddle flange – Where end loads will be exerted on a PE pipeline such as at line valve installations, puddle flanges are often used to absorb the force exerted on the valve gate when the valve is closed. The fitting is installed in the line, connected to the valve. The puddle flange is usually encased in concrete so that forces created by virtue of closing the valve are transferred through the flange to the concrete thrust block and the surrounding soil (Figure 6.3).

Stub flange joints – An alternative method of joining PE pipes to metal fittings, the stub flange is a drilled flange which is formed on one end of the PE pipe. Used with a backing ring, the flanged end of the pipe is joined to the flanged fitting using standard flange nuts and bolts.

General – Standard mechanical compression fittings such as couplings, adaptors and ferrule straps can also be used with PE pipes. PE is therefore a versatile product in terms of the jointing solutions that can be considered for use with this pipeline material.

Figure 6.3 Example of a puddle flange (Reproduced with permission. ©Radius Systems, www.radius-systems.co.uk. All rights reserved)

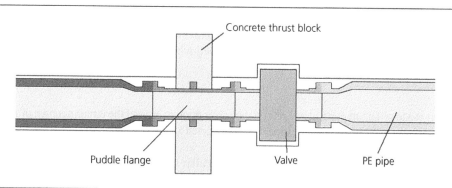

Concrete thrust block

Puddle flange · Valve · PE pipe

6.2.7.3 PVC pipes

Push fit mechanical joint – This is a male (spigot) to female (socket) flexible joint which uses a rubber ring gasket to provide the leak tight seal. The joint provides a small degree of angular deflection to accommodate some ground movement and to aid installation where consecutive pipes are not perfectly aligned. This is the most commonly used joint for PVC pipe systems.

Straight coupler – This is a mechanical compression fitting that is designed to join the plain or cut ends of pipes together using an internal sealing gasket to provide a leak-free tight seal.

Stepped coupler – Based on the same sealing principles as the straight coupler, it is designed to provide transition between pipes of different outside diameters and materials, and, therefore, permits the connection of PVC to ductile iron, cast iron or asbestos cement.

Flange adaptor – Based on the same sealing principals as the straight coupler, it will join the plain end of a PVC pipe to flanged fittings such as tees and sluice valves. It, therefore, enables PVC pipelines to simply incorporate control devices for network management.

6.2.8   Modes of pipeline failure

If all of the preceding considerations are properly accounted for in pipeline design and installation then network managers should have every confidence that their network is fit for purpose. However, inevitably, faults will occur due to poorly laid pipes, insufficient thrust restraint or inadequate external protection systems or as a result of third-party activities. The resulting damage will need to be located and repaired.

Failures can occur in a number of ways and the following paragraphs describe modes of failure and repair techniques.

6.2.8.1   Joint failure

Flanged joints and socketed joints on 'push-fit' pipes both require a rubber gasket to provide a watertight seal. It is important that the correct gasket is used for the application and that the gasket is fitted correctly. Equally important is the making of the joint so that bolts are tightened to the correct torque and in the right sequence in the case of a flange, and that pipe ends (spigot) are pushed home correctly in the case of a push-fit joint.

Failure to do this correctly can result in the gasket being forced out of the joint and a leak occurring at this location. If this type of failure is identified quickly the repair may be simply made by breaking and remaking the joint. However, if the leak is left undetected for a period of time then material can be lost from the pipe wall or the face of the flange which may require that an encapsulation fitting is needed to completely surround the joint. Alternatively, the joint might have to be cut out and replaced using flange adaptors and couplings combined with a cut section of pipe or using welded couplers in the case of plastic pipes.

6.2.8.2   Circumferential pipe breaks

A common mode of failure particularly in iron pipes is a fracture around the circumference of the pipe. Often caused by ground movement, the pipe wall is subjected to stresses which cause the pipe to fracture.

In most cases this can be rectified by using a repair collar or coupling, but if the fracture requires significant cut back to create an even face on both sections of pipe then two couplings and a section of cut pipe may be required to remake the joint. If the fractured pipe has deflected significantly through the course of the fracture then repair gangs must always check the integrity of any nearby joints as part of their repair works.

Early (1980s) medium density polyethylene (MDPE), or PE80, pipes were prone to rapid crack propagation (RCP) particularly when the crack initiated on the external wall of the pipe. In instances where external impact damage was caused to an MDPE pipe under pressure, the initial damage would propagate axially at extremely high speed. Larger diameter MDPE pipes were particularly prone to RCP and, as a consequence, MDPE pipes of DN250 and greater were de-rated. The PE pipe industry has, to a large extent, resolved the problem in their modern PE80 pipes and a water industry standard specification has been introduced to enable manufacturers to demonstrate that their modern PE pipes meet the necessary toughness and fatigue standards. In addition, a high-performance polyethylene (HPPE), or PE100, pipe has been introduced to the water sector which is not de-rated for RCP.

A UK Water Industry information and guidance note, IGN 4-32-18, provides guidance on the choice of pressure classifications or ratings for buried and above ground PE pipe systems (UK Water Industry, 2003).

6.2.8.3   Longitudinal pipe breaks
Longitudinal fractures occur along the length of a pipe and can be caused by a number of factors including ground movement, pressure surges and excessive mains pressure. By their nature they are expensive to repair since more ground has to be excavated to expose the extents of the fracture and in the case of uPVC (rigid PVC) and asbestos cement pipes usually result in a full pipe length being replaced from joint to joint.

Repairs are invariably completed by using a new length of pipe and suitable couplings although it is often necessary to cut off the socket end of the adjacent section of main in order to make the repair.

6.2.8.4   Pin holes
As the term suggests, pin holes are small breaks in a pipe wall usually caused when pipe walls are damaged or pitted due to corrosion so that the mechanical strength of the pipe is reduced. Pin holes can develop into larger breaks if they are not located and repaired quickly as the escaping water will quickly erode the pipe wall around the hole as it escapes under pressure (Figure 6.4).

Repairs to pin holes are usually accomplished by fitting a split repair collar around the pipe to cover the hole. Historically, experienced repair crews would often plug the hole first by using a Hazel twig or section of broom handle to stop the leak in order that the collar can be fitted.

6.2.8.5   Complications
Water main breaks will not always occur in convenient locations, therefore repair techniques for the types of failures covered previously may have to change because of the local environment or due to inaccessible sites. Water mains laid at significant depth or buried under buildings, in riverbeds, under major highways, breaks on encased bends or special fabricated fittings may test the inventiveness of the maintenance engineer. In many of these examples it is often easier to relay the main in order to bypass the break but, of course, that's not always possible.

Figure 6.4 Example of pinholes

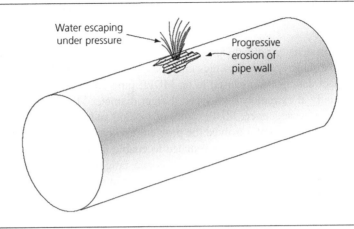

Water escaping under pressure

Progressive erosion of pipe wall

It is wise for a duplicated main to be incorporated in the design of a water network that needs to cross a major road or river or for pipes to be laid in ducts or tunnels so that pipe work can be withdrawn and replaced in the event of failure. In the case of special fittings, often the only solution is to weld the fitting in order to carry out a repair or to replace it like for like.

6.3.   Pumps

Pumps play a vital role in water supply systems. They provide or maintain the energy needed to carry water from the treatment works to the service reservoir within the distribution system and to the distribution mains and ensure the required amount of water can be delivered with sufficient pressure. By law, WSPs in England and Wales should ensure there is enough pressure to supply the top floor of a house under conditions of normal demand. The stipulated criteria or reference level of service – often referred to as DG2 (Ofwat, 2017) – are a pressure head of ten metres of water at the boundary stop tap, at a flow of nine litres per minute. WSPs are also required to maintain a minimum static pressure of seven metres of water in the communication pipe that supplies a property. These criteria do not apply to high-rise buildings for there are different requirements – for example, it is common to pump the water to a tank in the roof which then supplies the building.

Pumps operate primarily by adding pressure to the water to achieve a static lift and overcome the head losses that occur in the system. The main cause of head loss in distribution networks is pipe friction. However, other losses referred to as minor can occur at pipe junctions, bends and other locations where there are fittings such as valves or discontinuities. The major types of pump include the rotodynamic and positive displacement.

Positive displacement pumps are not normally used for water supply purposes these days. They are often of the reciprocating type and can be found at water and wastewater treatment works. Water is drawn into a cylinder through an inlet valve and released on the forward stroke of the piston through an outlet valve.

A rotodynamic pump imparts energy to the flow from the motion of a rotating component called an impeller at the centre of the pump (Figure 6.5). The volute is the casing that surrounds the impeller. The impeller blades are mounted on a shaft that is typically driven by an electric motor. Water

Figure 6.5 Centrifugal pump

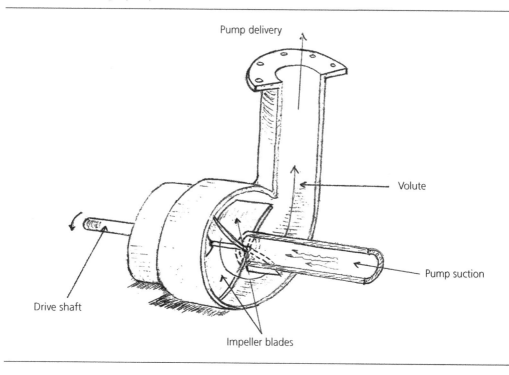

Pump delivery

Volute

Pump suction

Drive shaft

Impeller blades

enters the pump axially. The discharge from the pump can be tangential at the circumference of the casing or axial along the shaft. The pumps are thus referred to as centrifugal and axial flow, respectively. There is a third type known as mixed flow for which the discharge direction is intermediate.

In a centrifugal pump, the water entering the pump is forced outwards by the rotating impeller. Therefore, the water gains speed and moves into the volute where it slows down and so acquires extra pressure. The volute is snail shaped and collects and directs the flow to the outlet of the pump. Centrifugal pumps are versatile in that impeller design variations enable them to operate over a wide range in terms of the pressure head added and the volumetric flow rate and are, thus, widely used. Centrifugal pumps are suitable for pumping relatively small volumes of water against high heads. Multi-stage pumps, an arrangement with multiple identical impellers in series on a single shaft and driven by the same motor and in which water flows through the impellers sequentially, can be used to achieve even greater lifts.

The impeller of an axial flow pump operates in a similar fashion to a propeller in that the rotation of the impeller drives the water forward in the axial direction. Some of the increase in kinetic energy thus achieved is converted into a pressure increase. Axial flow pumps have a limited suction lift and should be submerged. They are utilised in situations in which the volumetric flow is large and the head required of the pump is low.

Mixed flow pumps combine both radial and axial flow as in centrifugal and axial flow pumps. Their range of operation in terms of flows and heads is intermediate between centrifugal and axial flow pumps.

The efficiency of a pump is the ratio of the power generated by the pump to the power supplied to the pump. The power generated is given by

$$P = \rho g Q H \tag{6.3}$$

where

      P = the power
      ρ = the density of the fluid
      g = the acceleration due to gravity
      Q = the volume flow rate
      H = the difference in total head between the inlet and outlet of the pump.

6.3.1   Performance characteristics

The head generated by a pump depends on the flow rate and, for a given pump, the relationship between the head and the flow defines its performance characteristic that can be shown graphically as a plot of the head against the discharge. Along with the head and discharge, the power required to operate a pump and its efficiency are primary considerations when selecting a pump to perform a particular duty. Self-evidently it is desirable to operate a pump at its maximum efficiency and as these parameters are all related to the discharge, it is advantageous to superimpose the plots showing their variations as illustrated in Figure 6.6.

The graphs in Figure 6.6 show the performance characteristics of the pump and are thus called performance or characteristics curves. The discharge corresponding to the maximum efficiency is referred to as the rated discharge. This is also called the design point, often quoted as the head and discharge at the point of maximum efficiency. It should be noted that the performance characteristics depend on the speed of pump (as shown in the next section). Also, the shapes of the curves depend on the type of pump (e.g. Potter *et al.*, 2017).

Figure 6.6 Characteristics curve for a centrifugal pump

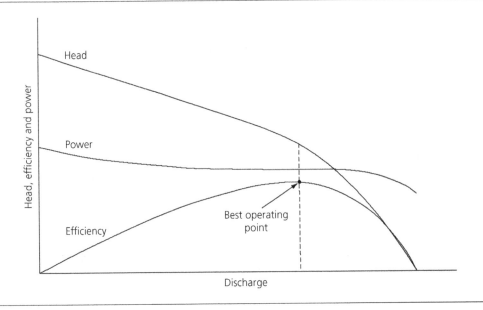

For a pipeline or pipe network, the head to be added by a pump increases as the flow rate increases because the pipe friction and other losses increase with the flow rate. The operating point is the point of intersection between the head-discharge curve of the pump and that of the pipe network (the system curve) as explained in the next section. In a design situation a key objective when selecting a pump is to make sure that the operating point is as close as possible to the maximum efficiency or rated discharge of the pump.

Pumps may be operated in series or in parallel. When operated in series, the water flows through each of the pumps in turn and thus the head generated by the pumps collectively is the sum of the heads developed by each of the pumps. The head-discharge curve for such a system is obtained by summing the heads for a given discharge. As the flow through the pumps is the same, pumps that operate in series should be identical or similar, to reduce the possibility of one or more pumps operating outside their normal operating range. For pumps that operate in parallel, the heads added by the pumps are identical while the overall discharge is the sum of the individual discharges. Accordingly, the head-discharge curve for the system is obtained by adding together the discharges for a given head. As the head across the pumps must be the same, pumps operated in parallel should be identical or similar, to reduce the possibility of one or more pumps operating outside their normal operating range.

Cavitation is an important design consideration for pumps. It is characterised by the formation and collapse of air bubbles and causes vibration and noise, reduces the efficiency of operation and damages impeller blades by erosion. Cavitation occurs when the pressure is less than the vapour pressure causing air bubbles to form, typically on the suction side of the impeller where the pressure is lowest. The air bubbles then collapse as they encounter a region of high pressure within the pump. Cavitation can be prevented by ensuring the pressure within the pump is kept above the vapour pressure.

The pressure above which cavitation will not occur is termed the required net positive suction head (NPSH) and corresponds to the minimum pressure that will raise water from the sump to the impeller. It depends on the flow rate and speed of the pump and is provided by the manufacturer. The NPSH is given by

$$NPSH = p/\rho g + v^2/2g - p_v/\rho g \qquad (6.4)$$

where
p = the pressure at the pump inlet
v = the velocity at the pump inlet
p_v = the vapour pressure
ρ = the density of the fluid
g = the acceleration due to gravity.

In terms of heads

$$p/\rho g = p_{atm}/\rho g - H_s - h_l \qquad (6.5)$$

where
H_s = the static lift in the suction pipe
h_l = the headloss (i.e. pipe friction and minor losses) in the suction pipe
p_{atm} = the atmospheric pressure.

To avoid cavitation, the actual NPSH should exceed the required NPSH.

6.3.2   Pump selection

When selecting a pump, one of the tasks accomplished early on in the process is to identify the type of pump that would be most suitable. This can be done with the help of the specific speed. The specific speed is defined as the rotational speed (rpm) required to develop a unit head (1 m) for a unit discharge (1 m³/s) – that is

$$N_s = NQ^{1/2}/(gH)^{3/4} \qquad\qquad (6.6)$$

where

N_s = the specific speed (rpm)
N = the impeller speed (rpm)
Q = the discharge at the point of maximum efficiency – the duty point (m³/s)
H = the head at the duty point (m)
g = the acceleration due to gravity.

Great care is required with regard to the specific speed, as the g in the denominator is often omitted and consistent units are not always adhered to.

Centrifugal pumps work best for specific speeds between approximately 2-16 rpm; for mixed flow pumps the range is approximately 7-29 rpm; and for axial flow about 27-76 rpm (Brandt *et al.*, 2017).

For a given pump, the effect of changing the rotational speed can be investigated using the following relationships often referred to as affinity laws. The affinity laws can also be used to compare two geometrically similar pumps

$$Q_1/Q_2 = N_1/N_2 \qquad\qquad (6.7a)$$

$$H_1/H_2 = (N_1/N_2)^2 \qquad\qquad (6.7b)$$

$$P_1/P_2 = (N_1/N_2)^3 \qquad\qquad (6.7c)$$

where

Q = the discharge
H = the head developed
P = the power required.

Pump selection and sizing go hand in hand with the design of the delivery pipe as the head H to be generated by the pump consists of the total static lift (i.e. the difference between the delivery and suction levels) and the pipe friction and minor head losses

$$H = H_s + h_l \qquad\qquad (6.8)$$

$$h_l = (\lambda L/D + \Sigma K)(v^2/2g) \qquad\qquad (6.9)$$

where

H_s = the total static lift
h_l = the head loss

λ=the friction factor
L=the pipe length
D=the pipe diameter
K=the minor loss coefficient – the summation includes all the minor losses including bends, fittings and so on
v=the velocity
g=the acceleration due to gravity.

A plot of H against the discharge yields the performance curve for the piping system which, when superimposed on the performance characteristics curves for the pump, identifies the operating point as the intersection of the head-discharge curves of the system and pump (Figure 6.7). The objective is to match the rated discharge (the duty point) and the operating point as closely as possible to ensure the pump operates at the highest efficiency possible and minimize the power required.

Alternative solutions based on different pump and delivery pipe systems may be arrived at in this way and the one with the smallest overall cost (i.e. capital and operation) selected.

Pumps in series or parallel may be adopted instead of a single pump. This may be because the design constraints cannot be met by a single pump or due to operational reasons. Pumps in series can achieve a higher head and offer greater flexibility in terms of the head developed. Pumps in parallel can achieve a larger discharge and offer flexibility in terms of the discharge because the number of pumps deployed can be varied since each pump can operate independently. In any case, adequate standby capacity must be provided to ensure continuity of supply in the event of a malfunction and allow for routine maintenance.

Figure 6.7 Pump characteristics and system curves. Option 1 is the preferred solution for the pump shown

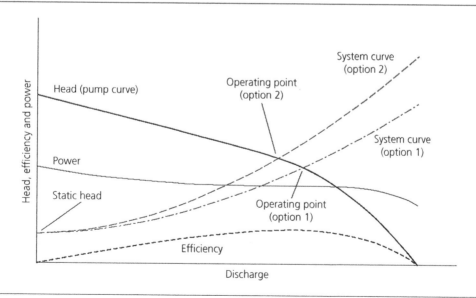

6.4. Valves
6.4.1 Control and operability of distribution networks

Earlier sections of this chapter have included information on pipe selection and the importance of installing the right pipe in terms of size and material. In essence the role of pipes (water mains) is to ensure that the volume of water within the distribution network is suitable to meet the needs of the customers that will use it. Having established that the network is capable of distributing the correct volume of water it is then necessary to provide the means by which to manage the network for both planned and unplanned maintenance and to take account of the types of conditions that will occur naturally within a hydraulic network.

Modern control rooms are able to interact with the distribution system to open and close valves and start and stop pumps to balance demands across the network. To do this with confidence, key installations are enabled to feedback information through supervisory control and data acquisition (SCADA) networks so flow, pressure and reservoir level data is visible 24 hours a day. SCADA and systems operations are discussed in greater detail in Chapter 9: 'Operation, maintenance and performance'.

This section will deal with the control valves that are present in a distribution network that provide its functionality and operability. It will also cover valves that are inserted into the network to permit distribution engineers and managers to interact with the network to create changes to the operating conditions. Additional information on the mathematical modelling of valves for distribution network analysis and simulation purposes is available in Tanyimboh and Templeman (2010).

6.4.2 Line valves

'Line valve' is a generic term that covers a variety of valves which have a primary purpose of changing flow volume within a pipeline. This can vary from the fullest extents of operation – fully open to fully close – or to a controlled, sometimes referred to as a 'throttled', position somewhere between fully open and fully close.

The following paragraphs describe the various types of valve in common use in the water industry together with an explanation of their functionality.

6.4.2.1 Gate or sluice valve

The purpose of the gate valve is merely to stop the flow of water in a pipeline, usually to permit maintenance activities to be undertaken. They are also used to separate areas of different pressure or to create water quality zones and district meter areas. The valve consists of a gate that moves vertically within a guide rail by operation of a screw thread. The screw thread connects to a stem which can be specified as either rising or non-rising depending on its intended use. Typically gate valves are offered with either hard or soft faces (sometimes referred to as resilient faced) to suit the preference of the water undertaking or to suit specific conditions. A gate valve is shown in Figure 6.8.

In high pressure applications on larger diameter mains, gate valves are often difficult to open from a fully closed position because of the pressure differential across the gate. To assist with this difficulty, larger gate valves can be purchased with a smaller diameter integral bypass that incorporates a small gate valve which can be operated to equalise pressure on either side of the larger gate to permit the main valve to be opened.

6.4.2.2 Butterfly valves

The butterfly valve comprises a disc that operates on a horizontal pivot around the centre (or offset from the centre) of the valve bore (Figure 6.9). The valve operates to any position between fully

Figure 6.8 Gate valve

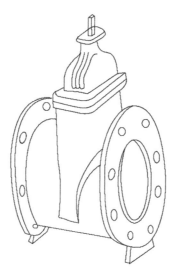

Figure 6.9 Butterfly valve

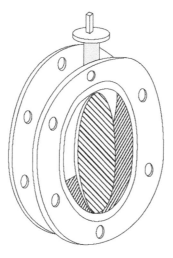

closed, perpendicular to the water flow, and fully open, parallel with the water flow, through a quarter turn. Like the gate valve, they are offered by most manufacturers with a metal faced or resilient faced option. Butterfly valves are typically used as flow control devices and can be operated by hand or by motorised actuation.

Butterfly valves are often used in water treatment works where flows need to be regularly changed and can be used as service reservoir inlet control valves linked with telemetry networks. While the butterfly valve offers excellent flow controllability, they are sometimes dismissed by distribution engineers because the pivot and gate occupy the pipeline even in the fully open position. This inevitably creates a head loss within the pipeline and restricts the passage of a swab during mains cleaning operations.

6.4.2.3   Eccentric plug valves

These valves are designed specifically for flow control. The control element is a profiled plug which pivots around a vertical stem. The plug rotates away from the seat as it begins to open reducing wear on the seat and exposing an increasing amount of the pipe bore as it moves towards its fully open position. The valve therefore provides high flow capacity with virtually a full bore when open.

Used increasingly for service reservoir inlet control and strategic network control, the valve can be operated manually by way of a hand wheel but mostly is configured with motorised actuators for remote operations by way of SCADA systems.

6.4.3   Air valves

When water mains burst or are fully/partially drained down for maintenance purposes, the water column is replaced by air. On refilling the main after the maintenance or repair operation, the system must be capable of exhausting the air out of the pipeline in order to avoid air locks or aerated water.

In the event of a planned or unplanned evacuation of water from a pipeline, air needs to be able to enter the pipe at a similar rate to the water escaping from it. If the network is not capable of achieving this then there is a risk that a vacuum can be created within the network which can lead to the main imploding or, at best, a potential weakness created that can lead to a future failure of the pipeline. When a distribution pipeline is sucking in air, other unwanted detritus can be drawn into the network. It is vital therefore that air valve chambers are free draining so water and associated debris cannot accumulate within the chamber.

Sudden starting or stopping of flows in a pipeline, due to, for example, the rapid closure of a valve or sudden stopping of a pump, can lead to a separation of the water column within the network. This in turn can lead to momentary negative pressures occurring in the pipeline followed by equally short-lived high pressures, referred to as transients. These undesirable events can create high stresses within a pipeline.

Air transfer in and out of pipelines can be managed in several ways but it is essential that pipelines are designed with appropriately sized and strategically placed air valves to manage the air movement to and from the distribution network. The importance of correct air valve placement and sizing cannot be over-emphasised.

Air valves should be located at high points on the distribution network where air will naturally accumulate. In addition, consideration should be given to installing air valves at gradient changes, both upwards and downwards, and at regular intervals along constant gradients.

There are a number of options to consider when selecting the most appropriate air valve for the location and dynamic network events the valve will need to be able to accommodate. Pipeline designers must consider this carefully before selecting a valve for each location. An example of an air valve is shown in Figure 6.10.

Figure 6.10 Single-orifice air valve

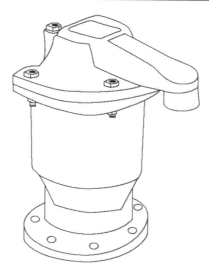

6.4.4   Automatic control valves

In the context of this chapter, the term automatic control valve refers to a valve that has been added to the distribution network to change flow or pressure automatically, therefore without manual or remote intervention.

There are many different control variations that can be deployed through the automatic control valve, limited only by the imagination of the end user. Additionally, the pilot rails and regulators can be designed to provide multiple control functionality such as pressure reduction and pressure sustaining within the same valve.

The following sections describe the principal control functions of the automatic control valve although it should be noted that this is not an exhaustive list of control functions that can be performed by the valve.

6.4.4.1   Pressure reducing valve (PRV)

The purpose of the pressure reducing valve is to maintain a constant pressure in the water distribution network downstream of the valve. PRVs are used extensively in networks characterised by hilly topography with gravity sources at high elevation and are installed to support pressure management initiatives for WSPs. The pilot regulator is sensing pressure variations in the downstream section of the valve created by changes in demand. This causes the pressure to change in the control space on the top of the main valve. The stem and disc then alter position relative to the seat in order to restore the downstream pressure to the desired level (Figure 6.11).

Many WSPs employ modulating devices with PRVs to alter (modulate) the pressure control set point to suit significant variations in demand within the downstream network due to, for example, industrial parks with large commercial water users. Electronic modulators can be deployed using timed set point changes or remote pressure sensing feedback. Alternatively hydraulic modulators, using an orifice plate to sense flow rate changes and to modulate the outlet set point, are becoming increasingly popular with many engineers.

Figure 6.11 A cut-away view of a Cla-Val globe-style automatic control valve (Reproduced with permission. © Cla-Val UK Ltd, www.cla-val.co.uk)

6.4.4.2   Pressure sustaining valve (PSV)

The pressure sustaining valve works to stop the pressure in the network upstream of the valve from dropping below a prescribed minimum level. A typical use for a PSV is on the inlet to a service reservoir where it is configured to prevent the demand into the reservoir from drawing down the pressure on the water main providing the feed, perhaps, where there are properties connected to the inlet main at high elevation. The pilot regulator senses the pressure in the upstream section of the valve. As the upstream pressure reduces towards its minimum setting, the regulator begins to close causing increased pressure in the control space, which causes the main valve to restrict the volume of water that passes through it.

6.4.4.3   Burst main valve (BMV)

BMVs are used where properties or sensitive environments are at risk of flood damage as a consequence of a burst main. The valve also serves to protect the network upstream of the valve from excessive drain down or de-pressurising due to the burst. The burst main valve is configured to sense a minimum downstream pressure that would be characterised by a burst in the network supplied through the valve. The regulator is configured to either fully or partially fill the control space to cause the main valve to fully or partially close to restrict the volume of water feeding the burst.

6.4.4.4   Altitude control valve

Altitude control valves (Figure 6.12) are used at service reservoir inlets. The regulator is configured to sense the maximum and minimum control levels (static back pressure) within the reservoir and operates the main valve at a linear or graduated rate to increase flow when the reservoir is low or reduce it as it approaches the top water level. The altitude valve has replaced the equilibrium ball valve at many sites as it is considered by many distribution engineers to be a more reliable form of inlet control.

Figure 6.12 A typical control configuration for altitude control valves (two-way flow) (Reproduced with permission. © Cla-Val UK Ltd, www.cla-val.co.uk)

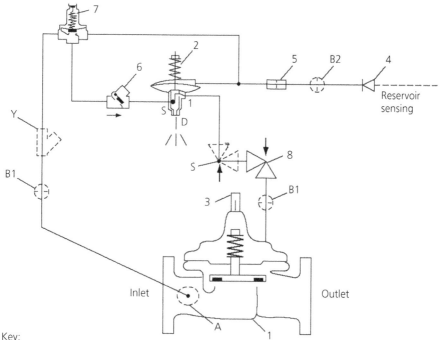

Key:
1, main valve; 2, altitude control; 3, valve position indicator; 4, bell reducer; 5, restriction assembly; 6, swing check control; 7, flow check control; 8, flow control (closing); A, strainer; B1 and B2, isolation valves; S, flow control (opening); D, check valve with isolation; Y, 'Y' strainer.

6.4.5   Non-return valves (NRV)

The non-return valve is a generic description for a valve that is referred to by various alternative names: reflux, check, swing check, one-way, recoil. The purpose of the NRV is to permit water to flow in one direction and to prevent water from flowing in the opposite direction (backwards) through the valve. Most prevalent at water pumping stations, the NRV is installed to prevent water passing back through the pumps when they are in their standby configuration. The designer needs to carefully consider several key design criteria when specifying non-return valves, not least of all the installed angle of the valve which may influence the correct operation of the valve, particularly where the gate relies on gravity to return it to its closed position.

There are numerous alternative designs for non-return valves including ball check, butterfly disc, dual plate and spring-assisted cone, to name a few. It is vital that the design engineer determines the critical parameters including the pressure drop across the valve, the flowing velocity of the water through the valve and the rate of deceleration of the water after the pump has stopped (e.g. Thorley, 1991). Deceleration is not easy to calculate, so simulation of the network using a computer model with surge analysis software would be advised to ensure that appropriate valve selection is made for the prevailing conditions. An example of a non-return valve (swing check) is shown in Figure 6.13.

Figure 6.13 Swing check valve

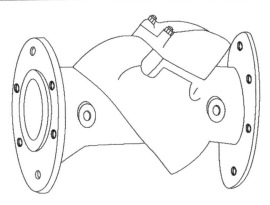

6.4.6 Ball float valves

We have already stated earlier in this sub-section that modern, automatic control valves are increasingly being deployed in water distribution networks to replace some of the traditional control methodologies. The ball float valve is one such device that has been relied upon for many years to manage flows into service reservoirs, tanks and water towers and, while it may have a more attractive alternative in the altitude valve, it continues to perform its function in many water distribution networks globally.

6.5. Service reservoirs

Water undertakings in England and Wales have a legal duty to maintain a wholesome, continuous and adequate supply of water to their customers. Therefore, the supplies must be uninterrupted and achieve the required standards with regard to quality, quantity and pressure. Self-evidently, the components (pumping stations and treatment works included) of water supply and distribution systems require regular maintenance and, to carry it out, some parts of the system have to be taken out of service. A similar situation with some elements of the system being unavailable arises when repairs are carried out. Superimposed on these 'abnormal' operating conditions is the inherent variability and uncertainty associated with water consumption from one day to the next, as discussed in Chapter 3: 'Water demand: estimation, forecasting and management'. This includes the diurnal variations as well as any underlying weekly patterns.

Water undertakings deal with these and many other eventualities, such as large flows for firefighting, using a variety of strategies a major component of which is storage of treated water in reservoirs close to the demand centre. Surface or underground tanks within the distribution system are commonly referred to as service reservoirs, whereas elevated tanks are often called water towers. A portion of the water in service reservoirs is held in reserve to deal with contingencies with the remainder being the variable storage that balances hourly fluctuations in demand. The contingencies may include major fires and source failures.

The principal function of service reservoirs is, thus, to improve the security and continuity of supply by providing additional storage. There are additional benefits that may be summarised as follows. The variable storage balances hourly fluctuations in demand and smoothes out the peaks. This has the advantage that it reduces the capacities of the water treatment and transmission (as opposed to distribution) facilities including pumps and trunk mains. Thus, the treatment and transmission facilities may be sized on the basis of the consumption that occurs on the maximum day, whereas the distribution system would be sized according to the maximum hourly consumption. Service reservoirs also simplify the operation and improve the efficiencies of treatment works and pumping stations. The treatment works can produce water at a constant rate (for longer periods). Similarly, the pumping need not track the daily consumption pattern closely but instead can be optimised to achieve a more steady operation and take advantage of any periods during which electricity is cheap. Service reservoirs also improve the pressures within the distribution system as discussed later in this section.

The common practice is to locate service reservoirs on high ground close to the demand centre to provide adequate pressure for the system served. Elevated tanks or water towers are often used as an alternative. A relatively central location will help avoid excessive pressure near the service reservoir and low pressure problems in the outlying areas. With respect to the main supply source, the service reservoir should be located beyond the main demand centre, to ensure the service reservoir and main source supply water from different directions to more easily cope with the peaks in demand. This arrangement should yield a more economical design (Siew *et al.*, 2016) compared to a service reservoir ahead of the demand centre. The adoption of two major supply directions in this way also enhances the reliability of the distribution system considerably. Other factors such as the aesthetics of the structure come into play and the final choice from among potential locations requires an economic and technical appraisal including hydraulic and water quality analyses (Siew *et al.*, 2016).

Service reservoirs are normally reinforced concrete structures. Steel tanks and prestressed concrete are also used sometimes, but the interested reader should refer to more specialised texts on the structural design and constructional aspects – for example, *BS EN 1992-3:2006*. A minimum of two compartments are required to help maintain continuity of supply as one compartment will remain operational if the other is taken out of service for maintenance, repairs and so on. For this reason, service reservoirs are usually rectangular in plan. They are normally covered to prevent contamination of the water. The Water Supply (Water Quality) Regulations 2000 is the statutory instrument set up to safeguard the quality of water in service reservoirs in England and Wales. It requires samples to be taken weekly from each service reservoir. In the UK, the Reservoirs Act 1975 as amended by the Flood and Water Management Act 2010 embodies the legal framework for the safety (design and construction included) of reservoirs that hold $25\,000\,m^3$ or more above the adjoining natural ground level. In England and Wales, under the Act, the Environment Agency and Natural Resources Wales (respectively) are entrusted with the responsibility to ensure that owners and operators of service reservoirs observe of the Act.

6.5.1   Balancing storage

Given the daily water consumption pattern and the water production rate, the storage needed to cope with the hourly fluctuations can be calculated using a mass balance approach on an hourly basis, as summarised in the following procedure. The calculations may be carried out using the average daily consumption based on the peak week, noting that different water undertakings apply peaking factors in slightly different ways (Surendran *et al.*, 2005).

1. Initial conditions:
   >   volume required $= 0$
   >   cumulative depletion $= 0$
   i.e. the reservoir is assumed full.

2. Surplus $=$ average hourly demand – an hour's demand

3. If 'surplus' is positive:
   cumulative depletion = max {0, (cumulative depletion – surplus)}
   If 'surplus' is negative, i.e. there is a supply shortfall:
   cumulative depletion = cumulative depletion – surplus
   volume required = max {volume required, cumulative depletion}

4. If the analysis is complete (24 hours):
   balancing storage = volume required
   Otherwise, return to step 2.

Table 6 .1 illustrates the calculations.

Table 6.1 Balancing storage calculations

Time	Demand	Surplus flow	Cumulative depletion
hours	m³/hr	m³/hr	m³
08:00	176	−16	16
09:00	198	−38	54
10:00	208	−48	102
11:00	236	−76	178
12:00	226	−66	244
13:00	227	−67	311
14:00	214	−54	365
15:00	189	−29	394
16:00	163	−3	397
17:00	163	−3	400
18:00	172	−12	412
19:00	183	−23	435
20:00	181	−21	456
21:00	168	−8	464
22:00	162	−2	**466**
23:00	155	5	461
00:00	134	26	435
01:00	105	55	380
02:00	89	71	309
03:00	85	75	234
04:00	88	72	162
05:00	92	68	94
06:00	96	64	30
07:00	130	30	0

Average demand = 160 m³/hour
Supply = Average demand (assumed)
Surplus flow = Supply – Demand
Balancing storage required = Maximum cumulative depletion = 466 m³

It is worth restating that there is a lot of uncertainty and randomness associated with water demands. These and other relevant factors (e.g. any reserve capacity at the treatment works), including the need for operational flexibility, should be allowed for. A balancing storage of about 25% of the average daily consumption is generally considered reasonable in the UK.

6.5.2   Contingency storage

The estimation of contingency storage is not straightforward and is highly system-specific as it is governed by the relevant risk factors mentioned earlier and any available additional contingency measures (e.g. alternative sources/service reservoirs from which some water could potentially be rerouted; any standby power generation capacity; the capacity of any duplicate major supply mains and so on). The monitoring and operational control practices of the water undertaking and the structure of the electricity tariff may also influence the overall amount of storage (balancing and contingency) provided. An overall capacity of up to 24 hours supply is generally considered reasonable in the UK.

Water quality deteriorates as the residence time increases (Grayman and Clark, 1993; Ghebremichael et al., 2008) and, along with the location and operating policy, should be considered when determining the overall storage volume. The overall age of water throughout a distribution system is very sensitive to the locations of service reservoirs and their operating policies as characterised by the diurnal water level fluctuations (Grayman et al., 1991). Other relevant factors are the shape and inlet and outlet arrangements. Mixing inside the service reservoir is very important and should be addressed at the design stage. Dead zones, with limited water exchange with the main bulk of the water in the rest of the reservoir, and short circuiting between the inlet and outlet should be avoided. These issues are best assessed with the help of detailed hydraulic and water quality models of the reservoir and distribution system. Chapter 7: 'Network modelling', and Chapter 9: 'Operation, maintenance and performance', contain more information on water quality (Seyoum and Tanyimboh, 2017).

The UK Water Supply (Water Quality) Regulations 2016 and 2018 require disinfection of all water from a treatment works. Chlorine is the most common disinfectant used to render pathogenic organisms in water harmless. A sufficient residual chlorine concentration should be maintained throughout the distribution system. However, chlorine reacts with organic and inorganic substances in water to produce a wide range of undesirable chlorinated compounds that are called disinfection by-products (DBPs). Therefore, the residual chlorine concentration decreases while, concomitantly, the concentrations of the DBPs increase with time. DBPs can cause taste and odour problems and are thought to constitute a potential health risk under certain circumstances (Brandt et al., 2017; Clark and Sivaganesan, 2002; Tanyimboh and Seyoum, 2021). Therefore, with respect to both DBPs and the chlorine residual, the quality of the water deteriorates with time. Self-evidently the overall age of the water in the system as a whole increases with the amount of storage provided. Water quality deteriorates with time and, thus, excessive storage beyond that required to safeguard the continuity of supply is best avoided.

6.6.   System integration
6.6.1   Regulation and monitoring
The European Drinking Water Directive 98/83/EC established obligations for member states in respect of water intended for human consumption. To discharge this responsibility, the UK

Government requires that all statutory water undertakers and licensed water suppliers in the UK operate under strict regulatory conditions. The conditions require that accurate and auditable reports are produced in order that regulators can monitor WSPs for compliance against their prescribed standards.

The Drinking Water Inspectorate (DWI) regulates water supplies in England and Wales and the Water Supply (Water Quality) Regulations 2016 and 2018 are the statutory instruments that dictate how water quality must be monitored. WSPs divide their regions into 'water into supply (WIS) zones' within which a mandatory sampling programme is undertaken to monitor and report their compliance levels. Results are provided to the DWI and are also reported to the Office of Water Services (Ofwat) as part of the 'June return' reporting programme.

Ofwat is the economic regulator for the water industry in England and Wales. They are responsible for ensuring that WSPs provide good service and value for money to their customers. Part of their remit is to monitor the performance and serviceability of the water supply infrastructure including leakage.

In order for WSPs to report their performance against the serviceability standards, District Meter Areas (DMAs) are established as a subset of WIS zones, primarily to aid the reporting on consumption and leakage levels in the water distribution network. A further definition is ascribed to water networks in order to monitor distribution system pressures. Pressure Managed Areas (PMAs) are subsections of DMAs and provide the means by which Average Zone Night Pressures (AZNPs) are monitored and reported to Ofwat (Figure 6.14).

Figure 6.14 Example of a Water into Supply (WIS) Zone (WIS – Water into Supply zone; DMA – District Meter Area; PMA – Pressure Managed Area; WTW – Water Treatment Works; PRV – Pressure Reducing Valve; BV – Boundary Valve; SR – Service Reservoir; M – Meter)

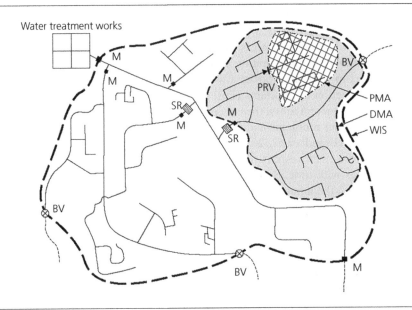

6.7. Surge control

Rapid changes in the flow rate or boundary conditions (e.g. line valve setting) in a distribution network can cause unsteady flow conditions often characterised by brief changes in pressure that can be detrimental, as stated in Section 6.3. For example, when the flow is stopped abruptly by the rapid closure of a valve, the water immediately upstream is compressed by virtue of its momentum and, as a result, the density of the water is raised with a concomitant increase in pressure. As the forward motion of the water in the pipeline upstream of the valve is progressively retarded and halted, a positive pressure wave is transmitted upstream at a speed or celerity that depends on the dimensions and material properties of the pipe. On reaching the upstream end of the pipe, depending on the boundary conditions, the wave is reflected back downstream as a positive or negative pressure wave which, in turn, will be reflected as a wave of the same type at the closed valve. In the absence of friction this cycle would continue indefinitely but in reality, the pressure waves die out quickly due to friction.

Conversely, a drop in pressure can occur immediately downstream of a pump that stops abruptly or a valve that closes rapidly. Thus, a negative pressure wave travels downstream to the end of the pipe. These unsteady flow conditions, known as transients or pressure surges, generate additional stresses that can cause damage to pipelines. Large increases in pressure can lead to bursts while pressure drops can cause pipes to collapse or fail subsequently due to fatigue. Sub-atmospheric pressures are undesirable as they may facilitate contamination of the water through open air valves.

Although transients are an inherent property of water distribution networks that cannot be eliminated entirely, they should be borne in mind and addressed at the design stage with the aim of keeping the pressure surges within the permitted bounds (as specified in codes of practice and standards). The analysis of transients is complex and requires special software. As mentioned previously, pressure waves are reflected or transmitted at the ends of pipe sections and where the material properties or dimensions of the pipes change. These interactions mean that the critical condition does not always correspond to the initial pressure surge. It can be shown (e.g. Potter et al., 2017) that the pressure generated by a sudden velocity change of Δv is given by

$$\Delta p = \rho c \Delta v \tag{6.10}$$

This is equivalent to

$$\Delta H = c \Delta v / g \tag{6.11}$$

where

$\qquad$ g = the acceleration due to gravity
$\qquad$ ρ = the density of the fluid
$\qquad$ c = the speed of the wave
$\qquad$ Δp = the changes in the pressure
$\qquad$ ΔH = the change in the head.

The speed (celerity) of the wave c depends on the pipe and fluid properties and is given by

$$c = \left[K / \rho (1 + DK/dE) \right]^{1/2} \tag{6.12}$$

where

K = the bulk modulus of the fluid
D = the internal diameter of the pipe
d = the pipe wall thickness
E = the pipe material Young's modulus.

The most important consideration regarding the attenuation of pressure transients is the speed with which changes in the flow take place. If a pressure wave is initiated at a valve, a time of 2L/c will elapse before the reflected wave returns to the valve, where L is the length of the pipe section or the distance from the valve to the boundary where the wave reflection occurs. The parameter 2L/c is important and is referred to as the pipe period. Thus, a valve closure that is completed within one pipe period is said to be rapid and the magnitude of the pressure transient it creates will be the maximum as given by Equation 6.10 (known as the Joukowski equation).

Conversely the surge will be less severe if the valve closure is slow, meaning it takes longer than one pipe period. The vital role that air valves, non-return valves and pressure relief valves play in controlling surge is explained in Section 6.4. Other surge control devices and measures are described below while more a basic characterisation is provided in Chapter 2: 'Basic hydraulic principles'.

The speed with which a pump stops is self-evidently important and should be considered. Increasing the time the pump takes to stop – for example, using a flywheel – will decrease the risk due to negative pressure. This, however, is not a straightforward decision to make.

A *pump bypass* permits water from the sump to enter the delivery pipe if low pressure develops in the delivery pipe. As mentioned in Section 6.4, a non-return valve in the delivery pipe prevents backflow. The bypass consists of a small diameter pipe with a valve that opens and then closes slowly.

In a similar way, a *feed tank*, with a non-return valve, releases water into a pipeline to restore the pressure if low pressure develops. A feed tank can be deployed downstream of a pump or at high points in a pipeline where accumulation of air and separation of the water column are likely. The feed tank should be designed to safeguard the quality of the water by preventing its contamination and stagnation and ensure sufficient water is available when required.

A *surge shaft* rises above the hydraulic grade line and dampens pressure fluctuations in the pipeline. It is connected to the pipeline and the water level in the shaft rises when the velocity is decreasing. On the other hand, when the flow accelerates, the surge tank augments the water in the pipeline and thus avoids negative pressures. The fact that a surge tank must rise above the hydraulic grade line is obviously a disadvantage as the location and construction cost are dictated by the topography.

Air vessels are part-filled with air under pressure and used at pumping stations. They release water when the pipeline pressure drops – for example, when a pump trips – to avoid negative surge. Water enters the vessels as a buffer against high positive pressures. *Accumulators* are a variant in which a flexible rubber membrane separates the air from the water to reduce the amount of air absorbed by the water. The air is replenished periodically.

REFERENCES

Brandt MJ, Johnson KM, Elphinston AJ and Ratnayaka DD (2017) *Twort's Water Supply*, 7th Edn. Arnold/IWA Publishing, London, UK.

Clark RM and Sivaganesan M (2002) Predicting chlorine residuals in drinking water: second order model. *Journal of Water Resources Planning and Management* **128(2)**: 152–161.

Ghebremichael K, Gebremeskel A, Trifunovic N and Amy G (2008) Modeling disinfection by-products: coupling hydraulic and chemical models. *Water Science and Technology: Water Supply* **8(3)**: 289–295. https://doi.org/10.2166/ws.2008.073

Grayman WM, Goodrich JA and Clark RM (1991) The effects of operation, design and location of storage tanks on the water quality in a distribution system. In American Water Works Association Research Foundation and American Water Works Association *Water quality modeling in distribution systems conference*. Denver, CO, USA.

Grayman WM and Clark RM (1993) Using computers to determine the effect of storage on water quality. *Journal of the American Water Works Association* **85(7)**: 66–77.

Ofwat (2017) *Delivering Water 2020: consultation on PR19 methodology*, Appendix 3: Outcomes technical definitions and Appendix to chapter 4: Delivering outcomes to customers. Ofwat, Birmingham, UK. www.ofwat.gov.uk/wp-content/uploads/2017/07/Draft-methodology-executive-summary.pdf [accessed 12.07.23]

Potter MC, Wiggert DC and Ramadan BH (2017) *Mechanics of Fluids*, 5th Edn. Andover: Cengage Learning.

Seyoum AG and Tanyimboh TT (2017) Integration of hydraulic and water quality modelling in distribution networks: EPANET-PMX. *Water Resources Management* **31(14)**: 4485–4503.

Siew C, Tanyimboh TT and Seyoum AG (2016) Penalty-free multi-objective evolutionary approach to optimization of Anytown water distribution network. *Water Resources Management* **30(11)**: 3671–3688.

Surendran S, Tanyimboh TT and Tabesh M (2005) Peaking factor based reliability analysis of water distribution systems. *Advances in Engineering Software* **36(11–12)**: 789–796.

Tanyimboh TT and Seyoum AG (2021) System-wide joint-dynamic-response approach to water quality evaluation of distribution networks with multiple service reservoirs and pumps. *SN Applied Sciences* **(2021)** 3:448.

Tanyimboh TT and Templeman AB (2010) Seamless pressure-deficient water distribution system model. *J. Water Management* **163(8)**: 389–396.

Thorley ARD (1991) *Fluid Transients in Pipeline Systems*. D&L George Ltd, Hadley Wood, UK.

UK Water Industry (2003) *The choice of pressure ratings for polyethylene pipe systems for water supply and sewerage details*. Water Industry Information and Guidance IGN 4-32-18. WRc plc, Swindon, UK. http://www.water.org.uk/home/member-services/wis-and-ign/current-documents-plastics-and-rubbers/ign-4-32-18c.pdf

FURTHER READING

BS EN 1992-3: 2006 Eurocode 2. Design of concrete structures. Liquid retaining and containing structures. British Standards Institute, London, UK

The European Drinking Water Directive. Council Directive 98/83/EC of 3 November 1998 on the quality of water intended for human consumption. *Official Journal of the European Communities* **L330**: 32–54

The Flood and Water Management Act 2010

The Reservoirs Act 1975, HMSO, London, UK

The Water Supply (Water Quality) Regulations 2016 and 2018(6.7b)(6.7c)

Dragan A. Savić and John K. Banyard
ISBN 978-1-83549-847-7
https://doi.org/10.1108/978-1-83549-846-020242009

Chapter 7
Network modelling

Dragan A. Savić
KWR Water Research Institute, Netherlands and University of Exeter, UK

Rob Casey
Thames Water, Reading, United Kingdom

Zoran Kapelan
Delft University of Technology, The Netherlands

7.1.   Introduction

This chapter provides a brief review of principles used for creating a computer model of water distribution network behaviour. Computer modelling or network modelling (also known as *network analysis*) will be discussed with different types of network analysis models covered, including steady-state, dynamic (extended period), water quality, transient and so on.

The chapter starts with an introduction to different types of models, before moving on to consider basic modelling principles, including complexity and uncertainty associated with network analysis. This is followed by the discussion of data sources for model building, focusing in particular on procedures for water consumption and demand assessment. The final part provides a discussion on model building for a particular purpose and the steps in model building, including model calibration and maintenance.

7.2.   Models

According to the Oxford English Dictionary, a model is 'a simplified mathematical description of a system or process, used to assist calculations and predictions.' As modern cities and their associated infrastructure, including water distribution systems, have grown over time, network analysis has become a crucial planning, design and operational activity. However, network analysis, like any other type of computer modelling suffers from the so-called 'garbage in, garbage out' syndrome. Often abbreviated as GIGO, this is a well-known computer axiom meaning that if invalid data is fed into a computer model, the output would also be invalid. Therefore, data for network modelling will also be discussed.

Although there are a number of possible classifications of mathematical models, this review will focus on *simulation models* and their use in water distribution system management (Box 7.1).

Over the last several decades, there has been a significant increase in the number of network analysis software packages both in the commercial and public domains. Coupled with increases in hardware power, it is now possible to simulate very large systems as all-mains models under complex dynamic scenarios. This increase in computer power coupled with software development has also led to a significant increase in the range of applications supported by modelling (Figure 7.1), moving from strategic and steady-state models confined to planning and engineering design departments, to all-mains extended period simulation (EPS) models used by various utility departments.

Box 7.1 Network simulation

A simulation is the process of using a network model that provides insight into the dynamics of a particular water distribution system and shows how it will behave over time and in various situations. For example, one might want to simulate variations of flow and pressure in a system over a 24-hour period to check for regulatory compliance. A model would be used and simulation runs performed to assess whether pressures satisfy minimum requirements under different demand scenarios.

Simulation is not only useful for testing and analysis when prediction or evaluation of system behaviour and management options is needed, but also for training of staff.

Figure 7.1 Model uses have expanded with increased computing power

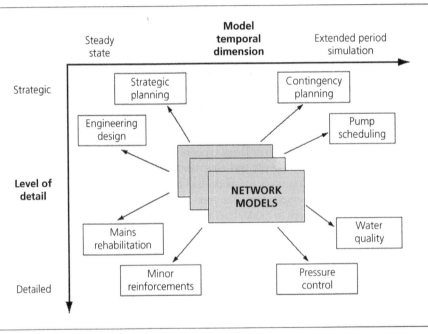

In addition, commercial software packages, such as WaterGEMS® (Bentley Systems), H₂ONET® (Innovyze), incorporate advanced graphical user interface (GUI) and geographic information system (GIS) capabilities to provide an accurate geographical representation of a network thus enabling simple communication of complex phenomena, even to non-specialists. Even freely available EPANET (Rossman, 2000) provides good modelling capabilities with a simple, but effective GUI (Figure 7.2). The latest version of the software, EPANET 2.2.0 (EPA, 2023), provides an updated and expanded open-source modelling tool that was a result of the involvement of a large community of researchers and practitioners.

The latest development in software delivery methods for water distribution system modelling embraces the software-as-a-service (SaaS) approach. As an example, the Qatium platform offers a software application that is made available to users over the Internet (Qatium, 2023). Users can access the software through a web browser, without the upfront costs of purchasing and maintaining

Figure 7.2 Epanet GUI

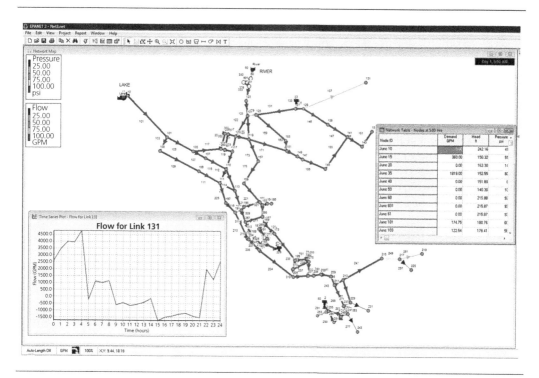

it. In addition, users can easily scale up or down their usage as needed. This is particularly important for small water utilities that often lack the resources to purchase desktop software or expertise to utilise it effectively in the management of their water systems.

7.2.1 Steady-state modelling

Early network models simulated only steady-state (snapshot) hydraulic conditions – that is, where all demand and operations are constant. Although conditions in a real water distribution system are always changing with time (e.g. demands, reservoir levels and so on) these models are useful when particularly important instantaneous conditions within a network need to be investigated. For example, a steady-state simulation would be performed to analyse peak demand conditions for pipe sizing, or minimum demand conditions important for investigating sediment deposition potential. Many hydraulic problems in a water distribution system are solved by a steady-state simulation, most often those relating to infrastructure design problems. Steady-state modelling serves also as a building block for more advanced types of modelling (e.g. extended period simulation, water quality analysis and fire protection studies). Once it is well understood, it is easier to grasp the concepts of these advanced modelling approaches.

Input data for steady-state simulation include all static network data (e.g. pipe and node characteristics) and fixed operational conditions (e.g. reservoir levels, valve statuses, pump flow rates and so on) for a particular time snapshot. The model calibration process often starts with considering a steady-state situation where nodes with reservoirs and pumps are removed from the model and flows are assigned to them as negative flows (inflow into the system). This allows for the flows through the system to be checked against measurements and errors corrected before calibration of the EPS model is attempted.

7.2.2 EPS modelling

In the 1970s, modelling capability was extended to include EPS models that could accommodate changing demand and operational conditions. These models are useful in determining the dynamic response of a network system – that is, how it behaves under varying demands, how a stressed system reacts in emergencies or how, like in the case of the Anytown problem (Section 7.5.1), the system could be strengthened most efficiently to provide sustainable service over a long period of time.

The established approach to performing EPS simulation computations involves steady-state analysis to determine pressure and flows in the system and a numerical integration technique to integrate reservoir flows in or out of the system to compute reservoir volume changes (Chapter 2). This type of simulation is extremely important for complex systems where important dynamic interactions occur involving varying reservoir water levels, pumps turning on and off, and pressure and flow control valve operation between different parts of the network (e.g. district meter areas (DMA) or pressure zones). An example of a dynamic output for pipe flow and node pressure is given in Figure 7.3.

Figure 7.3 Example output from an EPS model simulation run

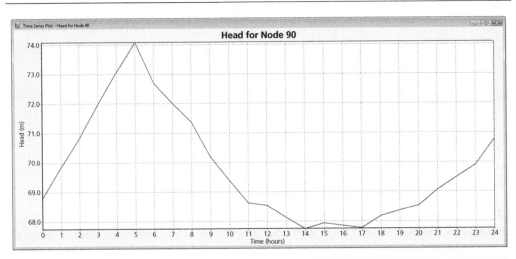

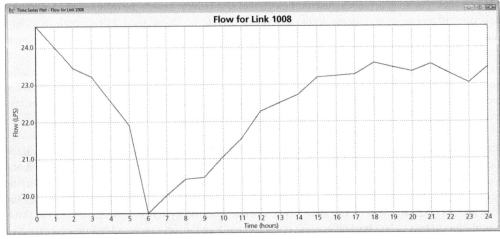

Input data for extended-period simulation models require not only the static data used in steady-state modelling, but additional data to describe time-varying demands, pump and valve characteristics (e.g. pump curves, valve losses and so on) and reservoir levels. Furthermore, the EPS simulation requires information on the computational time step and the duration of the simulation run. The decision about the time step and simulation duration should be based on the intended purpose of the model and should allow proper capture of equipment controls/functioning. For example, reservoir storage analysis should be performed at least over a 24-hour period to accurately capture the filling and emptying of the reservoir, with a time step that should not be longer than one hour to avoid missing an important change in status of valves or pumps – for example, when the reservoir gets full before the end of the computational time step. As for water quality modelling, the simulation is run for a sufficiently long period of time (e.g. several days) under a repeating pattern of source and demand inputs so that the initial water quality conditions, especially in storage reservoirs, do not influence the water quality predictions in the distribution system.

7.2.3   Water quality modelling

Around the early 1990s hydraulic modelling software was enhanced to include water quality modelling (EPA, 2005). This software enabled factors such as chlorine decay, source water mixing, contamination spread and water age to be modelled. The output of a water quality model can include the temporal and spatial distribution of a variety of constituents within a distribution system, including the fraction of water originating from a source, the age of water, the concentration of tracers (e.g. non-reactive constituents such as chloride or fluoride), the concentration of a reactive compound (e.g. chlorine or chloramine) and the concentration of disinfection by-products (e.g. trihalomethanes).

Although the modelling of chlorine decay and other chemical reactions is complex, key water quality parameters such as water age, source mixing and contamination spread can be successfully modelled using a standard hydraulic model which has been calibrated for flow and pressure. Water age models can often be used to gain some understanding of chlorine decay and bacterial growth issues within the distribution system, as high water age is often associated with these problems. With water quality models, it is important to use all main models to accurately predict water age, mixing and contamination spread.

The successful modelling of chlorine and other chemical reactions in the distribution system requires these parameters to be sampled throughout the network, which is a lot more expensive and difficult than standard pressure measurements. Also for chlorine decay, the chemical reactions are quite complex with reactions occurring both within the bulk fluid and at the pipe wall. However, chlorine decay models have been successfully built, calibrated and used to resolve water quality issues in many different countries. An example of chlorine residual level output for a particular node in a water distribution network is given in Figure 7.4.

Some modelling software suppliers have also included the capability to model sediment transport to predict possible water quality issues, such as brown water caused by significant velocity changes or flow reversals when sediment material is picked up and distributed around the network. These models can also be used to determine flushing routines to safely remove sediment.

However, once again much can be done with a standard hydraulic model to predict sediment-based water quality issues by identifying pipes with low velocities which are significantly increased or reversed following valve operations. As a general rule, it is assumed that pipes with a peak velocity

Figure 7.4 Chlorine residual levels at a node over a 24-hour period

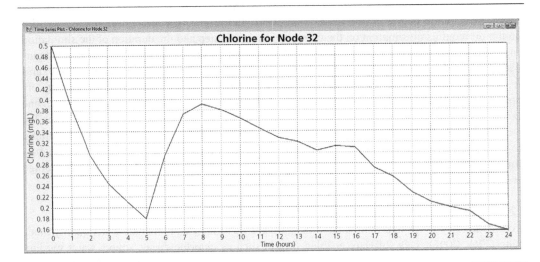

of less than 0.3 to 0.4 m/s are most likely to provide the highest risk in terms of sediment build-up. Furthermore, pipes with peak velocities above 0.6 to 0.8 m/s are most likely to be self-cleansing. This may vary from system to system and with different water sources, but local experience can be gained from flushing trials and turbidity monitoring.

With regards to flow reversals, the same rules regarding velocities as above will generally apply for sediment, however, where material is attached to the pipe wall (i.e. around the pipe as well as on the bottom) then the flow reversal may result in this material becoming detached into the bulk fluid.

7.2.4   Hydraulic transients

Pressure transients occur when there is a rapid change in the fluid velocity within a pipeline. These pressure changes are often generated by the rapid opening/closure of valves or switching/failure of pumps within the pipe system.

Prior to the 1970s, before computers were available to the design office, solutions to the complex equations that describe unsteady or transient flow conditions (Wylie & Streeter, 1978) were provided by graphical methods of analysis, such as those developed by Bergeron (1961). Analysis was usually confined to a single pipeline, which was sufficient to model surge issues on dedicated pumping mains.

The first computer programs, based on the method of characteristics, were only able to handle a few pipes, and hence a very simplified network. This coupled with the fact that the transient flow equations are different to the steady state or extended period simulations, and the analysis time period is much shorter, led to the separate development of transient programs from the basic steady-state hydraulic models.

However, the significant advances in available desktop computing power and analysis techniques have enabled surge analysis on much larger networks to be undertaken. This had led to some suppliers offering both extended period simulation and transient analysis facilities under a common GUI. This saves considerable time in model building, removes the need to transfer data from one package to another, and also enables the same modeller to undertake both the extended period simulation and the transient analysis (Box 7.2).

Box 7.2   Transient analysis

Transient analysis models are normally used to identify the extent of pressure surges generated by pump switching/failure or valve modulation within water networks. The model is then used to design surge relief measures such as air vessels, pump inertia and valve control.

The models can be calibrated using special transient pressure loggers that can record the pressure changes that occur over fractions of a second during a surge event.

7.3.   Basic modelling principles
7.3.1   Complexity

A water distribution system may be represented as a set of nodes and links. As more and more water utilities are building their models from GIS, tens or even hundreds of thousands of such elements can be included in a single model. Nowadays, the tendency is to build all-mains models, which then could be simplified if necessary – for example, for strategic decision-making (Box 7.3). A typical distribution system comprising urban and rural areas is shown in Figure 7.5. A close-up view of the smaller area showing details such as streets and buildings is shown in Figure 7.6.

Box 7.3   Occam's razor: complexity and parsimony

When building a model it is advisable that modellers use the well-known simplicity principle when deciding on the level of model simplification for a particular purpose. Popularly interpreted, this principle, also called Occam's razor (William of Occam, c. 1288 - c. 1348), states that *when there are two competing models that make exactly the same predictions [for a particular purpose], the simpler one is the better.* In other words, for the same level of predictive exactness (defined by the model intended use), the simplest model should be selected.

It is not always feasible nor necessary to include all elements and features in a model of a particular system and a decision on the level of detail the model has to be made to is often required – that is, what elements and features to include and what to omit (Obradovic and Lonsdale, 1998). This largely depends on the intended use of the model, as, for example, a model used for master planning (i.e. long-term development and rehabilitation planning) will have much less detail compared with a model used for operational optimisation (e.g. energy use optimisation), where detailed pump and valve characteristics play a crucial role. Attempting to include each individual service connection, tap, valve and every other component of a large system in a model could be an enormous undertaking without a significant impact on the model results (Walski *et al.*, 2003).

The process of model simplification in water distribution systems is called skeletonisation. It consists of selecting for inclusion in the model only the elements of the network that have a significant impact on the behaviour of the entire system. Choosing the degree of skeletonisation that is acceptable is one of the most difficult aspects of the modelling process and should depend upon the

Figure 7.5 A water distribution network

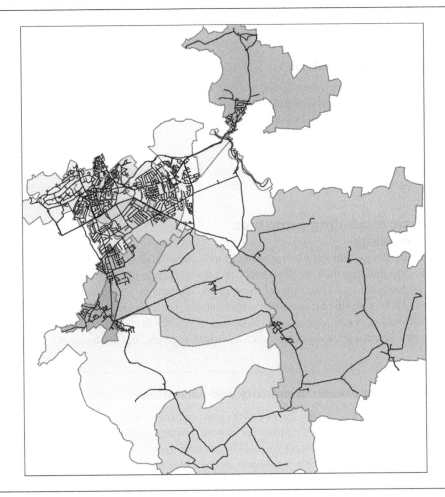

ultimate use of the model and the sensitivity of model results. Figure 7.7 shows an all-mains model and its skeletonised counterpart obtained by removing small-diameter pipes.

7.3.2   Uncertainty

Like the vast majority of mathematical models in other areas of engineering, water distribution modelling relies on deterministic approaches to describe the behaviour of a system. However, all real-life problems incorporate uncertainty in one way or another. Such contradiction between 'mathematical determinism' and 'natural uncertainty' can seriously affect the reliability of the results of modelling. A large number of problems in design, planning and management of engineering systems, including water distribution systems (WDS), require that decisions be made in the presence of various sources of uncertainty.

Different sources of uncertainties exist (Box 7.4) in the WDS modelling (Filion and Karney, 2002). Firstly, some of the WDS simulation model input variables are uncertain. The most frequently analysed are the uncertainties associated with water consumption and pipe roughness coefficients (Xu and Goulter, 1998). However, many other sources of uncertainty exist in real-life WDS, especially

Figure 7.6 Detail from the water distribution network

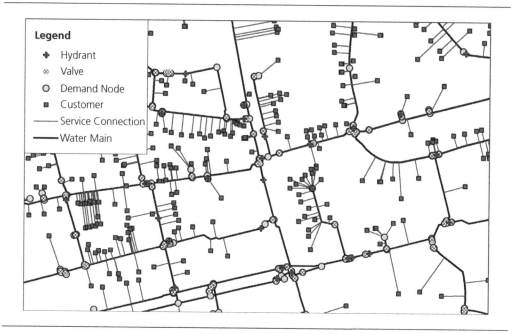

Figure 7.7 All mains model (a) and skeletonised model (b)

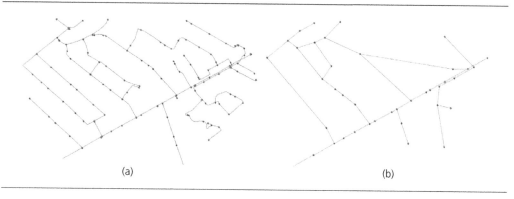

(a)   (b)

in systems that have been in use for many years (e.g. internal pipe diameter, status of some devices, basic characteristics of some devices (e.g. valve headloss curve or pump head-flow curve), network connectivity (are the two pipes at a 'junction' linked or not?) and so on). Secondly, WDS simulation models are not perfect – that is, they are only an approximation of the complex real-life WDS and their behaviour. For instance, in a WDS hydraulic model, demands are allowed to occur at network nodes only even though they are distributed along pipes in real-life networks. Thirdly, uncertainties originate from the uncertain environment in which WDS exist. This is especially true for the economic environment. For instance, the following uncertainties exist: uncertainties associated with the cost evaluation of different intervention (optimisation) options (e.g. structure of different cost models, various unit costs and so on), rehabilitation work budget uncertainties, macroeconomic uncertainties (e.g. discount rate used to calculate the net present value of all costs) and so on.

Box 7.4. Type of uncertainty

Two general types of uncertainty exist. The first is known as *reducible* (or *epistemic*) *uncertainty*. This uncertainty generally results from the lack of information about some aspect of the problem being analysed (e.g. the status of some valve in the WDS may not be known simply because that information is lacking; however, once the inspection is done, uncertainty can be reduced).

The second class of uncertainty is known as *irreducible* (or *aleatoric*) *uncertainty*. This type consists of fluctuations that are intrinsic to the problem being studied. Examples of this type are uncertainty associated with pressure and flow measurements.

In order to be modelled, uncertainties need to be characterised first. Several general ways exist to do this

(a) using probability theory – that is, probability density functions (Press *et al.*, 1990)
(b) using possibility theory – that is, fuzzy logic (Zadeh, 1965)
(c) using simple bounds on uncertain variables (Brdys and Chen, 1993).

Other approaches exist, too. Once characterised, uncertainties can be quantified by a number of different methods. These methods can be broadly classified into the analytical- and sampling-type methods. Examples of analytical methods include (Press *et al.*, 1990): the first-order second-moment model (FOSM), second-order second-moment model (SOSM) and so on. The sampling type methods include the Monte Carlo simulation (MCS) method and a large number of stratified sampling methods (e.g. Latin Hypercube method and so on). As a general rule, analytical uncertainty quantification methods are computationally faster but normally work under certain assumptions only (e.g. model linearity, independent, normally distributed uncertain variables and so on). The sampling type methods tend to be more general – that is, they are less restrictive (can handle the non-linear models and so on) but also much more computationally demanding.

7.4. Data for network modelling

An overview of the data collection, model building and calibration process is shown in Figure 7.8. The details behind this process are discussed in the following sections.

7.4.1 Water company data

Since the mid-nineteenth century, asset records have been created and archived by water utilities in the UK and worldwide. Paper maps were generally used to save records, with cartographic maps used as a reference source (e.g. Ordnance Survey mapping in the UK). Since the mid-1980s, most water utilities have made significant progress towards digitisation of their data records. Nowadays, water utilities possess large quantities of data derived from different sources (drawings, SCADA, asset and customer databases, work management systems and so on).

In general, the following data sources may be available within a water utility

- drawings (distribution network records, source and reservoir data and so on)
- pressure and flow data (manually logged)
- SCADA/Telemetry and IoT sensor data (e.g. reservoir levels, pump stations, or pressure and flow data – continuously logged)

Figure 7.8 An overview of the model building and calibration process

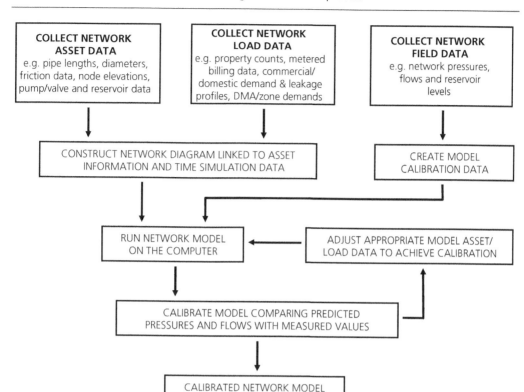

- work management data (e.g. job information based on work carried out as a result of mains burst, service pipe repair or work done at a pumping station)
- GIS data (information about above and below ground assets, e.g. treatment works, meter points, pipe age, pipe material and so on)
- historic performance data (asset failure data, e.g. pipe bursts/leaks, discolouration events, interruptions to supply, etc)
- customer contact data (customer phone calls made in relation to an issue/event on the system, e.g. no water, burst/leak, water quality and so on)
- water quality data (collected as part of the water quality testing/sampling process)
- energy use data (pump schedules, energy consumption and cost at pumping stations)
- customer meter readings (e.g. by way of a billing system).

These digital records are embedded within many organisational functions and are being used in a range of modelling and business scenarios, including network analysis. However, these data records are collected in different formats and often stored in different computerised systems (e.g. databases) requiring a number of different software packages and file formats for storing, editing, analysing and viewing data. For example, in a typical UK water company different software systems would be used for leakage and pressure monitoring, static asset data, GIS asset data, real-time pressure and flow data, billing data, network model data and so on.

Although data quality is generally thought to have improved over time, much of the data held on water company records should be checked for consistency and accuracy to avoid working with incorrect information – for example, in terms of location, size of pipe or pipe material. The problem is that, historically, the record drawings (required to be maintained by law) were either the design drawings, which were never modified to show 'as built' reality or, in some cases, simply schematics showing pipe centrelines. Finally, there is always the problem of unrecorded 'operational' enhancements which were not included in the system drawings. The last several decades have shown considerable investment in data collection and improvements in quality of record drawings with mobile IT allowing field workers to record what they actually encounter. However, caution is needed when this information is used for model building.

7.4.2   Water consumption and demand assessment

When speaking about water consumption and demand, it is commonly understood that consumption is the total amount of water used, whereas demand is the immediate rate of that consumption. The usual measure of consumption is the amount supplied per head of population – for example, in litres per capita per day (lcd), whereas the demand is measured in litres per second (l/s). Average per capita consumption across England and Wales in 2022/23 was 134 lcd (metered) and 173 lcd (unmetered), with water usage of households varying significantly from area to area. The difference can be caused by changes in behavioural routines as has been observed during the COVID-19 event (Abu-Bakar et al., 2021)). It is obvious that a good understanding of water consumption and demand estimates is a prerequisite for meaningful network analysis.

Metering is an essential component of water consumption assessment as it could provide valuable information about how much water is used, by whom, where and when. The primary metering points in a water supply system include production (bulk) meters, meters on discrete zones within the water distribution system (e.g. district metered areas) and customer meters. These meters are important for performing a water balance analysis aimed, mainly, at water loss management but this information should also be used for network analysis. For example, water consumption in a supply zone (measured by a production meter) should equal the sum of demands in its DMAs. Similarly, the total consumption in an area should be equal to the sum of demands in its supply zones.

It is clear that metering is essential for the effective management of a water distribution system. However, it is not always possible to achieve complete metering coverage, as, for example, in the UK where approximately half of the residential customers do not have permanent meters. On the other hand, production (bulk) water meters, DMA/pressure zone meters and meters for large industrial/commercial customers are a must and are almost universally available in all well-managed distribution systems. Information from these meters should be used for network analysis.

In order to use a model in the dynamic mode (i.e. for extended period simulation), information on temporal variations in water usage over the period being modelled are required. A typical diurnal diagram of water usage (Figure 7.9) at a district level can be constructed by use of statistical analysis. These need to be developed considering different demand conditions – for example, working against weekend days, or summer against winter days.

Diurnal demand curves should be developed for each major demand type or consumer class. For example, distinct diurnal curves might be developed for metered demand (commercial and industrial) and domestic demand (which is not always based on metered consumption). Leakage could represent another significant demand as it is a portion of the water put into the distribution and

Figure 7.9 A typical diurnal demand curve

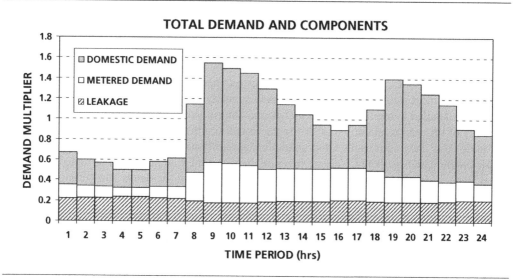

should be accounted for in modelling. Due to possible heterogeneity of water distribution systems that have developed haphazardly over time, it may even be necessary to allocate different levels of leakage to different parts of the system.

When modelling a water pressure zone, metered demand profiles are normally developed from standard metered profiles by consumer type (and large consumer meter profiles) and the leakage profile is determined from nightline and pressure measurements. Thus the domestic profile is often the remaining demand that makes up the total demand profile for the zone (Figure 7.9). For zones where the total demand can be measured for each DMA, it is, therefore, possible to construct a separate domestic profile for each DMA. However, this makes factoring up the model for future demands cumbersome and, often, a compromise is reached by using a common domestic profile so long as the model DMA profile is within 10% of the measured values.

7.5. Model building
7.5.1 Model purpose

Models are built for various purposes, with the aim of aiding analysis, design, operation or maintenance of water distribution systems. The first step in model building is to define its purpose, which in turn will affect the level of detail contained in the model. Model applications include master (strategic) planning, energy management, water quality analysis, regulatory reporting, emergency planning, water age analysis, planning of local distribution improvements, leakage analysis, resolution of system anomalies, surge (water hammer) analysis and so on.

Master planning or *strategic planning* involves various analyses to support the selection of capital improvement projects necessary to ensure the required level of service for the future. It uses long-term demand projections to assess the capability of the network to adequately serve its customers under a variety of conditions. The baseline master planning applications could include sizing of piping improvements, analysis of system pressures, storage capacity analysis, fire flow analysis and pump operation analysis. Due to uncertainty associated with future demand projections and population growth, it is not necessary to use a detailed all-mains model and a strategic (skeletonised) model is often employed. This type of model is normally constructed from an all-mains model with certain pipe sizes removed (e.g. below 200 mm dia.).

Operations management involves assessment of the current system operating conditions and planning of improved operations schemes. For example, pump operation optimisation (pump scheduling) for energy management means maximising pumping during periods of cheaper electricity. This type of analysis can identify potential savings on operational expenditure as pumping costs are one of the most significant investments made annually by a water utility. The model, which needs to be detailed in treating existing pumps and control valves, can also be used to assess the efficiency of the pumps and any need for replacement/refurbishment and the potential reduction in running costs that might result. Other modelling applications related to operations management include pressure management (to design pressure management schemes that contribute to a reduction in system leakage and burst incidents, extending the life of the assets and cutting pressure-related demand), fire flow analysis (investigation of the effect of exceptional demands and fire flow), storage facilities optimisation (to prevent excessive retention times and improve water quality), surge (water hammer) analysis and so on. These types of models may include all pipes above a certain diameter size (e.g. above 150 mm) and even some important pipes of smaller diameter where they are significant carriers.

Water quality modelling can be used for assessing system chlorination requirements (e.g. chlorine residuals) and simulation of pollution incidents/scenarios. For example, a water service provider needs to ensure that adequate chlorine residuals are maintained throughout the entire system in order to achieve the required level of protection, but also to avoid excessive levels of residual chlorine leaving the treatment plant that can cause taste problems, leading to customer complaints. A reasonable balance can be achieved by maintaining the chlorine levels leaving the treatment plant below the taste threshold and providing booster disinfection (i.e. adding disinfectant at some critical locations in the system) such that disinfectant residuals are maintained at a level greater than the minimum for public health. Accurate assessment of water quality through modelling can help the water utility reduce the probability of health incidents and customer complaints. A well-calibrated hydraulic model is a prerequisite for a good water quality analysis model. Hydraulic models are also used on their own (i.e. without the need for water quality modelling) to investigate water quality related issues, such as water age analysis, source blending (tracking of the hydraulic boundary between sources) or sediment modelling in a water distribution system. An example of such use is the simulation of mains flushing to facilitate the effective removal of the pollutant or sediment from the system helps minimise the disruption to normal supplies or identify customers affected by the flushing.

To illustrate some of the issues with planning, design and/or operations of water distribution systems, a simplified design problem is given next (Figure 7.10). The network, also known as the 'Anytown' water distribution system, was originally created as a challenging design benchmark for optimisation models (Walski *et al.*, 1987). The problem is simplified as it does not involve many of the features of real systems, such as, for example, multiple pressure zones, seasonal and local demand fluctuations, fiscal constraints, uncertainty of future demands, ageing pipe issues and complicated staging of construction. However, it has been demonstrated that the Anytown problem is complex and serves as an illustrative example exhibiting many real system features, such as locating and sizing new pumps and service reservoirs (Mays, 2000). The task is to determine the most economically effective design to reinforce the existing system to meet projected demands, taking into account pumping costs (OPEX) as well as capital expenditure (CAPEX).

Figure 7.10 A water distribution network design problem

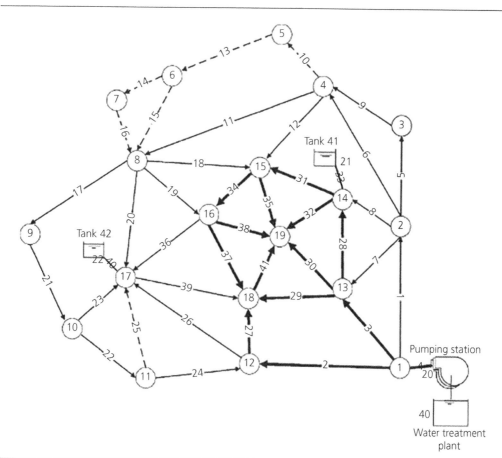

The town is formed around an old centre (solid lines representing pipes in this area) situated to the south-east of pipe 34 (Figure 7.10), where excavations are more difficult to undertake and consequently are more expensive. There is a surrounding residential area (the pipes represented by thin solid lines), with some existing industries near node 17 and a projected new industrial park to be developed to the north. Options include duplication (in a range of possible diameters) of any pipe in the system, addition of new pipes (dashed lines), selecting the operation schedule of pumping stations and provision of new reservoir storage at any location. Water is pumped into the system from a water treatment works by means of three identical pumps connected in parallel.

The design brief requirements are

(a) to maintain a minimum pressure of at least 28 m (or 40 psi, as in the original problem) at all nodes at average day flow as well as instantaneous peak flow (which is 1.8 times the average day flow)
(b) to meet fire flow requirements with a minimum pressure of 14 m (20 psi)
(c) to preserve the specified emergency volumes in reservoirs, which should empty and fill over their operational ranges during the specified average demand day

(d) to provide a certain level of reliability/redundancy in the network (e.g. by ensuring that each node is supplied by at least two pipes)

(e) to achieve optimum OPEX and CAPEX costs for the solution.

Additionally, one needs to consider water quality, as different design and operation plans will affect water quality in the network (which has not been considered in the original problem statement).

In order to address the design brief in a satisfactory manner, the designer has to consider installing new pipes, pumps and reservoirs or to consider existing pipes for cleaning and relining. Each of the 35 pipes could be considered for possible duplication or cleaning and relining, and six new pipes need to be sized (with ten new diameters available). The existing pump station could be upgraded or a new one built. Furthermore, the location of elevated reservoirs, overflow and minimum normal day elevation, diameter and bottom of the reservoir from minimum normal day elevation should also be found. All of the above design options need to be individually costed based on the type and size of the improvement and on the location of the particular network element (e.g. in the old centre or in the industrial zone) and the behaviour of the system tested against the above design requirements.

Even for this simplified problem, many alternative solutions are possible, each with its own advantages and disadvantages. Without network modelling, the task would be extremely difficult to complete and it would not be possible to assess the behaviour of the system until the proposed improvements were implemented. Therefore, modelling can save some valuable time and resources. For example, if a model of the existing system is available, a number of alternative solutions can be examined using the updated model (i.e. with new pipes, pumps and reservoirs added) in a short period of time.

To illustrate outputs and benefits of modelling two different solutions to the Anytown problem, with piping improvements and the reservoir locations and capacities, are given in Figure 7.11.

Both solutions satisfy the minimum pressure requirements and use the full reservoir operating range over the 24-hour simulation. However, the solution in Figure 7.11(a) has a total cost of about US $13M, a maximum water age value of about 35 hours (a surrogate for water quality, with the higher the age, the worse water quality is) and a resilience index value of 0.18 (the larger the value, the more reliable the solution). The solution shown in Figure 7.11(b) has a total cost of about US $17.3M, a maximum water age of about 34.5 hours and a resilience index value of about 0.20.

By closer inspection, the minimum cost solution needed one new reservoir at node 15, whereas the solution with the better reliability, but higher cost, needed one new reservoir at node 19 and relied on more pipes being duplicated, thus providing more capacity to supply demand even in the case of pipe failure in the system. The extended period simulation also provides additional information on how the two solutions behave – for example, they both have high fluctuation in reservoir levels (i.e. high turnover, thus contributing to better water quality) – but the solution in Figure 7.11(b) achieves a higher level of reliability with one of the reservoirs staying full for the most of the operating period (which, on the other hand, could contribute to water quality problems due to little turnover). Obviously, the ideal solution that satisfies all criteria to the fullest does not exist, but at least a compromise solution can be found based on detailed analysis afforded by modelling.

Figure 7.11 (a) minimum cost solution (b) highly reliable solution. (Modified with permission from Farmani *et al.* (2006). © IWA Publishing)

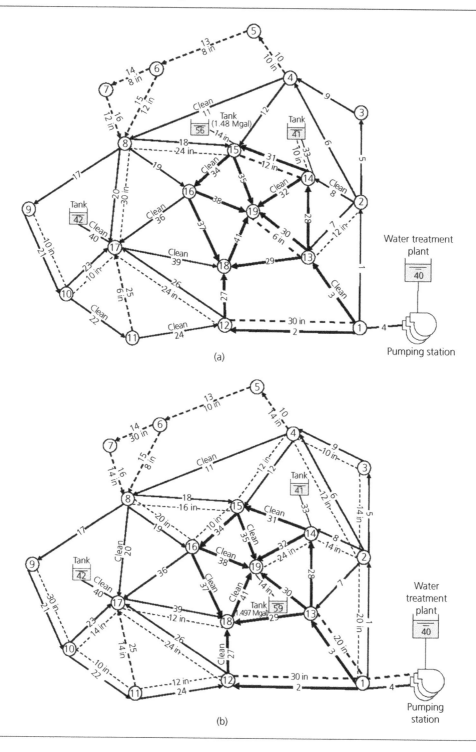

(a)

(b)

7.5.2 Data collection

Significant data need to be collected from various sources to build a good water distribution system model. The information comes from a number of sources – for example, paper maps, GIS, SCADA, work management systems and so on, often stored in a number of formats. It is important that all sources are thoroughly checked for accuracy and consistency before data is used in model building. For example, if GIS records are used to build the basic structure of the model, distribution network records drawings could be used to highlight any apparent anomalies, errors or omissions, prior to commencing model building. An example of a GIS map that was imported into the modelling software is given in Figure 7.12.

As can be seen from Figure 7.12(b), even if the automated GIS export into the network model is possible, the model created from GIS has a much greater level of detail than what is needed. For example, a number of adjacent nodes on short pipes could be removed (to create 'logical pipes') as they are not normally needed for most modelling applications. However, such alterations to the hydraulic model are not normally implemented in the GIS data, which poses a problem of model maintenance once the situation on the ground and consequently in the GIS changes.

Additionally, local knowledge of the structure of the system and its operations available from the personnel on the ground may have valuable insight to offer during the model building. It is also interesting to note that the latest trend is to use readily available data, such as from SCADA systems, and link it to models, although, like with GIS, we are still some time away from the situation where the data are automatically loaded into the water model.

As each network is represented as a set of nodes and links, data associated with each of these network elements need to be collected and entered into the analysis software. This, for example, means having the correct pipe diameters, in the proper locations and with appropriate connectivity to neighbouring pipes. Further information needed includes service reservoirs with their

Figure 7.12 (a) Water GIS map and (b) resulting model

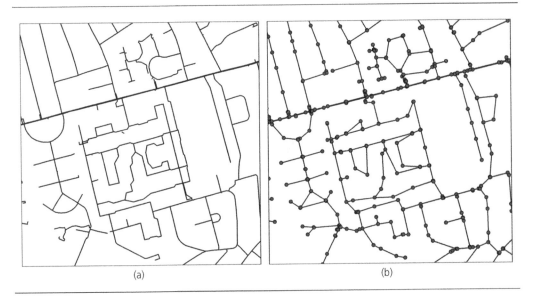

(a)   (b)

appropriate shapes and dimensions, pump and valve information (including settings and controls), elevation data for junctions and good data for demands (Speight *et al.*, 2009).

Nodal points, or simply nodes, indentify pipe junctions, locations of change of pipe characteristics (e.g. diameter, material), connections to reservoirs, pumps and valves, centres of demand, SCADA points and field measurement points. Topographic maps or digital elevation models (DEM) are commonly used for establishing ground elevations for each of the nodes (Figure 7.13). These can be verified by the global positioning system (GPS) measurements for important nodes – for example, field test points, reservoirs and pump stations.

All water demands are allocated at network nodes. This is an approximation of the real situation where service connections are distributed along the pipe (not normally at model nodes), but commonly utilised in network modelling software. The general rule for demand allocation is that nodal demands of the same type are lumped together in a demand area for a number of customers near a particular node. The demand along a pipe is split between the adjacent nodes so that the boundary of the two demand areas passes equidistantly between the nodes. Different methods of allocation could be employed, ranging from developing demand areas based on postcode information and different types of consumers to the assignment of geocoded customer meters to nearest nodes.

The pattern of demand chosen for each customer type (e.g. commercial/industrial, domestic and so on) should be determined either based on the field tests (e.g. for large metered users) or based on the local knowledge of the likely pattern of usage. Typically, a 30-minute or 1-hour time step is used for the demand pattern. However, it should be checked that true peak conditions are not lost due to coarse time steps. It should be pointed out that service reservoirs and valves are normally modelled as nodes. Detailed information on these elements is needed for building a realistic model of the network system.

Figure 7.13 DEM with a water distribution system

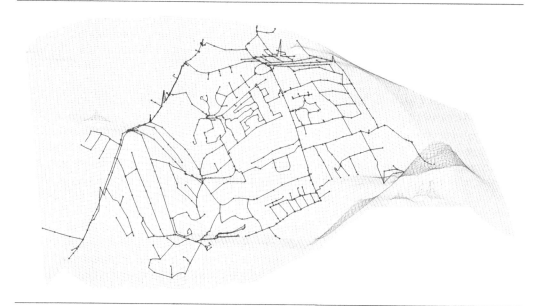

Link data, such as pipe diameters and roughness values, should also be entered into the modelling software. Pipe diameters are normally taken from the network drawings, GIS or asset databases, although care must be taken to use internal dimensions (i.e. the distance from one inner wall to the opposite) and not nominal diameters. Internal diameter may also change due to tuberculation – that is, the buildup of deposit (e.g. scale or rust) in a pipe. Depending on the software conventions, pumps are sometimes modelled as links or nodes. Pump tests are a preferred way of obtaining the most important information about the pumps – efficiency curves, necessary for the inclusion in the modelling software.

Nowadays, most modelling tools allow the use of both headloss equations – that is, Hazen-Williams and Darcy-Weisbach (Colebrook-White). Regardless of which equation is used, roughness values cannot be measured and need to be estimated through calibration. First estimates of pipe roughness values can be obtained from standard tables using available information on pipe material, size, age and internal pipe conditions. These values are then refined during the calibration process. Creating a well-calibrated model is more of an art than a science. Calibration is often limited to an operating range and a degree of accuracy. Hence calibrating certain node pressures against field-recorded pressures to within a specified precision under specific operating conditions does not mean that predictions of within one metre are guaranteed if the model is used to examine significantly different operating regimes. This is particularly true where high flows are involved, such as with rezoning, significant pumping regime changes, or exceptional demand conditions.

It is also worth remembering that measurement locations can only be used to calibrate the network upstream of the logging point; downstream the network remains un-calibrated. Furthermore, calibrating pressures to within pre-specified tolerance relative to the overall pressure does not guarantee a similar accuracy for flow at that point. Finally, attempting to calibrate a network at low demand periods (i.e. during the winter) or a network with a low overall hydraulic gradient is less likely to give accurate predictions if the network is better utilised with higher flows in the future. For example, shut valves at low flows are more likely to go undetected during the calibration process.

A high degree of interest in automated calibration has been shown in the past by researchers, but it has been considerably less covered by practitioners who mostly rely on trial-and-error procedures (Savic et al., 2009). However, the development of commercial modelling packages which include an automated (optimisation) calibration module in the available software (Wu and Clark, 2009) has the potential to change the attitudes of practitioners.

7.6   Model calibration

7.6.1   Introduction

A water distribution system model must be compared with field measurements on the real system to demonstrate that it adequately predicts the behaviour of the system under a range of conditions for an extended period of time, or, in other words, that it behaves similarly to the real system. This applies to both models simulating water quantity and quality, which are widely used by planners, water utility personnel, consultants and many others involved in analysis, design, operation or maintenance of water distribution systems. The iterative process of modifying the model parameters until the output from the model matches an observed set of data is called *calibration*. The goal of calibration is 'to reduce uncertainty in model parameters to a level such that the accuracy of the model is commensurate with the type of decisions that will be made based on model predictions' (EPA, 2005). Pipe friction characteristics and nodal demands are the most commonly encountered as calibration parameters.

Calibration methods are as old as network modelling; however, there are no general guidelines on how to calibrate a model nor there are standards for defining what a well-calibrated model is. Furthermore, the level of calibration effort required will depend on the intended use of the model, as water quality models would require a greater degree of calibration than models used for strategic analysis – for example, master planning.

In addition to the distribution system hydraulic models, water quality models are being used more often in everyday practice and, thus, have to be calibrated. Such a water quality model relies on the flow and velocity information calculated by the hydraulic model and, consequently, suffer if the hydraulic model is not properly calibrated (i.e. if inaccurate flow and velocity estimates from the hydraulic model are used). Tracer tests and online water quality data are starting to be used for water quality modelling applications. However, the cost associated with this more intensive field work is often a barrier to its implementation (Speight and Khanel, 2009).

7.6.2   Field measurements

Historically, most water utilities relied primarily on field tests for calibration of their hydraulic models but, with the advent of SCADA, this type of information is being used more and more to augment the field tests. Well-calibrated measurement equipment is a must for a credible model calibration exercise. This is achieved by proper maintenance and calibration of measuring devices that should be performed at regular intervals. The most important field measurements are simultaneous flow and pressure readings at selected nodes in the network. Additionally, monitoring of control valve operations, reservoir levels and major users' consumption meters should also be done, preferably at the same time. Finally, pump flows/heads should be monitored and actual pump characteristics developed as calibration should not rely only on pump manufacturers' curves, but on actually performed pump characteristics tests. The field test should be carried out for a period of time long enough to account for different demand/flow patterns. This will typically mean a full week, but could be extended if necessary.

Generally speaking, the more data that is used during calibration, the more confidence one can have in model predictions. However, there is a limit to the amount of field data that is available, often because of the costs that are involved with taking field measurements. Sampling design concerns the question of where to collect data during the field test. In other words, where should measurement equipment be placed in the network in order to collect data that, when used for the calibration of the model, will yield the best results. When considering the number and locations of measurement devices, the modeller should attempt to achieve the best trade-off between the field test cost and model prediction accuracy. This should avoid unnecessary expense in collecting redundant data – that is, data whose information contribution is already contained within another measurement.

7.6.3   Calibration approaches

Numerous calibration procedures for water network models have been developed since the 1970s. Generally, practical calibration methods may be grouped into two categories (Savic *et al.*, 2009).

(a) Iterative (trial-and-error) procedure models.
(b) Implicit models (or optimisation models).

Iterative calibration models are based on specifically developed, trial-and-error procedures (Rahal *et al.*, 1980; Walski, 1983; Bhave, 1988) where unknown parameters are updated at each trial using pressures and flows obtained from the simulation model. Once good agreement between predicted results and field data is obtained, the process stops. Implicit calibration refers to a situation where an automated optimisation technique is used with a simulation model (Savic *et al.*, 2009). The optimisation initially sets all input parameters and then, in each further iteration, updates them in order to minimise discrepancies between the model predictions and the corresponding field data. New parameters are then passed onto the simulation model, which in turn passes back the obtained model-predicted heads, flows, reservoir levels and so on. The optimisation tool employs an objective function to minimise the differences between measured and model-predicted variables. The development of commercial modelling packages, which include an automated (optimisation) calibration module in the available software (Wu and Clark, 2009), has the potential to improve the way practitioners perform calibration and reduce the cost of calibration.

Bayesian-based approaches have been developed to improve the calibration of WDS models (e.g. Kapelan *et al.*, 2007). These models improve upon the traditional parameter fitting approaches by being able to obtain, in a single model run, not only the calibration parameter values but also the information on errors associated with these estimates and the corresponding model predictions (e.g. pressures and flows). In this way, a modeller knows which model predictions to trust more (or less). The Bayesian approach also provides a natural framework for the maintenance – that is, for the periodic re-calibration of the simulation models.

7.6.4   Some calibration issues

Calibration of a WDS model is often an ill-posed problem. In practice, ill-posedness is typically manifested as non-uniqueness of the problem solution and it is usually a consequence of inadequate quantity and/or quality of measurement data, but may also be a consequence of an inadequate model parameterisation scheme used. A classical example of ill-posedness is the case of calibrating a WDS hydraulic model for unknown pipe roughnesses by collecting pressure measurement data at low-demand periods (i.e. during the winter) or in a network with a low overall hydraulic gradient. This is unlikely to give accurate calibration results because nodal pressures are simply not sensitive enough to the changes in pipe roughness values due to low head losses in the network (a consequence of low demands, i.e. low flows/velocities in pipes). The model calibrated this way is less likely to have good predictions if it is utilised with higher flows. Several improvements can be made to condition an ill-posed problem – for example, re-parameterisation of the WDS hydraulic model, provision of additional measurements under different conditions and/or incorporation of independent, prior information on calibration parameters (Kapelan *et al.*, 2004).

In existing calibration practice, the calibration problem is considered solved once a relatively good model fit is obtained. However, this fit is usually evaluated by the visual plots of model predictions against measurements and/or, in the best-case scenario, measured by some objective function value obtained. This, however, may lead to unreliable results and, in some cases, to incorrect conclusions (Kapelan, 2010). A more thorough assessment of the results obtained should be performed. In addition to the determination of parameter values which is conventionally done, the following could be evaluated (Kapelan, 2010).

(a) Check the well-posedness of the calibration problem (to determine whether the obtained solution is identifiable/unique and generally, is it meaningful at all).
(b) Check model fit using various statistics and analysis other than objective function value and graphs (to thoroughly evaluate model fit).

(c) Check parameter and model prediction uncertainties – that is, errors including residual analysis (to determine how reliable calibration results are).

Model calibration does not only provide a better model but can provide additional insights about the system that are gained through the calibration process (Ostfeld *et al.*, 2012). This can lead to the analysis of the network topology and sensor placement considerations as part of the sampling design problem.

The issue of sampling design is the one which is closely related to the problem of calibration. The objective of sampling design for calibration is to, in the general case, determine what (e.g. pressure and/or flow), where (which pipe/node?), how frequently (every 15 minutes, every minute and so on), for how long (days, weeks, permanently) and under what conditions (regular demand, fire fighting conditions and so on) to measure so that the data collected can be used to calibrate the relevant prediction model as accurately as possible. A number of different approaches have been developed in the past to achieve this (Yu and Powell, 1994; Bush and Uber, 1998; Kapelan *et al.*, 2003). The key is to understand the actual observability of the different parts of the network (i.e. pipes, nodes, and segments) based on the available sensors.

7.6.5   Model maintenance

Over time, the predictions of a calibrated model become less certain as changes to the real network occur. The accuracy of the model and its effective lifespan is normally determined by the amount of change in the network, making it difficult to set a universal number. These changes can usually be categorised into three groups.

(a) Network reconfiguration – changes to valve status (open/closed) or rezoning.
(b) Asset changes – new mains, pumping stations, PRVs and so on, plus mains renewal.
(c) Demand changes and development growth – changes in large consumers, new housing estates and mains.

Model maintenance is a means of maintaining the accuracy and extending the lifespan of a model to avoid rebuilding the model for every new project or application. New assets (e.g. pumping stations and mains) can be added to the existing model and a limited calibration carried on the new assets. Likewise, minor rezoning or new housing can be added to the existing model.

Unrecorded valve changes can be accommodated by a recalibration of the model and changes in commercial consumption by updating metered demands. However, after a period of time when these become significant, it is likely to be more economic to rebuild and recalibrate the model. The costs associated with maintaining a distribution system model may be more easily justified if it is used for a variety of applications by a water utility.

REFERENCES

Abu-Bakar H, Williams L and Hallett SH (2021) Quantifying the impact of the COVID-19 lockdown on household water consumption patterns in England. *npj Clean Water* **4(1)**: 1–9.

Bergeron L (1961) *Waterhammer in hydraulics and wave surges in electricity*. John Wiley & Sons, NY, USA.

Bhave PR (1988) Calibrating water distribution network models. *Journal of Environmental Engineering, ASCE (American Society of Civil Engineers)* **114 (1)**: 120–136.

Bush CA and Uber JG (1998) Sampling design methods for water distribution model calibration. *Journal of Water Resources Planning and Management, ASCE* **124(6)**: 334–344.

Brdys MA and Chen K (1993) 1. Joint state and parameter estimation of dynamic water supply systems with unknown but bounded uncertainty. *Integrated Computer Applications in Water Supply*, B. Coulbeck, (ed.). Research Studies Press, Brookline, MA, USA, vol. 1, pp.335–355.

EPA (2005) *Water Distribution System Analysis: Field Studies, Modeling and Management, A Reference Guide for Utilities*. US Environmental Protection Agency, Cincinnati, OH, USA.

EPA (2023) EPANET 2.2.0: An EPA and Water Community Collaboration. https://www.epa.gov/sciencematters/epanet-220-epa-and-water-community-collaboration (last accessed on 29/05/2024).

Farmani R, Walters GA and Savic DA (2006) Evolutionary multi-objective optimization of the design and operation of water distribution network: Total cost vs. reliability vs. water quality. *Journal of Hydroinformatics* **8(3)**: 165–179.

Filion YR and Karney BW (2002) Sources of error in network modeling. *Journal of AWWA (American Water Works Association)* **95(2)**: 119–130.

Kapelan Z (2010) *Calibration of Water Distribution System Hydraulic Models*. Lambert Academic Publishing, London, UK, p. 284.

Kapelan Z, Savic DA and Walters GA (2007) Calibration of WDS Hydraulic Models using the Bayesian Recursive Procedure. *Journal of Hydraulic Engineering, ASCE* **133(8)**: 927–936.

Kapelan Z, Savic DA and Walters GA (2004) Incorporation of prior information on parameters in inverse transient analysis for leak detection and roughness calibration. *Urban Water* **1(2)**: 129–143.

Kapelan Z, Savic DA and Walters GA (2003) Multi-objective sampling design for water distribution model calibration. *Journal of Water Resources Planning and Management, ASCE* **129(6)**: 466–479.

Mays LW (2000) *Water Distribution Systems Handbook*. McGraw-Hill, New York, USA.

Obradovic D and Lonsdale P (1998) *Public water supply*. E & FN Spon, London, UK.

Ostfeld A, Salomons E, Ormsbee L *et al.* (2012) Battle of the water calibration networks. *Journal of Water Resources Planning and Management* **138(5)**: 523–532.

Press WH, Flannery BP, Teukolsky SA and Vetterling WT (1990) *Numerical Recipes: The Art of Scientific Computing*. Cambridge University Press, Cambridge, UK, p. 702.

Qatium (2023) The Water Management Platform. https://qatium.app/ (last accessed on 06/01/2023).

Rahal CM, Sterling MJH and Coulbeck B (1980) Parameter tuning for simulation models of water distribution networks. *Proceedings of the Institution of Civil Engineers* **69**: 751–762.

Rossman LA (2000) *EPANET 2, Users Manual*. US Environmental Protection Agency, Cincinnati, OH, USA.

Savic DA, Kapelan Z and Jonkergouw P (2009) Quo vadis water distribution model calibration? *Urban Water Journal* **6(1)**: 3–22.

Speight V and Khanal N (2009) Model calibration and current usage in practice. *Urban Water Journal* **6(1)**: 23–28.

Speight V, Khanal N, Savic DA, Kapelan Z and Jonkergouw P (2009) *Guidelines for developing, calibrating and using hydraulic models*. Water Research Foundation, Denver, CO, USA. (In press.)

Walski TM, Brill ED, Gessler J *et al.* (1987) Battle of the network models: epilogue. *Journal of Water Resources Planning and Management, ASCE* **113(2)**: 191–203.

Walski T (1983) A technique for calibrating network models. *Journal of Water Resources Planning and Management, ASCE* **109(4)**: 360–372.

Walski TM, Chase DV, Savic DA *et al.* (2003) *Advanced water distribution modeling and management.* Haestad Methods Press, Waterbury, CT, USA, p.751.

Wu ZY and Clark C (2009) Evolving effective hydraulic model for municipal water systems. *Water Resources Management* **23**: 117–136.

Wylie EB and Streeter VL (1978) *Fluid Transients.* McGraw Hill, NY, USA.

Xu C and Goulter IC (1998) Probabilistic model for water distribution reliability. *Journal of Water Resources Planning and Management, ASCE* **124(4)**: 218–228.

Yu G and Powell RS (1994) Optimal design of meter placement in water distribution systems. *International Journal of Systems Science* **25(12)**: 2155–2166.

Zadeh L (1965) Fuzzy sets. *Information and Control* **8(3)**: 338–353.

Dragan A. Savić and John K. Banyard
ISBN 978-1-83549-847-7
https://doi.org/10.1108/978-1-83549-846-020242010

Chapter 8
Design of water distribution systems

Seneshaw Tsegaye
Florida Gulf Coast University, USA

Mohamed Mansoor
Jacobs, UK

Harrison Mutikanga
Uganda Electricity Generation Company, Uganda

Kalanithy Vairavamoorthy
International Water Association, UK

8.1.   Introduction

Water distribution systems (WDSs) vary from simple to complex. The main objective of all water systems is to supply safe water for the cheapest cost. These systems are designed based on least cost and enhanced reliability considerations, and design principles should satisfy both hydraulic and engineering requirements. The hydraulic requirements include pressure, velocity, sufficient flow, minimum operational cost and so on, and the engineering requirements involve selection of durable materials, system component configuration, ease of access to components and so on.

This chapter begins by considering the design objectives of WDSs. It describes performance indicators that can be used to drive WDS design. The chapter then introduces optimisation techniques and how they can be applied in planning and design of conventional and intermittent WDSs. This is followed by the discussion of decentralised WDSs, focusing on clustering model and procedures for optimising the boundaries of decentralised WDSs. The chapter includes a discussion on the need to develop WDSs that can cope with future uncertainties and change requirements. To complement this, the chapter concludes by providing a brief look at the impacts of the COVID-19 pandemic on the operation of the WDSs and highlights the need for future preparedness. An alternative approach to build resiliency in the wake of the COVID-19 is also discussed.

8.2.   WDSs requirements

8.2.1   Design objectives

The primary purpose of a WDS is to provide good potable water to the public in sufficient quantities and pressure at all times. A WDS must, therefore, be properly designed to sustainably meet the objective at the least cost possible. A well-designed WDS should minimise operational costs, supply water for fire protection and provide sufficient level of redundancy to support minimum level of service during emergency conditions (i.e. power loss or water main failure). Achieving these objectives requires acquisition of basic information about the users, including historical water usage, population trends, planned growth, topography and availability of water resources, to name just a few. This information can then be used to plan and design the WDS to provide sufficient water of good quality and adequate pressure.

Generally, most of the developed countries operate conventional fully pressurised WDS, where water is available 24 hours a day. Many countries, especially in the developing world, are experiencing intermittent water supplies (IWS). Nearly a billion people receive water from networks that operate intermittently. Approximately 41% of the piped supplies in middle and low income countries operate intermittently (Vairavamoorthy et al., 2008). Therefore, intermittent systems are a reality. In most situations, the nature of intermittency can range from a few hours a day to almost all day (Taylor et al., 2018) and, in some countries, water is supplied for only a day every week or sometimes every fortnight (Coelho et al., 2003). The IWS are not always caused by water scarcity (due to global change pressures), but factors like economic scarcity and technical scarcity could force the continuous supplies to operate intermittently (Totsuka et al., 2004).

Design objectives of an intermittent system differ from those of a conventional fully pressurised system. The quantity of water received by the customer depends on the system pressure as opposed to the conventional supplies where consumption is not affected by the pressure in the system. Due to the intermittent nature of the network, water quality is a critical factor as contaminants are more likely to ingress into the network during low pressures, in addition the household storage associated with IWS also contributes to the deterioration of the water quality (Coelho et al., 2003). Quantitative microbial risk assessments have shown that intermittent water supplies cause 17 million infections annually (Bivins et al., 2017). Therefore, the primary objective of the intermittent systems is to provide an adequate and equitable supply of acceptable quality at a low cost.

8.2.2   Performance indicators and levels of service

WDSs often represent a large proportion of capital investment in water supply. This investment is buried in the ground and deteriorates with time. Because of the key role played by the WDS, it is critical that performance indicators (PIs) are established to evaluate its performance towards delivering the required levels of service.

Three major performance criteria are used to evaluate a WDS: *adequacy, serviceability and efficiency*. Adequacy refers to the delivery of an acceptable quantity and quality of water to the customer. Serviceability (reliability or dependability) measures the ability of the distribution system to consistently deliver an acceptable quantity and quality of water, while efficiency measures how well resources such as water and energy are utilised to produce the service. The task of measuring and evaluating performance is accomplished by performance assessment systems through well-defined PIs. One of the key requirements of a good PI is that it must be clearly defined, with a concise meaning, and easy to understand, even by non-specialists – particularly by consumers. The level of service provided should be looked at from three perspectives

- the regulator's perspective
- the customer's perspective
- the water supplier's perspective.

The levels of service PIs provided by the WDS are classified as structural, hydraulic, water quality and customer satisfaction. The PIs corresponding to each performance criterion are as follows (Deb et al., 1995; Coelho and Alegre, 1998; Ofwat, 2009)

- **Adequacy**: Pressure, flow, water quality, customer satisfaction/ complaints, response time to customer complaints
- **Serviceability**: Service interruptions, inoperable valves and hydrants, main breaks, water quality violations
- **Efficiency**: Leakage, metering functionality, pumping efficiency.

The indictors used for measuring the appropriate levels of service vary according to the local conditions. In England and Wales, the Water Act 1989 and the Water Industry Act 1991 established a regulatory framework which is based on the definition of quantifiable levels of service. In the UK, the service level indicators are defined and regulated by the economic regulator, The Water Services Regulation Authority (OFWAT). For instance, the level of service indicator for pressure is assessed in terms of the number of connected properties that have received, and are likely to continue to receive, pressure below the reference level when demand for water is at a normal level. The reference level of service is defined as 10 m head of pressure at a boundary stop tap with a flow of 9 l/min (Ofwat, 2009).

8.2.2.1 Adequacy

The adequacy of WDS is measured in terms of how well the customers of the system are served. Hydraulic performance measures relate to the delivery of adequate supplies of water and are measured in terms of pressure and flow. Pressure in a WDS is measured in terms of the number of customers at risk of being supplied at pressure levels lower than the reference level. Water quality performance measures are assessed by the number of water quality violations that the system experiences. Customer satisfaction can be measured by surveys of customers, or in terms of complaints, response rate to complaints or system resiliency (how quickly the system recovers from failure to meet the customer's needs).

In general, when the term adequacy is used, it is implied that the water supply is a fully pressurised continuous system. However, in the context of intermittent supplies, the adequacy needs to be looked at in conjunction with equity. Due to the pressure dependent nature of the supply, not all consumers receive equal amount of water, which is further exacerbated by the leakage in the system. In an intermittent system, customers do not receive an adequate supply, rather the quantity of water received depends on the water availability and system pressure. Therefore, customers who are nearer the source tend to get a higher share of the water than those are away from the source. Hence, maintaining equity of supplies is essential. However, equity does not necessarily mean each customer receives equal amount of water, rather it could be that each household getting a similar portion of the required demand.

8.2.2.2 Serviceability

Serviceability is a measure of the consistency of service that customers of the system experience and is related to how well WDS assets are managed now and in the future. The structural performance measures include service interruptions, mains bursts, and hydrant and valve functionality. In terms of dependability, the water quality performance measure of interest is water quality violations of extended duration. Supply interruptions are measured in terms of number of customers affected by interruptions to supply lasting longer than the reference duration, without adequate warning and appropriate justification by the water service providers.

Serviceability measures need to be modified when applied to intermittent systems. Consistency of service could be determined by availability of service to customers, in this context, rather than interruptions, the availability measure looks at the frequency of water availability. The water quality aspects are of great concern in intermittent systems and can be identified by the breaches in acceptable levels of water quality parameters in the network after the first flush.

8.2.2.3 Efficiency

The efficiency of a WDS is measured in terms of three structural performance measures. The leakage-energy nexus measures how well water resources are utilised. Leakage, a significant component of water losses, represents a major waste of a precious resource, but the benefits of reducing leakage need to be balanced against the costs of finding and fixing leaks or replacing mains. Impact of leakage is more critical in intermittent supplies as this reduces the supplies to the consumer. The operation of the intermittent supplies is driven by leakage rather than demand after a certain demand threshold is satisfied, in other words any additional water supplied to the network is lost to leaks (Taylor, *et al.*, 2018).

The excessive use of energy for pumping can be a symptom of distribution system problems (undersized mains, tuberculation, main breaks, leakage, worn-out impellers etc.). Customer metering under-registration due to inaccurate meters does not promote efficient use of water and result in substantial revenue loss to the water service providers. All these may result in increased capital and operating costs, which may affect customers as tariff increases.

8.2.3   Basic design principles – conventional systems
In order to design WDS, key basic principles must be adhered to and these include the following (Mays, 2000)

- water demand assessment and projections (design period and peak factors)
- storage and water balancing requirements
- analysis of the WDS using hydraulic modelling techniques
    - laws of hydraulics (energy and continuity equations)
    - application of pipe head loss formulae (Colebrook-White or Hazen-Williams)
- optimisation of the design.

8.2.4   Basic design principles – intermittent systems
Design principles for both conventional fully pressurised and intermittent systems are similar, but the latter slightly differ from the former in demand allocation and network analysis. The design criteria are determined based on the customer preferences on the timing of supply, duration of supply, pressure at outlet, type of connection, location of connections (standpipes) and so on. This is termed as 'people-driven levels of service (Vairavamoorthy *et al.*, 2002). Therefore, in order to design an intermittent WDS, the following principles must be adhered to.

- Water demand assessment and projections are based on secondary network demand analysis. Secondary network being a collection of consumers supplied from a point in the network (design period and peak factors, duration of supply, timings of supply).
- Storage and water balancing requirements.
- Analysis of WDS using pressure dependent hydraulic modelling techniques
    - laws of hydraulics (energy and continuity equations with pressure dependent functions)
    - application of pipe head loss formulae (Colebrook-White or Hazen-Williams).
- Optimisation of the design.

8.3.   Optimal design of WDSs
Design of WDSs is accomplished by trial and error methods of different alternative scenarios using a network solver. Because of the complex interactions between components, identifying changes to improve a design can be difficult even for small systems. This approach may not guarantee an optimal design of the WDS. This section highlights research efforts being made to obtain optimal designs of WDSs. This section highlights the basic concepts in optimal designs of WDS.

8.3.1   Problem formulation

Optimisation of a WDS is often viewed as the selection of the least net present value (NPV) combination of component sizes and settings such that the criteria of demands and other design constraints are satisfied (Zecchin et al., 2005). However, many real world decision-making problems need to achieve several objectives such as minimise risks, maximise reliability, minimise deviations from desired levels, minimise cost and so on (Savic, 2002), and it is not always easy to attribute monetary values to these objectives. The monetary values usually include the costs for construction, operation and maintenance. According to Mays (2000), the main constraints are supplying the desired demands with an adequate pressure head being maintained at withdrawal locations. Also, the flow of water in a distribution network and the nodal pressure heads must satisfy the governing laws of conservation of energy and mass. Problem formulation is normally the most important part of the optimisation process. It involves the selection of design variables, constraints and objective functions.

8.3.1.1   Design variable

A design variable for an optimisation problem is any quantity or choice directly under the control of the designer. It involves many forms, as WDSs comprise many components and performance criteria – for example, the selection of diameters for all the pipes, pump types and locations, the sizing and locating of tanks, valve pressure settings and valve locations.

8.3.1.2   Constraints

A constraint is a condition that must be satisfied in order for the design to be feasible, and constraints can reflect resource limitations, user requirements or bounds on the validity of the analysis models. The constraints in the WDS optimisation problem could be specified to include minimum and maximum allowable pressures at each demand point, minimum and maximum velocity constraints for each of the pipes, and water quality requirements. Further constraints may be added for materials and for different rehabilitation alternatives (cleaning, relining or both) (Walski et al., 2003).

8.3.1.3   Objective functions

An objective function is a numerical value that is to be minimised or maximised. The optimal design of a water distribution network is often viewed as a least cost optimisation problem. However, there are other possible objectives such as network reliability, redundancy and water quality that can be included in the optimisation process. Conventionally, water engineers have treated these problems as single objective optimisation problems instead of multi-objective ones. This type of optimisation is useful to provide decision makers with insights into the nature of the problem, but usually it cannot provide a wide range of alternatives that trade different objectives against each other (Savic, 2002).

Once the design variables, constraints, objectives and the relationships between them have been chosen, the optimisation problem can be expressed mathematically – for example, the mathematical formulation of the WDS optimisation model for finding the least cost combination of pipe sizes can be stated as

$$\min f(D_1,...,D_{npipe}) = \sum_{k \in npipe} C(D_K, L_K) \tag{8.1}$$

$$\text{subject to} \sum Q_{in} - \sum Q_{out} = Q_e \tag{8.2}$$

$$\sum_{k \in Loopl} \Delta H_k = 0, \forall l \in NL \qquad (8.3)$$

$$\Delta H_k = H_{ul\,snode,k} - H_{dl\,snode,k} = \omega \frac{L_k}{C_k^{\beta} D_k^{y}} Q |Q_k|^{\beta-1}, \forall k \in npipe \qquad (8.4)$$

$$H_{iN} \geq H_{min\,iN}, \forall iN \in NN \qquad (8.5)$$

$$D_k \in [D], \forall k \in npipe \qquad (8.6)$$

where

$D_1,...,D_{npipe}$ are npipe discrete pipe diameter decisions selected from the set of commercial pipe sizes [D].

$C_k(D_k, L_k)$ is cost of pipe k with diameter D_k and length L_k.

Equation 8.2 represents conservation of mass for each node where Q_{in} and Q_{out} are flow into and out of the node respectively, and Q_e is external inflow (negative) or demand (positive) at each node.

Equation 8.3 represents the conservation of energy around a loop where ΔH_k denotes the head loss in pipe k and NL are the total number of loops in the system.

The head loss in each pipe is the head difference between connected nodes (Equation 8.4) where C is the Hazen-Williams roughness coefficient (the Darcy-Weisbach equation can also be used).

Equation 8.5 requires the total node pressure H for any node iN (where the total number of nodes is NN) is equal to or greater than a pre-specified minimum pressure H_{min}.

Objective functions for the intermittent supply are the same as conventional supply except that the demand is pressure dependent (Vairavamoorthy, 1994). In intermittent systems demand is pressure dependent, hence $Q_e = f(p)$, where $f(p)$ is the pressure dependent demand function and p is the pressure at the outlet or node.

8.3.2 Application of multi-objective optimisation to WDSs

The design of a WDS is often viewed as a single-objective, least-cost optimisation problem with pipe diameters acting as the primary decision variables. The problem of choosing the best possible set of network improvements to make with a limited budget is a large optimisation problem to which conventional optimisation techniques are poorly suited. The exponential growth of the problem size with the increase in the number of discrete decisions persuades designers to use multi-objective optimisation approaches in WDS design. Three major improvements of the multi-objective optimisation approach are identified by Cohon (1978).

(a) A wider range of alternatives is usually identified when a multi-objective methodology is employed.
(b) Consideration of multiple objectives promotes more appropriate roles for the participants in the planning and decision-making processes – that is, 'analyst' or 'modeller' (who generates

alternative solutions) and '*decision maker*' (who uses the solutions generated by the analyst to make informed decisions).

(c) Models of a problem will be more realistic if many objectives are considered.

Optimisation techniques are tools to develop useful information for the decision makers. However, single-objective models require that all design objectives must be measurable in terms of a single fitness function which depends on prior ordering of objectives (i.e. a weighting scheme). Thus, single-objective approaches place the burden of decision-making on the shoulders of the analyst. However, multi-objective optimisation allows decision makers to assign relative values based on trade-off curves between different objectives (Savic, 2002).

An example of a dual-objective trade-off curve obtained in a single run of an optimisation routine (Kapelan *et al.*, 2005) is shown in Figure 8.1. This curve defines an efficient frontier – that is, the minimum network design cost against the level of system robustness achieved or, equivalently, the minimum robustness that can be attained for various levels of investment. Several points can be noted here – for example, that due to the discrete nature of the network design problem, the solutions on the trade-off curve are represented as discrete points and to reach a robustness level of, say, about 80%, the minimum investment cannot be less than US $45 million. Furthermore, the decision maker could also benefit from the knowledge that increasing the robustness from 80% to 99% would require an increase in investment of almost US $15 million.

In recent decades, the focus of optimisation for WDSs has also shifted from the use of traditional optimisation methods, such as linear programming, decomposition methods, nonlinear programming and the use of heuristics derived from nature (HDN), such as genetic algorithms (GA), simulated annealing (SA) and, more recently, ant colony optimisation (ACO). These optimisation techniques encourage the implementation of multi-objective formulation in the planning and design of WDSs.

Figure 8.1 Example of a dual-objective trade-off curve obtained in a single run of an optimisation routine. (Data from Kapelan *et al.* (2005))

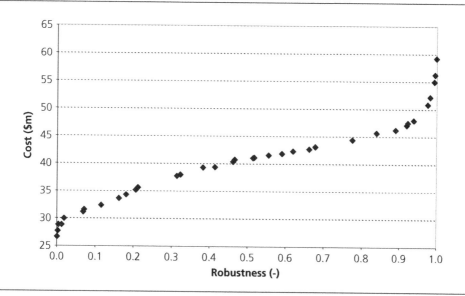

Multi-objective optimisation techniques have been successfully applied to the planning, design and management of WDS – for example, Savic *et al.* (1997) have used a multi-objective optimisation technique for pump scheduling. The pump scheduling problem is formulated as two objectives: the minimisation of energy costs and the minimisation of pump switches. These multi-criterion methods provide a choice of trade-off solutions from which a decision maker can select a suitable one to implement. Hence, this approach is used to find a spread of good, trade-off solutions with respect to all objectives and achieves remarkable reductions in operation costs by optimising the pump scheduling problem.

Improvement in a WDS performance can be achieved through replacing, rehabilitating, duplicating or repairing some of the pipes or other components (pumps, tanks etc.). Halhal *et al.* (1997) implemented a multi-objective approach (capital cost and benefit as dual objectives) to enhance the performance of an existing water distribution system. Pipes are considered in improving performance by

- increasing the hydraulic capacity of the network through cleaning, relining, duplicating or replacing existing pipes
- increasing the physical integrity of the network by replacing structurally weak pipes
- increasing system flexibility by adding additional pipe links
- improving water quality by removing or relining old (cast iron) pipes.

A study by Zecchin *et al.* (2005) developed parametric guidelines for the application of ACO algorithms to WDS and found this approach to outperform other HDNs for two well-known case studies. Zecchin *et al.* also highlighted the high requirement for calibration in using HDNs because of the drawbacks associated with their searching behaviour, and, hence, performance is governed by a set of nondeterministic parameters. Babayan *et al.* (2007) also use two objectives (cost of the network design/rehabilitation and probability of network failure) to formulate the problem associated with stochastic WDS design. The problem is solved using a multi-objective GA after converting it to an equivalent, simplified deterministic optimisation problem. The design method was tested and successfully applied to a case study.

8.3.3 Example applications of optimisation for WDSs performance, least cost, equity and reliability design

8.3.3.1 Application of optimisations to WDSs design

Multi objective optimisation has been applied by the Regional Municipality of York (Canada) to develop the water supply for the area that arose due to an expected rapid development and doubling of population over a 35-year period taken from 1996 as the base. This enables the optimal selection of feasible solutions from the options of different infrastructure configurations (Morley *et al.*, 2001). Given the complexity of the problem, the GA approach was applied to a calibrated hydraulic model to carry out extensive optimisation of the regional infrastructure requirements that were needed to meet the demands in 2031. The optimisation applied the common processes used for similar problems elsewhere. These included

1. a GA with an objective function defined on a set of decision variables – for example, pipe diameters, reservoir storage requirements and so on
2. a calibrated model of the system to simulate its hydraulic behaviour and to ensure that continuity and head-loss equations are satisfied at all times (hard constraints)

3. a penalty term to penalise insufficient level of service (soft constraints) – for example, pressures at nodes, imbalance of reservoir flows and so on.

The GA was programmed to ensure that the overall solution reflected cost effectiveness on a whole-life cost basis. It took into account expected operating and maintenance costs as well as capital costs. Failure to meet a set of performance criteria (e.g. acceptable pressure ranges and storage requirements) was penalised during the optimisation process.

The optimal result proposed the construction of 85 new transmission mains added to the existing 750 mains in the system. Six new pumping stations were proposed and three old pumping stations were identified for expansion with a total of 42 new pumps. The optimised plan also recommended three pumping stations to be decommissioned. Seven new ground reservoirs and elevated tanks were proposed (total volume of 78 million litres) with two existing elevated tanks identified for decommissioning.

This enabled the authority to realise considerable savings; for instance the phased optimal solution for 2011 would cost US $102 million instead of the previous manual solution cost of US $156 million – a saving of US $54 million or 35%. The estimated cost for the 2031 planning horizon was found to be around US$138.88 million. The size of the problem solved by the GA which is estimated to be 1×10^{357}, demonstrates its superiority in dealing with complex, nonlinear and discrete optimisation problems. It is not surprising that the impressive improvement over the manual solution has been achieved by the optimisation approach.

8.3.3.2 Determining the optimal level of service

Customer expectations of the reliability of water supply are likely to rise over time, leading to a need to improve the distribution system to reduce interruptions and ensure adequate pressure. For example, in the UK, there has been increased concern about ensuring reliable supplies, following rising concerns about security and the prolonged loss of supplies to 138 000 households in Gloucestershire in 2007, as a result of flooding. This requires additions to the system to reduce the number of communities which are dependent on a single mains link from the network. However, the scale and pace of improvements needs to be balanced against other potential service improvements, such as improving the taste and odour of water supplies and the need to ensure customers' bills are affordable.

To review water prices in UK for 2009, water companies balanced service improvements and changes in bills by establishing customer willingness to pay for improvements, established through surveys using choice experiments. This involved presenting customers with alternative packages of services and bills, covering all the main potential areas of improvement to water and sewerage. Customers' choices of the best package allowed estimation, through statistical analysis, of the willingness to pay for improvements in each of the service measures. An example of Severn Trent Water company's results for service measures affected by network performance is shown in Table 8.1.

These willingness to pay results are used in optimisation modelling. For example, for a project to improve resilience of the distribution system, its benefits were assessed in terms of:

■ the probability of water supply interruption before and after the scheme
■ the numbers of customers affected.

Table 8.1 Level of performance and customers willing to pay (Data from Severn Trent Water Business Plan (2009))

Service measure	Current performance	Potential change	Household willingness to pay	Non-household willingness to pay
Interruptions – number	11500	7500	£7.26	£1.76
per year	11500	3500	£14.53	£3.51
Low pressure – number of	10000	15000	–£8.17	–£1.38
customers at risk	10000	5000	£8.17	£1.38
	10000	2000	£8.17	£2.21

This gives an average number of customers affected by interruptions per year, which is multiplied by the willingness to pay to give a total benefit from the scheme. The NPV of the benefits was then compared with the NPV of costs (including the cost of carbon) to determine whether there were net benefits from the project and, therefore, whether it should be included in company plans.

8.3.3.3 Optimisation for the rehabilitation of WDSs

Improving the performance of WDS involves a high cost while available financial resources are limited. Thus, there is a need for the development of optimal rehabilitation plans. Many objectives can be incorporated in rehabilitation decision models – for example, work by Cheung et al. (2003) demonstrates a multi-objective optimisation for rehabilitation of WDS recognising the multi-objective nature of the problem. The problem formulation is done based on three-objective network rehabilitation function such as minimisation of cost and pressure deficit and maximisation of hydraulic benefit, considering various combinations of rehabilitation choices. The individual objectives are expressed in Equations 8.7 through 8.9 as follows.

Minimise cost

$$F_1 = \sum_{l=\chi} c_l L_l + \sum_{k=\pi} c_k L_k \qquad (8.7)$$

where

l is the index of the pipes to be rehabilitated (cleaned or left unaltered)
k is the index of the new pipes (replaced or duplicated)
χ is the set of alternatives related to the pipes requiring rehabilitation
π is the set of alternatives for new pipes
L is length of the pipe
cl are rehabilitation unit costs
ck are unit costs of new pipes.

The decision problem corresponds to the identification of pipes to be added in parallel or as a new pipe.

Minimise pressure deficit

$$F_2 = \sum_{i=1}^{LC} \max(H_j - H_{j\min})_i \quad j=1,2,...,nn \qquad (8.8)$$

where

the pressure deficit is the sum of maximum nodal deficits on the network for each demand pattern, and

j is the index of nodes

nn is the total number of nodes in the system

Hj is the energy

$Hjmin$ is the required minimum energy at node j.

LC denotes the number of demand patterns considered (three demand patterns are investigated: peak, average and minimum demands).

Finally, a modified hydraulic benefit formulation is used as the third objective function. It is quantified as the difference between the pressure deficiencies in the network before improvement (*DEFO*) and after improvement (*DEFP*) represented by each solution found which is calculated by

$$DEFO / DEFP = \gamma \sum_{j \in \chi} \left| H_{min} - H \right| Q_j$$

Maximise hydraulic benefit

$$F_3 = DEFO - DEFP \tag{8.9}$$

where γ represents the specific weight of water

χ is the set of nodes related to the energy below minimum required energy at node j

H is the energy

H_{min} is the required minimum energy at node j

Q is the demand at node j.

For intermittent systems, $Q = f(p)$, where $f(p)$ is the pressure dependent demand function and the p is the pressure at node.

The study employed an elitist multi-objective evolutionary algorithm, called the Strength Pareto Evolutionary Algorithm (SPEA), to generate a series of non-dominated solutions. The method uses these elite solutions to participate in the genetic operations along with the current population in the hope of influencing the population to steer towards good regions in the search space. The rehabilitation analyses were conducted on a simple hypothetical network. The paper demonstrated the advantage of SPEA methods in visualising trade-offs (costs, pressure deficit and hydraulic benefit) and choosing a satisfactory solution; and showed the benefit of multi-objective optimisation technique to the decision-making of water distribution systems problems. Moreover, the importance of choosing appropriated recombination and mutation operators for reading a stable Pareto front is suggested by the authors in improving results using multi-objective evolutionary algorithms for WDS.

8.3.3.4   Optimisation of equity in intermittent WDSs

Due to the nature of intermittent water distribution systems, the customers generally do not get equal amount of water or required demand. The inequity in supply is caused by the variation in pressure between the nodes. There are multiple approaches to introduce equitable supply to customers. Ilaya-Ayaza et al. (2017) approached the issue by using integer linear programming and multi criteria analysis to minimise the peak flows by producing an optimal supply schedule that reorganises the peak demand times of different sectors. Effah et al. (2018) optimised the equity and cost by minimising the deviation of equity between nodes. The objective function for the optimisation is expressed as

$$DE = Min\sum_{i=1}^{n} \left|\left(\%Qav - \%Qs\right)\right|$$

(8.10)

Subject to

$$\sum_{i=1}^{n} V_i = A_w$$

(8.11)

where

DE is the deviation in equity (equity is defined as the quantification of actual volume supply to the nodes and any deviation in supply among the nodes)

Qav is the percentage average volume of water supplied to all customer nodes or the ratio between the amount of water received and water required at nodes

Qs is the volume of water supplied to the customer node expressed as a percentage

A_w is total availability water for supply

V_i is the volume of water at node i.

n is the number of nodes in the system.

The study employed hydraulic modelling and opsonisation tools, namely Epanet and GANetXL, which execute the optimisation runs and allows to analyse the result of the Pareto-optimal solutions visually.

8.4.   Decentralised WDSs

With increasing global change pressures, it has become obvious that the current practice of urban water supply system is not sustainable. In order to cope with these challenges, different authors have proposed decentralisation of urban water systems. While there is great consensus on the benefits of decentralised water systems (Weber *et al.*, 2007; Bieker *et al.*, 2010; Cornel *et al.*, 2011; Böhm *et al.*, 2011), one of the challenges for cities to make this transition is a lack of clear guidance on defining the optimal scale and boundary of decentralised WDSs. A new clustering methodology and tool (Tsegaye *et al.*, 2020a) to decentralise WDSs into small and adaptable units is presented.

8.4.1   Methodology for clustering WDSs

This section presents a methodology for clustering WDSs based on two major optimisation principles

(a) minimisation of the distance from water source to water users, and
(b) maximisation of the homogeneity within the cluster.

The clustering methodology uses location of water sources (surface water, groundwater, and stormwater collection points), topography/ digital elevation model, spatio-temporal population growth and associated demand, land use characteristics and socio-economic status of the area under consideration as an input parameter. These parameters are used to define source-demand distance and intra-cluster demand, and topographic homogeneity of the study area (Herrera *et al.*, 2010). Figure 8.2 shows the steps used to determine cluster boundaries.

8.4.2   Minimisation of source-demand distance

The first part of the clustering method involves prior grouping of spatially distributed available water sources, such as surface water, groundwater reclaimed water and stormwater. This method

Figure 8.2 The proposed method for clustering WDS (Tsegaye *et al.*, 2020a)

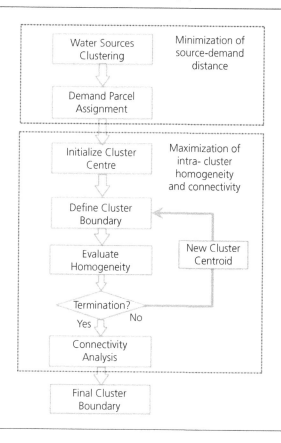

involves grouping water sources and determining their group centre such that the effort required for supply storage is minimised. Then each demand parcel is assigned to one source group centre such that the distance between source and demand parcel (grid cell) is minimised. The Euclidean norm minimisation approach is used to minimise source-demand distance.

Source allocation is a demand assignment problem where demand parcels are assigned to the nearest source centre. The method employs a minimisation of the sum of Euclidean norms within the cluster. The sum of Euclidean norms is used to determine the membership of parcels (demand) based on the shortest distance to the water source centres. This increases the compactness (Dopp, 2011) and reduces the cost of pipe networks and the energy needed for pumping long distances. Compacted networks with closer proximity also increase resource efficiency by reducing leakage that would be higher in large centralised systems.

Given a set of parcels (representing the study area) with dimension vector $P = \{P_1, P_2, \cdots, P_n\}, P \in \mathbb{R}^N$. Euclidean norm defines $\|P\| = (P * P)^{\frac{1}{2}}$, *if* $N - 1$ *then* $\|P\| = |P|$, the absolute value of $P \cdot \|P\|$ is the Euclidean norm of P that is used to measure the distance between points (Nachbar, 2009).

Suppose $P = (X, Y) \in \mathbb{R}^2$ and the source centres are defined by $C = (X_1, Y_1) \in \mathbb{R}^2$, then the shortest distance from the source to the parcel is determined using Equation 8.12.

$$min \, \|P\| = \sqrt{(X_1 - X)^2 + (Y_1 - Y)^2} \qquad (8.12)$$

Given the Euclidean norm of each parcel (from each source centre), distance minimisation is performed using Equation 8.13. Then each parcel will have membership (to the source centre) based on the minimisation of Euclidean norms. The membership defines grouping of similar parcels which are assigned to the same source centre. The Euclidean norm minimisation algorithm is shown in Figure 8.3.

$$\min d_{(P,P^C)} = \sum_{K=1}^{C} \|P\| = \sum_{k=1}^{C} \sqrt{\sum_{j=1}^{n} (P_j - P_j^C)^2} \qquad (8.13)$$

Where $d_{(P, \, P^c)}$ is the Euclidean norm from the source centres, $\|P\|_{min}$ is the minimum Euclidian norm of each parcel from source centres, P is an attribute which is described by parameters where the variation needs to be minimised (i.e. location and elevation parameters).

The movement of water is based on an absolute distance which depends on the link (pipe) layout and pressure distribution; this requires hydraulic simulation of the whole network. However, to simplify the clustering process, the minimisation of the Euclidean norm is employed by using the relative distance based on the coordinate of demand parcels and supply centres. Once the parcels are assigned to the source centre by the minimising Euclidean norm principle, the membership values are used in the maximisation of cluster homogeneity.

The size of a cluster could be used to pre-determine an initial number of clusters or source groups and could be changed during the process of clustering – for example, considering an area with eight water sources and 121 demand parcels (each representing a 100 m by 100 m area), the pictorial representation of the demand parcel assignment to the nearest source centre is shown in Figure 8.4.

8.4.3 Maximisation of intra-cluster homogeneity and connectivity
Maximisation of intra-cluster homogeneity allows clustering the parcels so that parcel attributes within a cluster are closely related to one another (Herrera *et al.*, 2010). Three major parameters can be considered in the clustering process. These are membership (determined by

Figure 8.3 Basic minimising Euclidean norm algorithm

Minimising Euclidean norm algorithm

(a) For the given C source centres, the Euclidean norm of a parcel is determined with respect to their parameter $P = \{P_1, P_2,..., P_n\}$, yielding the distances $d_{(p, \, pc)}$.
(b) Given the set of Euclidean norms $\{d_1, d_2,...,d_c\}$ for each parcel, the total cluster Euclidean norm is minimised by assigning a parcel to the nearest source centre.
(c) Steps (a) and (b) are repeated until all parcels are assigned to the closest source centre (then a membership will be assigned to each parcel based on the source centre to which they belong).

Figure 8.4 Assignment of parcels to the source centre. (X and Y are location parameters, Z is elevation above sea level (asl), Qd is parcel demand, Qs and Qg are capacity of local water sources and group source, respectively) (Tsegaye et al., 2020a)

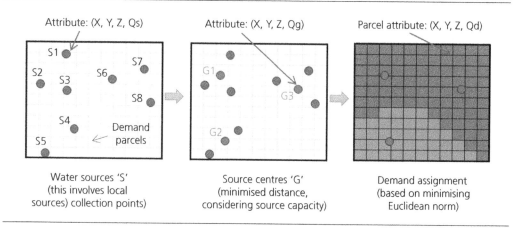

| Water sources 'S' (this involves local sources) collection points) | Source centres 'G' (minimised distance, considering source capacity) | Demand assignment (based on minimising Euclidean norm) |

Euclidean norm minimisation), topography (elevation of the parcels) and spatio-temporal demand distribution (determined from the population distribution, land use and socio-economic parameters). The clustering process involves the grouping of similar parcels. An inter-cluster homogeneity is used as a measure of similarity between parcels and a K-means optimisation technique is employed to maximise intra-cluster homogeneity by minimising the total cluster variance with respect to the mean value. A connectivity analysis is used to ensure the linkage of parcels within a cluster.

Given a set of parcels p representing the study area {X1, X2,..., Xp}, where each parcel has n-dimensions (i.e. topography, elevation), K-means clustering aims to partition the parcels (p) into K clusters (K≤p) within an assigned data-set S {S1, S2,...,S_k}. For the given cluster assignment-A that involves K groups, the total cluster variance is minimised through minimisation of the sum of the squares of Euclidean norm for all clusters using Equation 8.14. Figure 8.5 shows the basic K-means algorithm used in clustering water supply systems.

$$A = arg \min_{S} \sum_{i=1}^{K} \sum_{X_j \in S_i} \left\| X_j - \mu_i \right\|^2 \tag{8.14}$$

$$\mu_i = \frac{1}{N_i} \sum_{Xj \in Si} X_j \tag{8.15}$$

where
   A is cluster assignment
   K is the number of clusters
   N_i is the number data-set assigned to S_i
   μ_i is mean of parcels in cluster S_i and is calculated using Equation 8.15 (Al-Saleh et al., 2009).

Figure 8.5 Basic K-means algorithm

Basic K-means algorithm

(a) Initialization of K means $\{\mu_1, \mu_2, ..., \mu_k\}$ where each mean is defined by d-dimension vector (n-parameters)

(b) Given an initial set of K means, the algorithm assign parcels to the closest mean so that the total variance is minimised with respect to the mean

(c) Calculate a new mean to be the centroid of the cluster

(d) Repeat steps (a) and (b) until the assignments do not change

---

Intra-cluster parcel connectivity analysis needs to be done to avoid the possibility of detaching parcels of the same cluster in different spatial locations. Intra-cluster parcel connectivity, defined as the linkage of a parcel within a cluster, is used to check whether a parcel of one cluster is located in another cluster. Given the membership of parcel 'p' defined as $P_{(m,n)}$ and neighbourhood parcels as $P_{(n\pm1,m\pm1)}$, if parcel $P_{(m,n)}$ of one cluster neighbours two or more parcels from another cluster and has only one neighbour from its own cluster, the evaluation of the minimum Euclidean norm of the parcel $P_{(m,n)}$ is performed with respect to the neighbouring cluster centroid and is re-assigned to the closest one.

In addition, the periphery parcels, which do not have many neighbours, are merged to the nearest cluster. This connectivity analysis alone does not guarantee the existence of cluster members in another spatial location. One can use the smallest recommended size of cluster and/or the smallest demand that a cluster should supply to decide on merging isolated parcels to the neighbouring cluster. An isolated parcel group will be kept as an independent cluster if the demand it supplies is greater than the required minimum size/ demand within the cluster. However, a parcel group that does not satisfy the above mentioned condition will be merged to the neighbour cluster. The decision of which cluster to combine will be made by evaluating the minimum Euclidean norm value with respect to the centroid of neighbouring clusters.

8.4.4   Case study application of the clustering methodology

The clustering methodology was applied to a real case study in Arua, Uganda. Arua is located in the Northern Region of Uganda. The Aura municipality is one of the fastest growing municipalities in the country. According to the statistical abstracts of the Uganda Bureau of Statistics (UBOS, 2011), the population of the Arua municipality was 59 400 in 2011, with the population around the periphery of the municipality reported to be 49 893. Figure 8.6 shows the predicted spatial extent of Arua in 2032 and the output of source-centre identification process. Once the groups are identified, the X, Y coordinate and supply capacity (Qs) are used to calculate source-centres.

Once a source centre was identified, the discretised square parcels (150 m by 150 m) were assigned to the source centres. Each parcel has a location, topography and demand attribute. This stage used the location attribute (X, Y) coordinate of parcels and the centroid of the available sources as inputs to minimise the source-demand location for each parcel. Given the Euclidean norm of each parcel (from the 7 source centres), the distance minimisation was performed using Equation 8.13. Given the input parameters, a K-means algorithm (Equations 8.14 and 8.15) was applied to maximise the intra-cluster homogeneity and determine the cluster boundaries for the study area.

Figure 8.6 (a) Available groundwater and surface water sources and their groups; (b) water-source centres, S_i (Tsegaye *et al.*, 2020a)

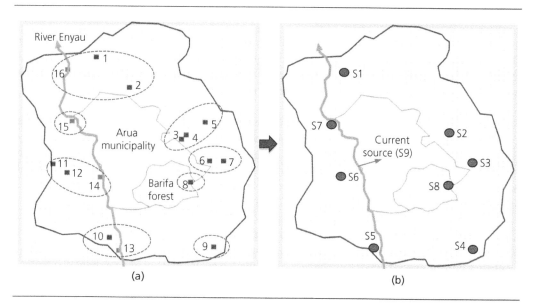

The parcel connectivity analysis was performed, refining the boundary and merging parcels of one cluster which are located in a different cluster. A group merging was performed if a cluster/group was too small. Groups with a size less than 20% of the maximum cluster size were distributed to the neighbouring cluster to avoid large variation in cluster size. However, a recommended cluster size and/or the smallest demand that a cluster should supply were used to decide whether to merge isolated parcels. Figure 8.7 (a) shows the final cluster boundaries after isolated neighbouring parcels were re-distributed, and (b) shows the final cluster boundary with source centres for the case study area.

Figure 8.7. Maps showing cluster boundaries (a) after re-distributing small/disconnected parcels, and (b) with source centres (green circles) (Tsegaye *et al.*, 2020a)

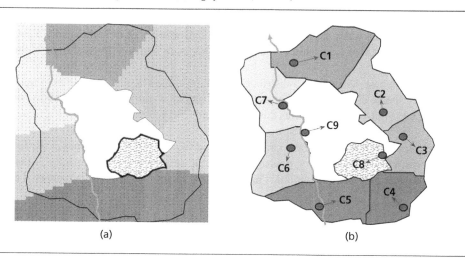

8.5. Planning of WDSs under uncertainty

Traditional planning of WDS development and management has been based on deterministic assumptions when future conditions are not known with certainty.

Projections of future global change pressures are plagued with uncertainties and one of the major challenges that water service providers face is designing WDSs with incomplete information concerning the future global change pressures. Based on the possibility of returns above or below expected, uncertainty has a good and a bad side (Bernanke, 1983).

The level of investment for these systems is often very high (measured in billions of dollars) while decision-making has large, long lasting consequences and is fraught with risk and uncertainty (Savic, 2005). Planners have been dealing with uncertainty and risk associated with global change pressures for the past few decades and contribute a significant shift away from the traditional design principles towards design of WDS under uncertainty (Babayan *et al.*, 2005; Babayan *et al.*, 2007; Giustolisi *et al.*, 2009.; Vairavamoorthy, 2008; Tsegaye *et al.*, 2020b).

8.5.1 Global change pressure affecting the future design of WDSs

WDSs are facing a range of dynamic global change pressures such as climate changes and variability, population growth and urbanisation, changes in public behaviors and socio-economic conditions, ageing and deterioration of buried infrastructure, technological development, governance and privatisation and so on. The major pressures include climate change, population growth and urbanisation, and ageing and deteriorating of water distribution infrastructure (Khatri and Vairavamoorthy, 2007).

In order to develop sustainable urban water systems, one must recognise the global change pressures. Hence there is a need for us to pay attention to these changes in the context of how these systems will be designed and operated in an ever-changing environment. However, investment decisions are still one of the major challenges for WDSs which are performing in an inevitable changing environment.

8.5.1.1 Climate change

Climate change will affect the availability of water by interrupting the water cycle process. Although the regional distribution is uncertain, precipitation is expected to increase in higher latitudes, particularly in winter. Potential evapotranspiration (ET) rises with air temperature. Consequently, even in areas with increased precipitation, higher ET rates may lead to reduced run-off, implying a possible reduction in renewable water supplies. The frequency and severity of droughts could also increase in some areas as a result of a decrease in the total rainfall, more frequent dry spells and higher ET.

The periods of highest demand on a distribution system in the UK and many other countries are during hot, dry spells, due to demand for uses such as garden watering. Figure 8.8 shows an example from the UK of how demand varies during the year.

With hotter, drier summers the peaks can be expected to increase. This increases the likelihood that customers will experience low pressure and where service reservoir capacity is inadequate the supply of water may completely fail. This is likely to require investment in increased capacity.

8.5.1.2 Population growth and urbanisation

Population growth and urbanisation is one of the world's major challenges in developing water distribution infrastructure. The numbers and size of the cities in the world are increasing due to

Figure 8.8 Demand variation during the year (Data from Seven Trent Water (2009))

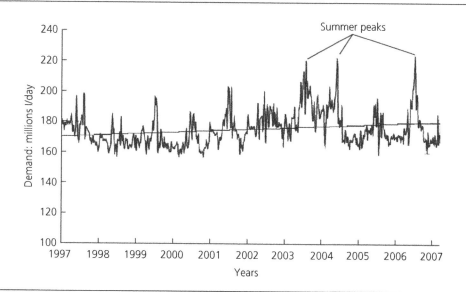

the higher rate of urbanisation. In 1950, New York City and Tokyo were the only two cities with a population of over ten million inhabitants. By 2015, it is expected that there will be 23 cities with a population over ten million, 19 of which will be in developing countries. In 2000, there were 22 cities with a population of between five and ten million. Cities in developing countries are already facing enormous backlogs in infrastructure and services, and confronted with insufficient water supply, deteriorating sanitation and environmental pollution.

Population growth and rapid urbanisation will create a greater demand for water while simultaneously decreasing the ability of ecosystems to provide more regular and cleaner supplies. Moreover, a rapid increase in built-up areas disturbs the local hydrological cycle and environment by reducing the natural infiltration opportunity. In addition, uncertainty in prediction of population growth and urbanisation may cause variation in water demand which will affect system operation and create many severe problems – for example, high risk will be associated with a network which conveys insufficient water to meet people's living needs and industrial consumption. Under these uncertainties related to population growth and urbanisation, we need to think about increasing system ability to deal with unpredicted global drivers.

8.5.1.3   Aging and deterioration of infrastructure systems
In some cities worldwide, the concept of asset management in water distributions has been neglected for many years. A significant proportion of WDSs are old and malfunctioning due to deterioration. This is also associated with the lack of records and data about the location and condition of the infrastructure, and lack of efficient decision support tools and management of the infrastructure (Misiunas, 2005). Deterioration of WDSs will result in high rates of water losses and higher chances of infiltration and exfiltration of water. This will create a higher risk of drinking water contamination and outbreak of water-borne diseases (Vairavamoorthy *et al.*, 2008).

The escalating deterioration of WDSs also threatens our ability to provide safe and sufficient drinking water services for current and future generations. Deterioration is an exogenous factor which is coupled with many uncertainties. Therefore, this highlights the need for effective decisions against these drivers and the uncertainty associated with them.

8.5.2   Design of WDSs under uncertainty

The design of WDSs is confronted with a dilemma. On the one hand, it is foreseeable that uncertain future drivers will change the basic conditions of a WDS itself during their long operational life span. A WDS is affected by several uncertain future global drivers discussed above. As a consequence, during its life span, fundamental alternations of the basic conditions are expected. On the other hand, long-lasting decisions for the planning of urban water systems have to be made, even if it is expected that the bases for these decisions will change. Hence it is necessary to design a WDS that can adapt to the expected future global changes. Moreover, the possible future developments caused by global change are associated with several uncertainties.

Additionally, there are uncertainties in the input data of the present predicting models and additional uncertainties occur during the downscaling of the models on local level – for example, after construction had begun of a large project to expand water supplies in the Skane region in Sweden, water consumption unexpectedly stopped increasing. There has been nearly no growth in consumption for ten years in the region and has multiple causes: a decreased birth-rate, environmental regulations to reduce industrial water use, higher prices for water and restrictive government development policies. This history of water development for the Skane region highlights two major factors associated with the failure of the project: prediction of future uncertainty drives and long-term planning. This illustrates the hazard of conventional planning based on deterministic growth forecast when a long period is considered (Erlenkotter *et al.*, 1989). In most cases, such an uncertain shift in future water consumption may lead to an unnecessary investment and/or cause system performance problems. Therefore, the design of WDSs that operate in an uncertain world requires decisions to be made against a background of various sources of uncertainty.

The recognition of future uncertainty in both design requirements and the operating environment is the most important issue in WDS planning and management and is a significant shift away from traditional practice (Hassan and de Neufville, 2006). Babayan *et al.* (2005) considered the uncertainty associated with water demand when predicting the behaviour of a system. The research focused on the design of a WDS with minimum cost, while meeting the pressure requirements in terms of given a robustness level under uncertain demand. A stochastic WDS design methodology is used to obtain robust and economic solutions for water distribution network design, where the robustness of the network is defined as its ability to provide an adequate supply to customers despite fluctuations in some or all of the design parameters.

Babayan *et al.* (2007) developed a multi-objective optimisation approach to formulate the problem associated with a stochastic (i.e. robust) WDS design under uncertain (future water consumption and pipe roughness) variables. The problem formulation is based on two parameters – for example, minimisation of cost of the network design/rehabilitation and probability of network failure. The most uncertain parameters, future water consumption and pipe roughness, are considered as an independent variable with pre-specified probability density function. The problem is solved using GAs after converting it to an equivalent, simplified deterministic optimisation problem. The methodology was tested and compared using the well-known problem in the literature of New York tunnels reinforcement and showed that neglecting uncertainty in the design process may lead to serious under-design of water distribution networks.

Giustolisi *et al.* (2009) proposed a procedure for robust design through a multi-objective (minimisation of design cost, maximisation of WDS robustness) approach, considering nodal demands and pipe roughness as uncertain variables. The research follows a two-step design procedure for computational efficiency: a deterministic design (i.e. constrained least-cost design procedure) as the first step and then the deterministic solutions, as an initial population, to solve the robust design problem multi-objectively, implementing the minimisation of design costs and the maximisation of WDS robustness as objective functions. This research is a significant achievement in the design of WDSs under uncertainties.

Both infrastructures (i.e. WDSs) and living organisms struggle to survive in an ever-changing environment. It could be beneficial for WDS designers to replicate how living things function and interact with their environment. According to Darwin's theory of evolution, individuals or species that are better equipped to adapt to changing environments tend to live longer. Similarly, systems that are better equipped to adapt to changing environments will outlast more-rigid organisms or systems (Saleh *et al.*, 2003). Thus, if a system is to be designed for an extended design life and value delivery, the ability to cope with uncertainty and change has to be embedded in the system. In general, the design of and investment decision for WDSs should consider the inevitable future alterations and look for ways of adding value from them. One fundamental way of doing this is by designing for flexibility.

8.6. Introduction to flexible designs for WDSs

One of the challenges that water service providers face is to design WDSs with incomplete information concerning future global change pressures. Relevant strategies are needed for coping with both negative (worse than expected) as well as positive (better than expected) outcomes associated with uncertainty (Dean, 1951). Three basic strategies have been identified by de Neufville (2004), including reducing the uncertainty in the system, increasing system robustness and increasing system flexibility

Flexibility is proposed as a key property to be embedded (Schulz *et al.*, 2000) in high-value assets, particularly if they are to be designed for an increased or longer design life. It has also been cited by many scholars as a key goal for dealing with uncertainty in the design of complex systems (de Neufville, 2004; Saleh *et al.*, 2001). As postulated by Silver and de Weck (2007) and Zhao and Tseng (2003), increasing the flexibility of a system provides a potential solution to deal with uncertainties acting on systems which are required to adapt/evolve to future stages. Scholtes (2007) also recognises flexibility as way to transform a risk associated with uncertainty in to an opportunity.

Uncertainty is not always a negative to be mitigated; it can also be a positive to be exploited (de Neufville, 2004). Based on the possibility of returns below or above expected, uncertainty has a good and a bad side (Bernanke, 1983). Flexibility offers an opportunity to exploit the benefits of uncertainty and to enhance the ability to act or to respond to the future change requirements in a cost-effective manner. However, system flexibility has not yet received sufficient attention in the design of WDSs.

In recent years, flexibility has become a key concept in many fields such as manufacturing, software engineering, architecture, finance and so on. The theoretical background and definition of flexibility have been discussed by many researchers. However, very little effort has been made to define it formally and clearly for WDSs – for example, Allen *et al.* (2001) define flexibility as the ease of changing the requirements of the system with a relatively small increase in complexity (and rework). Saleh *et al.* (2001) define the flexibility of a design as 'the property of a system that allows it to respond to changes in its initial objectives and requirements – both in terms of capabilities and attributes – occurring after the system has been fielded – that is, is in operation, in a timely and cost-effective way'. Shah *et al.* (2008) characterise flexibility as 'the ability of a system to respond to potential internal or external changes affecting its value delivery, in a timely and cost-effective manner'. Fricke and Schulz (2005) also define flexibility as a 'system ability to be change easily. Changes from external have to be implemented to cope with changing environments'.

Lack of a clear definition of flexibility is the major problem in addressing its distinct features. Most of the confusion about the concept of flexibility comes from the subtle distinction between systems features. In general, three major problems in the existing flexibility theory can be identified.

(a) Incompatibility of the existing definition of flexibility between one system and another.
(b) The lack of description for a measure of flexibility or the ability to rank different designs according to their flexibility.
(c) The overlap of the concept of flexibility with other properties describing change (changeability, adaptability, robustness and so on).

Adaptability characterises the ability of a system to adapt itself towards changing environments (Fricke and Schulz, 2005). Adaptation is an internally initiated change, while flexibility is externally initiated (Shah *et al.*, 2008) and Robustness is the property of a system which allows it to satisfy a fixed set of requirements, despite changes occurring in the environment or within the system itself (Saleh *et al.*, 2001).

According to Upton (1994), constructing a definition of flexibility is not straightforward, since definitions are often coloured by a particular situation or problem. There is, therefore, a need to recast the existing definition of flexibility to suit urban water systems (i.e. WDSs). Based on its general definitions, flexibility can be defined for the field of urban water management. Flexibility is the ability of urban water management as follows: the ability of urban water systems to use their active capacity to act or to respond to relevant alterations in a performance-efficient, timely and cost-effective way. This definition covers most of the important characteristics of flexibility, such as to overcome alterations, the capacity for change and the characteristics of the change process, and the metrics of flexibility such as the costs of change, the duration of change and system performance.

8.6.1 Designing for flexibility

In order to design flexible WDSs that have the capability to cope with future alterations and to make the most of uncertainities, the following basic features should be considered.

- Drivers: flexibility to *what* and *when?*
- Option: *what* flexibility and *where* to embed?
- Level: *how* much flexibility?

These basic features can be addressed through four major stages: uncertainty modelling; option identification and/or system alternatives; flexibility generation and evaluation; and decision-making. Similar approaches have been followed in the design of flexible systems in different disciplines (Ramirez, 2002; Cardin and Neufville, 2008; Shah *et al.*, 2008) – for example, Shah *et al.* (2008) developed a three 'D' (dice, design and decision, and discounting) concept in response to the common problem of uncertainty facing designs. Dice represents the uncertain future within which the engineering solution will be delivering benefit. Designs and decisions represent designers' control over current design choices and, as the design allows, over choices in the future, in response to the resolution of uncertainty. Discounting is used to represent comparison between future benefits and costs associated with subsequent contingent decisions.

With respect to addressing the objectives of designing flexible WDSs, the three steps listed above can be followed. Uncertainty modelling addresses the issue with respect to *flexibility to what and when?* It identifies and quantifies the key sources of uncertainties in WDSs. The distribution of uncertain parameters should also be limited to the number of possible future states. Option identification addresses the issues of *What flexibility and where to embed?* in the system. It attempts to identify the 'best' sets of options in a WDS that most likely offer better lifetime flexibility in the uncertain environment. The flexibility generation and evaluation stages analyse different system alternatives for the described future states. Flexibility generation and decision-making stages should attempt to address the issue related with *How much flexibility?* Uncertainty description, option identification and flexibility generation in designing of flexible WDSs are discussed briefly below.

8.6.1.1   Uncertainty description and modelling

Future conditions will certainly differ from past sequences. A statistical analysis of recorded sequences and the stochastic generation of various possible future sequences are done to account for this. Since the statistical characteristics are themselves uncertain, there is no assurance that generated sequences are representative of the range of sequences that might occur in the future (Beard, 1982). However, the description and characterisation of these unknown conditions is the most important factor in the design of flexible WDS. The capacity of uncertainty to be resolved in future is usually understood as the characteristic that allows it to generate value (Ramirez, 2002). Uncertainty is identified as a key element of flexibility. It creates both risks and opportunities in a system and it is with the existence of uncertainty that flexibility becomes valuable (Nilchiani, 2005).

Uncertainties can be modelled by a number of different methods. These methods are classified into analytical and sampling approaches. The analytical approaches include first order second moment, second order second moment approaches and so on, and the sampling methods include Monte Carlo simulation. The sampling type methods are much more computationally demanding. The choice of the particular method depends on the information available and none of the methods give a precise result (Nilchiani, 2005).

8.6.1.2   Option identification

In finance, an option is defined as the 'right, but not the obligation' to perform an action. The action could be expanding, contracting, deferring or abandoning. The key feature of an option is that cost of exercising the option of using our right to perform an action. It is in this respect that an option has value (de Neufville, 2001). Real options are options that relate to physical assets rather than to financial instruments. Real options can be categorised as those that are either 'in' or 'on' projects. In exploring flexibility in engineering systems, it is also identified as flexibility 'in' and 'on' a system. Flexibility 'in' a system is a technical aspect of the design that enables the system to adapt to its environment, while 'on' a system relates to management decisions without altering technical components (de Neufville, 2002) – for example, the flexibility to defer WDS expansion for a specific phase is non-technical and is flexibility 'on' a system. Most of the sources for flexibility 'on' a WDS are well known. These sources include investment deferral, multistage deployment, expansion and so on.

The design of flexible systems which have the ability to contend with an ever-changing environment often needs an identification of flexible options in the system. The identification of potential flexible options has been discussed by several researchers. Cardin and Neufville (2008) define flexible design options as a physical component enabling flexibility 'in' a system. Several techniques have been used to identify the flexible options in a system, such as change propagation analysis (Eckert *et al.*, 2004), the sensitivity design structure matrix (Kalligeros, 2006), the interview method (Shah *et al.*, 2008) and so on. However, the appropriateness of the methods depends on the type of system and the source of uncertainty and needs to be explored.

8.6.1.3 Generation and valuation of flexibility

Generating flexibility in a system design is an investment problem where a premium has to be paid so that an option that can be exercised later. The investment decision depends on the trade-off between the cost of capturing the options and the expected benefit that may arise from future uncertainties. The estimation of the value of flexibility has three major elements (de Neufville, 2002):

- estimation of the loss associated with the system without flexibility
- calculation of the value of the flexible options
- identification of the strategies for exploiting the options, to permit the best use of the flexibility built into the system.

The generation and valuation of flexibility should consider the changes that result in change in value delivery. Changes and responses to changes can be measured in the form of costs and monetary or non-monetary benefits. The criterions for generating and measuring of flexibility can be deduced from the definition of flexibility (Eckart *et al.*, 2010).

- *Capability for change*: The capability for change indicates for which uncertain future states a change of the system is possible.
- *Performance of system*: Flexibility should guarantee that future alterations have minor impacts on the system performance.
- *Costs of change*: Due to the long operational life span of WDSs, the costs for several changes in the whole life span of systems should be considered. The life cycle costs should consider the cost of planning, development and management of a system, the cost of embedding flexibility 'in' or 'on' a system and damage costs associated with missing/delayed action on altering conditions.
- *Duration of change*: The duration of change is the period which is required to adapt the system to new requirements. This matrix may not be a significant factor as the time is usually sufficient to react to alterations in a WDS.

Different techniques can be used to value flexibility and identify a strategy that effectively hedges risks and ensures optimality in terms of the criteria listed above, such as discounted cash flow (DCF), decision analysis (DA) and real options analysis (ROA). DCF-based approaches (such as NPV, internal rate of return and the payback period) require the analyst to estimate the net cash flows during the design life. These methods presume one line of action from time zero and evaluate the project when fixed. These techniques do not consider the dynamics of the project in the valuation (Arboleda and Abraham, 2006). ROA and DA are better suited to deal with uncertainty. Both of these techniques view projects as processes that management can continually modify in light of changes in the environment. They promote a change from a deterministic type of valuation to a dynamic planning process, which encourages flexible designs that effectively deal with the uncertainties in the environment (Ramirez, 2002). According to Nilchiani and Hastings (2007), the choice of the valuation method depends on how well the methodology can capture a specific type of uncertainty in a system.

The valuation of flexibility should also be coupled with a decision-making process. Flexibility-based decision-making should look at the 'best' set of alternatives (systems) that deliver good life-time benefits. This is achieved through comparing with a non-flexible system.

8.7. Preparing for outbreaks: implications on resilient WDSs

Recent outbreaks have disrupted the operation of our critical infrastructures, our global economy and public health in numerous ways, impacting efficiency and prosperity across all sectors (Neal, 2020; Gude and Muire, 2021; Renukappa et al., 2021). The purpose of this section is, therefore, to discuss the impact of uncertainties associated with outbreaks on WDSs and water utilities, and to recommend further studies for innovative processes and strategies for design and operation during outbreaks. The section critically evaluates the issues underlining the impacts of pandemics on service and operation and highlights the need for preparedness. An alternative approach to build resiliency in the wake of Pandemics such as the COVID-19 event is discussed.

8.7.1 Water and the pandemic

A major component of all urban water systems is distribution, which constitutes approximately 80–85% of the total cost of a water-supply system. Traditionally, WDSs are designed using the 'worst case scenario,' or 'robustness' to improve system reliability. Such deterministic assumptions are historically inaccurate and require a new design approach that recognises uncertainties and offers more adaptability. Due to inherent variability in water consumption levels and values of demands at nodes, uncertainty analysis for resilient WDSs design has received much attention in the past. Although interest in applying resiliency to cope with uncertainty has been growing in the past few decades, the COVID-19 pandemic has further spurred its framing for water supply systems.

Access to adequate drinking water is necessary, especially during outbreaks. Most of our existing water systems are not sufficiently resilient to deal with such pandemic that create bulk shift in quantity, quality, timing and distribution of demand – for example, the recent COVID-19 pandemic highlights the need for revisiting the way how we design and operate our water systems. As nations grappled with the immediate health and economic consequences of the pandemic, the operation of water utilities faced notable but manageable challenges. From sudden shifts in water demand patterns and disruptions in supply chains to the urgent need for enhanced safety protocols, the pandemic tested the resilience and adaptability of water distribution systems. In addition, it has exacerbated pre-existing challenges such as the rising costs of treating emerging contaminants, intermittency of water supply and financial insecurity. While most water utilities have kept the water running to their customer, the revenue losses have caused delays or cancellations to capital projects and routine maintenance that may impact their ability to supply safe, affordable water into the future (Cooley et al., 2020).

Many other industrial sectors depend upon water supplies as well as wastewater treatment services for their daily operations. It is almost impossible to carry out routine operations in some sectors such as chemical and hazardous materials production, hospitals, manufacturing and some commercial facilities without these essential services (USDHS, 2014; NIAC, 2016). Figure 8.9 shows the dependence of various industry sectors on water supplies – for example, hospitals provide critical services and are more dependent on water supplies.

Figure 8.9 Potential functional degradation of the critical industry due to lack of water supplies (Gude and Muire, 2021)

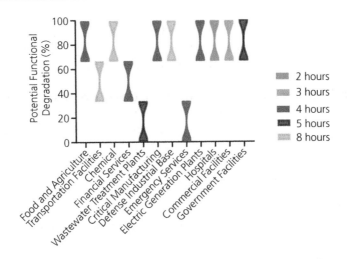

8.7.2 Impacts of the stay-at-home orders and lockdowns on water consumption patterns

The impact of shelter-in-place, business shutdowns and stay-at-home orders shifted demand pattern and consumption in many cities around the world. Recent studies (Cooley *et al.*, 2020; Makino *et al.*, 2021) show that the stay-at-home orders and business shutdowns increased residential demand and decreased non-residential (i.e. commercial, industrial and institutional) demand. The net effect of these changes varies from community to community, depending on the relative proportion of residential and non-residential water uses and the major economic sectors in the community.

Changes in consumption and diurnal water demand pattern have affected utility water operational conditions and water quality. Knowledge of accurate demand variations over a 24-hour period is essential for effective management of water systems. The hourly demand variation is represented by diurnal demand curves. The diurnal demand curves are employed in determining adequate sources of supply, pumping facilities, reservoirs and the transmission/distribution facilities. The shelter-in-place and stay-at-home orders affected the typical diurnal water demand pattern that was based on large water use in the morning and in the evening. Despite the change in the total consumption, users behave differently during stay-at-home orders (Makino *et al.*, 2021). These disrupt the typical diurnal demand pattern and result in an altered peak demand and/or shifted peak hours.

Some of these impacts are short-lived, generally limited to the period when stay-at-home orders are in place and businesses closed. However, there could be longer-term impacts if, for example, unemployment remains high, people continue to work from home, or there are deeper changes to the economy (Cooley *et al.*, 2020). A thorough study is crucial to fully understand the scope and duration of these impacts on water distribution systems. Such research will aid in developing smart water distribution systems that can adapt to changing demand patterns and disruptions.

Improving these systems will enhance the operational resilience of utilities, enabling them to better manage variability and uncertainty while maintaining reliable and efficient water service.

Makino *et al.* (2021) conducted a study on the impact of COVID-19 on water service provides in Latin America. Most of the utilities in this study reported increases in residential water consumption when compared to the consumption levels reported in 2019. However, non-residential water consumption decreased significantly during the strict lockdowns imposed in various South American countries, because businesses such as offices, retail and other commercial establishments were forced to close for extended periods of time or allowed staff to work from their homes.

These changes in residential and non-residential water consumption generally offset each other, which resulted in only slight increases in total water consumption. Water service providers experienced financial difficulties, primarily due to changes in consumption patterns that led to lower average tariffs. As consumption patterns shifted, residential consumption increased while non-residential consumption decreased. These changes in residential and non-residential water consumption affected the utilities' average water tariffs because utilities usually charge higher tariffs on non-residential customers. As a result, reductions in non-residential water consumption translated into lower average tariffs. When compared to pre-pandemic levels, residential consumption was up by 3.1 per cent on average from March to May 2020. Non-residential consumption was down by 14.2 per cent, while some service providers registered decreases of up to 73.8 per cent. Total water billed only saw small changes, increasing by 1.8 per cent, on average (Makino *et al.*, 2021).

8.7.3 Impact of stay-at-home orders and lockdowns on water quality

The increase of water stagnation time (water age) in distribution system causes the formation of disinfection by products, intensification of corrosion, nitrification, re-growth of microorganisms, biofilm formation, raising the temperature and colour creation, and changes in taste and odour of water. These changes could pose serious threats for the consumers' health (WHO, 2014).

As society resume normal activities following stay-at-home orders and lockdowns, it is important to recognize that the coronavirus may not be the only health concern people face upon returning to the office. The pandemic has raised some new water-quality concerns in need of special attention. One of these relevant to municipal water systems is the potential for health risks associated with stagnant water inside building plumbing (Gleick, 2020). Studies have clearly shown that stopping water in indoor building pipes has a significant effect on the quality of drinking water (Lautenschlager *et al.*, 2010; Wang *et al.*, 2012; Pepper *et al.*, 2014).

The full extent and duration of pandemic related water quality impacts are still not fully understood. As buildings begin to reopen, it is important to take prompt action to ensure the safety of the water supply and to improve the operational resilience of utilities. Thoughtful management of these issues will help mitigate risks and support the continued health and well-being of communities.

8.7.4 Considerations for Enhancing WDS Resilience for Pandemics and Beyond

Even though the extent or duration of the pandemic, range of impacts on different water utilities, or effectiveness of efforts to mitigate these impacts is not clear yet, there are several key aspects that water utilities could possibly consider (Cooley, 2020). Figure 8.10 outlines considerations that can be made before, during, and after a pandemic to build resilience against its impacts. By examining these aspects, utilities might strengthen their ability to handle future disruptions and ensure continued service reliability.

Figure 8.10 Key considerations for improving water utility resilience in response to pandemics

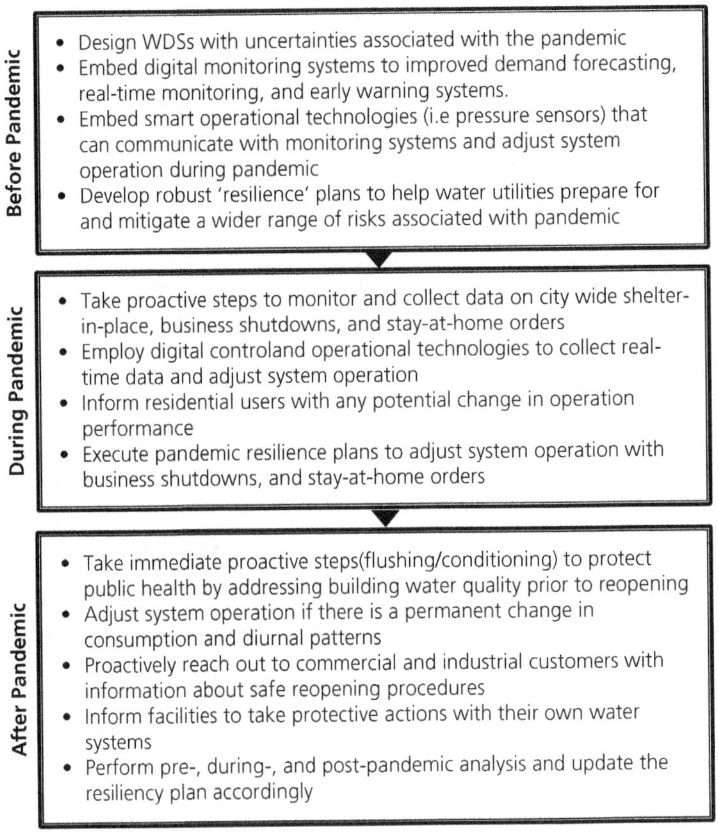

Before Pandemic
- Design WDSs with uncertainties associated with the pandemic
- Embed digital monitoring systems to improved demand forecasting, real-time monitoring, and early warning systems.
- Embed smart operational technologies (i.e pressure sensors) that can communicate with monitoring systems and adjust system operation during pandemic
- Develop robust 'resilience' plans to help water utilities prepare for and mitigate a wider range of risks associated with pandemic

During Pandemic
- Take proactive steps to monitor and collect data on city wide shelter-in-place, business shutdowns, and stay-at-home orders
- Employ digital controland operational technologies to collect real-time data and adjust system operation
- Inform residential users with any potential change in operation performance
- Execute pandemic resilience plans to adjust system operation with business shutdowns, and stay-at-home orders

After Pandemic
- Take immediate proactive steps(flushing/conditioning) to protect public health by addressing building water quality prior to reopening
- Adjust system operation if there is a permanent change in consumption and diurnal patterns
- Proactively reach out to commercial and industrial customers with information about safe reopening procedures
- Inform facilities to take protective actions with their own water systems
- Perform pre-, during-, and post-pandemic analysis and update the resiliency plan accordingly

REFERENCES

Al-Saleh MF and Yousif AE (2009) Properties of the standard deviation that are rarely mentioned in classrooms. *Austrian Journal of Statistics* **38(3)**: 193–202.

Allen T, Moses J, Hastings D *et al.* (2001) *ESD Terms and Definitions*. Massachusetts Institute of Technology, Engineering Systems Division, MA, USA.

Arboleda CA and Abraham DM (2006) Evaluation of flexibility in capital investments of infrastructure systems. *Engineering Construction and Architectural Management* **13(3)**: 254.

Babayan A, Kapelan Z, Savic D and Walters G (2005) Least-cost design of water distribution networks under demand uncertainty. *Journal of Water Resources Planning and Management* **131(5)**: 375–382.

Babayan A, Savic D and Walters G (2007) Multiobjective optimization of water distribution system design under uncertain demand and pipe roughness. *Topics on system analysis and integrated water resource management* **161**: 1–15.

Beard LR (1982) Flexibility – a key to the management of risk and uncertainty in water supply. *Optimal Allocation of Water Resources*.

Bernanke BS (1983) Irreversibility, uncertainty, and cyclical investment. *The Quarterly Journal of Economics* **98(1)**: 85–106.

Bieker S, Cornel P and Wagner M (2010) Semicentralised supply and treatment systems: integrated infrastructure solutions for fast growing urban areas. *Water Science and Technology* **61(11)**: 2905–2913.

Bivins AW, Sumner T, Kumpel E *et al.* (2017) Estimating infection risks and global burden or diarrheal disease attributable to intermittent water supply QMRA. *Environmental Science and Technology* **51(3)**: 7542–7551.

Böhm HR, Schramm S, Bieker S *et al.* (2011) The semicentralized approach to integrated water supply and treatment of solid waste and wastewater – a flexible infrastructure strategy for rapidly growing urban regions: the case of Hanoi/Vietnam. *Clean Technologies and Environmental Policy* **13(4)**: 617–623.

Cardin MA and de Neufville R (2008) *A Survey of State-of-the-Art Methodologies and a Framework for Identifying and Valuing Flexible Design Opportunities in Engineering Systems.* Massachusetts Institute of Technology, Cambridge, MA, USA.

Cheung PB, Reis LFR and Carrijo IB (2003) Multi-objective optimization to the rehabilitation of a water distribution network. In *Advances in Water Supply Management* (Maksimovic C, Butler D, and Memon FA (eds.)). Balkema, Rotterdam, pp. 315–325.

Coelho ST, James S, Sunna N, Abu Jaish A and Chatila J (2003) Controlling water quality in intermittent supply systems. *Water Science and Technology, Water Supply* **3(4)**: 119–125.

Coelho ST and Alegre H (1998) *Performance indicators for water supply and wastewater drainage systems.* ICTH report, LNEC, Lisbon, Portugal. (In Portuguese.)

Cohon JL (1978) *Multiobjective Programming and Planning.* Academic Press, New York, USA.

Cooley H, Gleick PH, Abraham S and Cai W (2020) *Water and the COVID-19 Pandemic: Impacts on Municipal Water Demand.* Pacific Institute, Oakland, CA.

Cornel P, Meda A and Bieker S (2011) Wastewater as a source of energy, nutrients and service water. *Treatise on Water Science* **3(92)**.

de Neufville R (2001) Real options: dealing with uncertainty in systems planning and design. *5th International Conference on Technology Policy and Innovation.* Delft University of Technology, Delft, the Netherlands.

de Neufville R (2002) *Architecting/Designing Engineering Systems Using Real Options.* Engineering Systems Division Internal Symposium, Massachusetts Institute of Technology. Cambridge, MA.

de Neufville R (2004) Uncertainty management for engineering systems planning and design. *Engineering Systems Monograph.* Cambridge, MA, USA.

Dean J (1951) *Capital Budgeting-Top-Management Policy on Plant, Equipment, and Product Development.* Columbia University Press, NY, USA.

Deb AK, Hasit YJ and Grablutz FM (1995) *Distribution System Performance Evaluation.* AWWA (American Water Works Association) Research Foundation, AWWA, Denver, USA.

Dopp K (2011) *Legislative Redistricting-Compactness and Population Density Fairness.* Available at SSRN 1945879.

Effah EA, Memon FA and Bicik J (2013) Improving equity in intermittent water supply systems. *Journal of water supply: Research and technology – AQUA* **62(8)**: 552–562.

Eckart J, Sieker H and Vairavamoorthy K (2010) Flexible urban drainage systems. *Water Practice and Technology* **5(4)**.

Eckert, C., Clarkson, P. J., and Zanker, W. (2004) Change and customisation in complex engineering domains. *Research in Engineering Design* **15(1)**: 1–21.

Erlenkotter D, Sethi S and Okada N (1989) Planning for surprise: Water resources development under demand and supply uncertainty I. The general model. *Management science* **35(2)**: 149–163.

Fricke E and Schulz AP (2005) Design for changeability (DfC): Principles to enable changes in systems throughout their entire lifecycle. *Systems Engineering* **8(4)**: p.342.

Giustolisi O, Laucelli D and Colombo AF (2009) Deterministic versus stochastic design of water distribution networks. *Journal of Water Resources Planning and Management* **135(2)**: 117–127

Gleick PH (2020) *Water and the Pandemic: Reopening Buildings After Shutdowns: Reducing Water-Related Health Risks*. Pacific Institute, Oakland, CA, USA. https://pacinst.org/publication/reopening-buildings-after-shutdowns/

Gude VG and Muire PJ (2021) Preparing for outbreaks – implications for resilient water utility operations and services. *Sustainable Cities and Society* **64**: 102558.

Halhal D, Walters GA, Ouazar D and Savic D (1997) Water network rehabilitation with structured messy genetic algorithm. *Journal of Water Resources Planning and Management* **123(3)**: 137–146.

Hassan R and de Neufville R (2006) *Design of Engineering Systems under Uncertainty via Real Options and Heuristic Optimization*. Massachusetts Institute of Technology, unpublished paper.

Herrera M, Canu S, Karatzoglou A, Pérez-García R and Izquierdo J (2010) *An approach to water supply clusters by semi-supervised learning*. 5th International Congress on Environmental Modelling and Software – Ottawa, Ontario, Canada July 2010.

Ilaya-Ayza AE, Benitez J, Izquierdo J and Perez-Garcia R (2017) Multi-criteria optimization of supply schedules in intermittent water supply systems. *Journal of computational applied mathematics* **309**: 695–703

Kalligeros K (2006) *Platforms and Real Options in Large-Scale Engineering Systems*. Massachusetts Institute of Technology, Cambridge, MA, USA.

Kapelan ZS, Savic D and Walters GA (2005) Multiobjective design of water distribution systems under uncertainty. *Water Resources Research* **41**: W11407–W11421

Khatri KB and Vairavamoorthy K (2007) Challenges for urban water supply and sanitation in the developing countries. In *Symposium on Water for a Changing World – Enhancing Local Knowledge and Capacity*, Delft.

Lautenschlager K, Boon N, Wang Y, Egli T and Hammes F (2010) Overnight stagnation of drinking water in household taps induces microbial growth and changes in community composition. *Water research* **44(17)**: 4868–4877

Makino M, Serrano H, Flores B and Hurtado J (2021) *Building Financial Resilience. Lessons Learned from the Early Impact of COVID-19 on Water and Sanitation Service Providers in Latin America*. Report of the Water Global, Practice, World Bank

Mays LW (2000) *Water Distribution Systems Handbook*. McGraw-Hill, NY, USA

Misiunas D (2005) *Failure monitoring and asset condition assessment in water supply systems*. Lund University, Department of Industrial Electrical Engineering and Automation, Lund, Sweden.

Morley MS, Atkinson RM, Savić DA and Walters GA (2001) GAnet: genetic algorithm platform for pipe network optimisation. *Advances in Engineering Software* **32(6)**: 467–475.

Nachbar J (2009) Basic properties of Euclidean Norm. *Economics* **511**.

National Infrastructure Advisory Council (NIAC) (2016) Water sector resilience final report and recommendations. https://www.cisa.gov/sites/default/files/publications/niac-water-resilience-final-report-508.pdf [Access date: Apr 15].

Neal MJ (2020) COVID-19 and water resources management: Reframing our priorities as a water sector. *Water International* **45(5)**: 435–440.

Nilchiani R (2005) *Measuring Space Systems Flexibility: A Comprehensive Six-element Framework*. Massachusetts Institute of Technology, Cambridge, MA, USA.

Nilchiani R and Hastings DE (2007) Measuring the value of flexibility in space systems: A six-element framework. *Systems Engineering* **10(1)**: 26–44.

Ofwat (2009) *Service and delivery-performance of the water companies in England and Wales 2008-09 report Supporting information*. Ofwat, London, UK.

Pepper IL, Rusin P, Quintanar *et al.* (2004). Tracking the concentration of heterotrophic plate count bacteria from the source to the consumer's tap. *International Journal of Food Microbiology* **92(3)**: 289–295.

Ramirez N (2002) *Valuing Flexibility in Infrastructure Developments: The Bogota Water Supply Expansion Plan*. Massachusetts Institute of Technology, Cambridge, MA, USA.

Renukappa S, Kamunda A and Suresh S (2021) Impact of COVID-19 on water sector projects and practices. *Utilities Policy* **70**: 101194.

Saleh JH, Hastings DE and Newman DJ (2001) Extracting the essence of flexibility in system design. MIT, Cambridge, MA, pp. 59–72.

Saleh JH, Hastings DE and Newman DJ (2003) Flexibility in system design and implications for aerospace systems. *Acta Astronautica* **53(12)**: 927–944.

Savic D (2002) Single-objective vs. Multiobjective optimisation for integrated decision support. In *Integrated Assessment and Decision Support*, Rizzoli AE and Jakeman AJ (eds.), *Proceedings of the First Biennial Meeting of International Environmental Modelling and software Society*, Lugano, Switzerland, pp. 7–12.

Savic D, Walters G and Schwab M (1997) Multiobjective genetic algorithms for pump scheduling in water supply. In *Evolutionary Computing* (Corne D, Shapiro JL (eds.)). AISB EC 1997. Lecture Notes in Computer Science, vol 1305. Springer, Berlin, Heidelberg, pp. 227–235.

Savic D (2005) *Coping With Risk And Uncertainty In Urban Water Infrastructure Rehabilitation Planning*. School of Engineering, Computer Science and Mathematics, University of Exeter, Exeter, UK.

Scholtes S (2007) *Flexibility: The Secret to Transforming Risks into Opportunities*. [Online] http://www.eng.cam.ac.uk/~ss248/publications/BusinessDigest.pdf [Access date: Nov 12]

Schulz AP, Fricke E and Igenbergs E (2000) Enabling changes in systems throughout the entire life-cycle – key to success? *10th annual INCOSE conference*. Minneapolis, MN, USA.

Shah NB, Viscito L, Wilds J, Ross AM and Hastings D (2008) Quantifying flexibility for architecting changeable systems. MIT System Design and Management Thesis Seminar, Massachusetts Institute of Technology, Cambridge, MA, USA.

Silver MR and de Weck OL (2007) Time-expanded decision networks: A framework for designing evolvable complex systems. *System Engineering-New York* **10(2)**: 167.

Taylor J, Slocum AH and Whittle AJ (2018). Demand satisfaction as a framework for understanding intermittent water supply systems. *Water Resource Resarch* **55(7)**: 5217–5237.

Tsegaye S, Missimer TM, Kim JY and Hock J (2020a) A clustered, decentralized approach to urban water management. *Water* **12(1)**: 185.

Tsegaye S, Gallagher KC and Missimer TM (2020b) Coping with future change: Optimal design of flexible water distribution systems. *Sustainable Cities and Society* **61**: 102306.

Totsuka N, Trifunovic N and Vairavamoorthy K (2004) Intermittent urban water supply under water starving situations. *Proceedings of 30th WEDC international conference on people centered approaches to water and environmental sanitation*. WEDC, Loughborough University, Loughborough, UK, pp. 505–512.

Uganda Bureau of Statistics (UBOS). *Mid-Year Projected Population for Town Councils*. UBOS, Kampala, Uganda, p.2011.

Upton DM (1994) The management of manufacturing flexibility. *California Management Review* **36**: 72.

USDHS (2014) *Sector resilience report: Water and wastewater systems*. National Protection and Programs Directorate. Infrastructure Sector Assessment, Report of Office of Cyber and Infrastructure Analysis (OCIA), USA.

Vairavamoorthy K (2008) Managing water in the city of the future. In *Water and sanitation for all* (pp. 43–54). Water Management Academic Press.

Vairavamoorthy K and Elango K (2002) Guidelines for the design and control of intermittent water distribution systems. *Waterlines* **21(1)**: 19–21.

Vairavamoorthy K, Gorantiwar SD and Mohan S (2008) Intermittent water supply under water scarcity situations. *Water International* **32(1)**: 121–132.

Vairavamoorthy K, Akinpelu E, Ali M *et al.* (2001) *Guidelines for Design of intermittent water distribution systems*. South Bank University London, UK.

Vairavamoorthy K (1994) *Water distribution networks: Design and control for intermittent supply*. PhD thesis Imperial College of Science Technology and Medicine, London, UK.

Wang H, Edwards M, Falkinham III JO and Pruden A (2012) Molecular survey of the occurrence of Legionella spp., Mycobacterium spp., Pseudomonas aeruginosa, and amoeba hosts in two chloraminated drinking water distribution systems. *Applied and environmental microbiology* **78(17)**: 6285–6294.

Walski TM, Chase DV, Savic DA *et al.* (2003) *Advanced Water Distribution Modeling and Management*. Haestad Press, Waterbury, CT, USA.

Weber B, Cornel P and Wagner M (2007) Semi-centralized supply and treatment systems for (fast growing) urban areas. *Water Science and Technology* **55(1-2)**: 349–356.

World Health Organization (WHO) (2014) *Water safety in distribution systems*. WHO, Geneva, Switzerland.

Zecchin AC, Simpson AR, Maier HR and Nixon JB (2005) Parametric study for an ant algorithm applied to water distribution system optimization. *IEEE transactions on evolutionary computation* **9(2)**: 175–191.

Zhao T and Tseng CL (2003) Valuing flexibility in infrastructure expansion. *Journal of infrastructure systems* **9(89)**.

Dragan A. Savić and John K. Banyard
ISBN 978-1-83549-847-7
https://doi.org/10.1108/978-1-83549-846-020242011

Chapter 9
Operation, maintenance and performance

Joby Boxall
University of Sheffield, UK

Neil Dewis
WSP, UK

John Machell
University of Sheffield, UK

Ken Gedman
Stantec UK, Leeds, UK

Adrian Saul
Emeritus Professor, University of Sheffield, UK

Frank van der Kleij
Stantec UK, Bristol, UK

Adam Smith
Yorkshire Water, UK

Nathan Sunderland
Yorkshire Water, UK

9.1.   Introduction

Water distribution systems develop in piecemeal fashion around available sources to meet customer demands and public health requirements. While the infrastructure from inception by the Romans rarely remains, most UK systems readily date back to the Victorian era. Driven by different needs over the years, each system has been built from a variety of materials, using different design and construction techniques and a whole host of piecemeal layouts. In addition, little focus was given to the individual source water quality characteristics of the water they conveyed or of the treatment process, which was often poor. As a consequence, such poor quality water has caused the system materials to deteriorate at different rates by way of a number of physical chemical and biological mechanisms. This fragmentary development, system complexity and asset deterioration in the face of a changing consumer base and environmental conditions are the primary causes of many of the current operational, maintenance and performance challenges that face the water industry today.

Recent regulatory and customer-driven expectations, especially for continuity of supply, leakage and water quality, have challenged the industry to be more efficient and the introduction of reward and penalty mechanisms over the last ten years have aimed to incentivise the industry and drive innovation.

The challenge to maintain and develop the system to be more resilient in the face of climate change while maintaining an attractive investment for shareholders continues to be of primary concern for the industry. In a climate of escalating cost the industry is, and must continue, changing from a reactive to a proactive management style, and employ new technologies and methods to reduce water losses and optimise energy use, while always ensuring safe high quality water. The focus is now on developing long term strategies, underpinned by adaptive planning, to enable rapid response to external factors (climate change, population growth etc.) along a cost-efficient road-map making the best use of existing legacy infrastructure.

In 2019, every water company in England agreed a 'Public Interest Commitment' (PIC) that reflected the public purpose of the water company, setting five demanding long-term goals including a pledge to reach net zero on operational emissions by 2030. The water industry, through the PICs and subsequent development of route maps to deliver long-term change, is committed to take its part in tackling the threat of climate change and limiting the rise of global temperatures.

This chapter begins by describing the development of water distribution pipe networks to provide a clean supply of wholesome drinking water, to ensure public health, through to the more recent operational drivers associated with water quantity – for example, reduced leakage, minimal energy use and service quality (e.g. a reduced number of customer contacts). It is this piece meal development, rather than idealised design, as presented in the last chapter, that is often the primary cause of many of these operational, maintenance and performance challenges that face the water industry.

A focus has also been given to an overview of the regulations and standards, which range from a need to meet stringent regulatory standards to the softer measures of customer expectations and customer orientated care. The main body of the chapter is structured around the operation and maintenance cycle under the 'MAIDE' (monitoring, analysis, interventions, decision and evaluations) concept of five core elements. Although, more commonly in use now is the PALMM approach (prevention, awareness, location, mitigation and mend). Under these headings the relevant issues, techniques and approaches are discussed with reference to both the quantity of water, including loss of supply and leakage, and the quality of water. The argument is developed, firstly by reference to historical tried and tested methodologies, through to the latest developments and approaches to optimise system performance, operation and maintenance.

9.2. Historic development of networks and regulation

The primary function of a distribution network is to supply wholesome water to the point of use, whenever required, as efficiently as possible. In the UK, relatively simple systems were expanded at the time of the industrial revolution to meet the needs of populations as they grew, and in the 1960s over 65% of the distribution network in the UK had been constructed. These systems were constructed using basic engineering principles that transferred water from a source to the point of delivery with the inclusion of storage to meet peak demands and some resilience incorporated into the system to ensure the security of supply to strategic locations.

Since this time, significant further piecemeal development of the networks has taken place (approximately 30% of the system), driven by changing demand due to population growth and the need to improve the network to meet regulatory standards. In this latter respect, four notable acts have contributed significantly to the progressive design and development of the distribution networks: The Water Act 1945, The Water Act 1973 and The Water Act of 1989.

The Water Act 1945 focused largely on providing a blueprint for the development of water distribution systems with guidance and clarity on how to modify and expand the existing network. However, the Acts of 1973 and 1989 had a greater impact on the strategies for operation and maintenance of the distribution network and resulted in a change in the structure of the industry.

The Water Act 1973 brought together the Regional Water Authorities that allowed greater connectivity of the disparate piecemeal systems. This was followed by the Water Act 1989 which heralded the last significant organisational change to the water industry in England and Wales, due to the privatisation of the water companies bringing financial resources previously denied under the public sector borrowing requirement regime but accompanied by far tighter financial discipline from the capital markets. It had a huge influence. It is clear, therefore, that the current operation and maintenance of water distribution systems is complicated and has been influenced by the many different design criteria, regulatory standards, and construction methods and materials. However, it may be argued that, possibly, the most fundamental issue associated with the operation and maintenance of the systems is that they are underground and that, if they work well, the concept of 'out of sight out of mind' has too often been applied.

9.3.   Monitoring

The operation and management of water distribution systems produces an inherent conflict in that there is a need to provide a continuous supply of water to meet demand while at the same time delivering wholesome and high-quality water to the customers' taps. Despite the initial purpose of the water supply system to ensure public health, network design and evolution is a compromise between quantity and quality of supply – for example, emphasis on looped systems for resilience in continuity of supply can increase residence times and, hence, risk water quality and the need to ensure the delivery of high demands for firefighting, and has led to systems that may be considered to be oversized for water quality. Such networks contain significantly more water than is required to meet the usual daily demands. Systems contain significant redundancy, helping to ensure the continuity of supply under most event scenarios – for example, such systems are heavily meshed with numerous loops and valves, which facilitates maintenance and repair whereby, at the time of a malfunction event (e.g. bursts, discoloration etc.), small numbers and lengths of pipe, and hence customers, may be isolated to facilitate repair.

Such over engineered systems also have an advantage in that they are adaptable to meet the changing requirements of customers, to accommodate population growth and the capacity to supply new developments. In contrast, it is known that the quality of water deteriorates with time and, hence, to maintain water quality, it would be preferable to have smaller diameter pipes that supply only the peak demands plus minimal headroom for events and network developments. These smaller pipes produce higher flow velocities, thereby reducing the residence time and age of the of water within the system. This reduces the potential for the water to deteriorate.

It should be noted that it is feasible to optimise the operation and the management of over engineered systems with redundancy, such that areas of low velocity and high residence times are minimised (Prasad and Walters, 2006). The ongoing resilience and service measures have given rise to a complex dynamic between redundancy, which allows for mitigation of failure against consolidation to drive efficiency, and taking a risk based approach to service failure, including automation of network valving and rapid operational response to events such as supply of water by tankers in emergencies.

To balance the conflict between the quantity and quality issues, the water industry in England and Wales is regulated by two independent bodies.

- Ofwat, the Water Services Regulation Authority, is responsible for economic regulation, flow and pressure standards and policing for flow and pressure.
- The Drinking Water Inspectorate (DWI) is responsible for all aspects of water quality.

These two bodies are effectively responsible for the regulation of the operation and maintenance activities that water companies undertake. Additionally, there are four other bodies that influence the performance, targets and operational and maintenance techniques that are used by water companies.

- *The Consumer Council for Water* in the UK was set up in 2005 to provide a strong voice for water and sewerage consumers in England and Wales. Their job is to make sure the consumers' collective voice is heard and to ensure that the consumers are fully engaged at the heart of the water industry's thinking and decision making.
- *Water UK* is the association that represents all UK water and wastewater companies and suppliers at a national and European level. They provide a framework for the water industry to engage with government, regulators, stakeholder organisations and the public, and actively seek to develop policy and improve understanding in areas that involve the industry, its customers and stakeholders.
- *Water Regulations Advisory Scheme* (WRAS) is primarily concerned with the materials and products that are used in the water supply industry. They have a significant influence on the products and tools available for the operation and maintenance of water supply systems.
- *The Environment Agency (EA)* was established on 1 April 1996 under the Environment Act 1995. The EA has wide ranging responsibilities across the water industry in England primarily on safeguarding water resources and protecting the environment. This includes overseeing the actions water companies take to secure public water supply – that is, minimise water losses from distribution networks. Its equivalent body in Wales is *Natural Resources Wales (NRW)*.

9.3.1 Quantity

Ofwat regulate the 17 regional water companies in England and Wales. Of these, 11 companies provide both potable water and sewerage services, while six companies supply only potable water. In addition, there are a number of other companies who provide regulated water and sewerage services in small, typically, new development sites within the regional water companies.

Ofwat has many functions and these include

- ensuring that the companies provide customers with a good quality and efficient service at a fair price
- monitoring the companies' performance and taking action, including enforcement, to protect consumers' interests
- setting the companies challenging efficiency targets.

The Guaranteed Standards Scheme (GSS) within the Water Supply and Sewerage Services Regulations 2008 sets out the requirements for water companies to maintain a minimum static pressure head of seven metres in the communication pipe serving the premises supplied with water. Whereas, the Water Industry Act 1991 Part III Chapter II 'Means of Supply' Section 65 requires that there is a pressure of not less than ten metres head (1 bar) at the external stop tap of a property *and* that the flow should not be less than nine litres per minute within the property. Ofwat require companies to report the number of properties that receive pressure lower than this standard annually and companies compile a list of the number of properties at risk of low pressure. This level of service does not, however, override the statutory duty to deliver a continuous supply of water at a pressure that is able to reach the upper floors of properties.

In addition, companies have to provide a continuous supply of water, without interruption. Under the GSS regulations, a company must automatically make a payment to affected customers if the supply is not restored by the time given on the notice for planned work or within 12 hours for unplanned interruptions in an emergency (24 hrs due to a leak or burst on a strategic main).

As part of the annual service delivery reporting, each company in England and Wales is required to report on a number of common performance and asset health performance commitments to Ofwat as part of the annual return reporting process. There are several metrics relevant to the operation and maintenance of network

- leakage
- mains burst
- Event Risk Index (Water Quality)
- properties at risk of receiving low pressure
- customer contacts about water quality (DWI)
- Compliance Risk Index (DWI).

The performance indicator Customer Minutes Lost (CML) is a metric for supply interruptions that calculates the performance of water companies in terms of the average number of minutes lost per customer for the whole customer base for interruptions that lasted three hours or more. This common performance measure covers planned and unplanned interruptions and there are no exclusions for asset failure caused by third parties. In reality, the measurement of flow and pressure delivered to every customer is not feasible, particularly as the measurement of flow is generally complex and requires a permanent installation of equipment. Pressure is more readily measured and most operational staff carry suitable devices to manually obtain pressure data from any customer tap, fire hydrant or other accessible tapping point on a distribution system. Such manually obtained spot sample pressure data is usually not recorded formally as it is used primarily to inform current local operational decisions.

For regulatory and more strategic management of the system, both flow and pressure are continuously measured at strategic points. In practice, therefore, the current status of the system at each individual customer's tap is unknown and, hence, water companies usually only become aware of the quality of service at an individual premises through customer contacts. Therefore, failure to meet the regulatory standards or the delivery of a poor quality of service is only reported to the company once it has occurred.

There have, however, been significant developments towards the creation of smart networks through the increased level of sensor deployment (especially pressure loggers and acoustic noise logging devices) that facilitates more proactive management of water supply networks. Smart networks, when optimised, introduce enhanced levels of insight into water network performance and thereby enable early warning alarms, support a quicker decision making process to mitigate against interruptions, reduce leakage run times and overall improve the level of service to customers. Smart water network solutions integrate data from multiple sources and typically present it in a single visualisation platform, which informs asset and operational decision making through machine learning. A key additional component of the introduction of a smarter network is the installation of smart customer water meters. The level of metering in England and Wales is increasing with a greater number of smart water meters being installed.

Most companies invest significant effort in policies and procedures for handling and avoiding customer contacts as well as categorising, including both quantity and quality issues, and recording them for future analysis. The importance and weighting placed on customer contacts reflects the UK attitude of water supply as a service industry.

Water supply systems consist of treatment, storage and distribution assets joined and controlled by pumps, valves and other operational components (Chapter 5). For management and reporting purposes, these systems are broken down into hierarchical areas comprising production management zones, distribution management areas (DMA) and sub-DMAs created, for example, for pressure control (Chapter 5). Flow and pressure data are collected from all these areas in order to provide operational and regulatory data. The layout of a typical network and of the typical locations at which flow data is recorded are shown in Figure 9.1.

Flow is usually monitored in trunk mains, at the inlets and outlets of service reservoirs, pumping stations, some large diameter distribution mains, DMA inlets and exports, at some sub DMA inlets and exports, and major consumer premises. Figure 9.2 shows an example flow pattern measured at DMA level. Pressure is also measured at these locations and at critical pressure monitoring points inside DMAs, specifically selected at locations where the first indications of pressure problems may be anticipated to occur. Common practice for such pressure monitoring is usual to select the highest point. However, this is not necessarily the most sensitive location to change occurring in the system, such as leaks and burst.

Methodologies have been developed to identify 'optimal' spatial resolution for sensor location in order to capture event data irrespective of where in a DMA the event occurs (Farley et al., 2008). Analysis of such data can help to determine where the event actually took place. Mounce et al. (2010) present the online application and verification of an automated event detection system, based on artificial intelligence self-learning algorithms. An example event detection using this technique is shown in Figure 9.2.

For hydraulic design and monitoring purposes and, more recently, the need to accurately determine leakage levels, manufacturers have developed flow and pressure instruments to a high level of sophistication. Off-the-shelf instruments are capable of measuring flow and pressure to an accuracy of 0.1% and even 0.01% if the instruments are rated correctly for purpose and manufacturers' installation procedures are strictly followed. The installation, calibration and maintenance requirements of flow instruments are far more onerous when compared to that for pressure equipment, so it is simpler and more cost effective to collect pressure data than flow data. However, irrespective

Figure 9.1 Division of a distribution system into DMAs, including main flow meter locations (Reproduced with permission from Morrison (2004). © IWA Publishing)

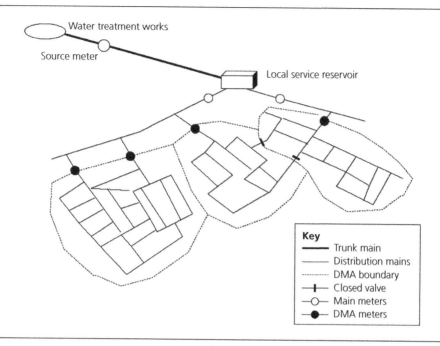

Figure 9.2 Example flow–time series plot and event detection using an artificial intelligence (AI) technique (Redrawn with permission from Mounce *et al.* (2010). © ASCE)

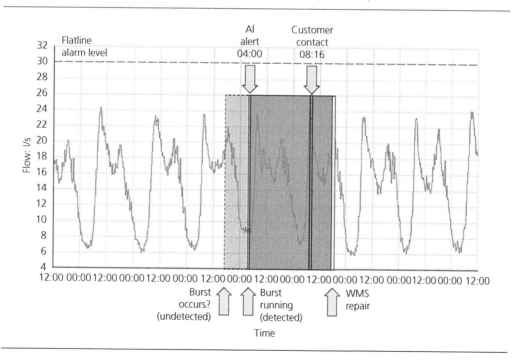

of the type of instrumentation used, it is recommended that regular testing and calibration of the instrumentation is made to ensure that accuracy of measurement is achieved.

Continuous sampling from fixed instruments produces a time series of data that represents all the changes in the measured parameters at the point in the system at which the measurement is being made, over the duration of the measurement period. The unofficial industry standard for the collection interval of both flow and pressure data, at most locations, is 15 minutes. The basis and justification of this is somewhat obscure, but it does provide a reasonable balance between the volume of data and definition of daily patterns. However, ideally, the frequency of data collection should be different for different applications. A good representation of the overall pressure dynamics within a network can be observed at 15-minute intervals, but clearly this is insufficient for rapidly changing events – for example, the shape and amplitude of pressure transients cannot be resolved with data points less than a hundredth of a second apart. Flow dynamics can also be well represented by 15-minute data points but higher frequencies allow a much better understanding of the system hydraulics – for example, it is feasible to assess the different contributions to the overall flow from different types of demand such as domestic and industrial.

The output of most flow measurement instrumentation is an electronic pulse per unit of flow. The number of these pulses are generally counted over, and then divided by, the data resolution period. Hence most flow data is in the form of time averaged values. This 'smoothes' the data but, in doing so, much valuable data may be lost. Pressure instrumentation varies, some provides an instantaneous value while others average values. It is important to know which prior to making use of the data (Mounce *et al.*, 2012).

The data that is recorded at permanently installed flow and pressure monitors is subsequently required for analysis. To facilitate this, most flow and pressure instrumentation is, or can be, equipped with a memory such that it can store data. This data can then be obtained either manually at the site or remotely by using communication technology. Pressure and flow instrumentation at large installations, such as pumping stations and service reservoirs where power is available, have commonly been connected to centralised online data storage by way of SCADA systems, and such connectivity facilitates regular data transfer. More recently the diminishing costs of automated data transfer by way of mobile data networks, low power radio and through 'internet of things' networks is allowing all types and volumes of recorded data to be transferred from a large number of monitors, many at remote locations in the distribution system.

The emergence of low-cost solutions for permanent sensor application in distribution network are opening up new opportunities for managing performance. Pressure and transient pressure loggers can now be located in multiple locations in DMAs, identifying assets and operations which potentially cause stress on the network. Furthermore, the additional sensors deployed in the network are now underpinning the use of near real time hydraulic modelling or digital twins, providing insight into asset performance, network connectivity anomalies and more accurate measurement of service impacts.

For many applications and analysis, it is possible to consider water distribution operation to approximate to steady state, or a sequence of steady states snapshots. In Chapter 2, hydraulic transients were introduced – instabilities of rapidly fluctuating pressures caused by water flowing in pipes not being able to transition instantly from one state to another. Figure 9.3 shows how a system

Figure 9.3 Increasing levels of information revealed by increasing the rate of data acquisition (Data courtesy of University of Sheffield)

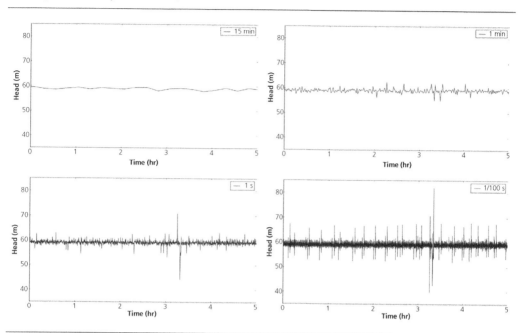

that would be considered steady state, from 15 minute resolution data, becomes increasingly more complex as the resolution of data is increased to 100 Hz, revealing the occurrence of transients. As data capacities increase, the collection of such data from operational systems is increasing and, henc,e appreciation of how widespread transients are. Transients are caused by any and every sudden change – from pumps, valves, demand and so on.

Due to the high wave speed of transients, pressure data around 100Hz is sufficient to capture the occurrence of transients, and hence associate them with any impacts. The wave forms of transients are impacted by every feature of the system they pass, hence there is potential to extract information from the wave forms, such as leak detection (Shucksmith *et al.*, 2012). This is either attempted through inverse modelling approaches or data analysis, requiring far higher data rates to fully characterise the details of the wave forms.

The great and faster the change the bigger the instability (transient) that will occur. This is commonly considered due to change in velocity, as in Equation 2.35, but can also be due to the release of trapped pressure (Collins *et al.*, 2012). An important common misunderstanding is that the size of the transient (the change in pressure) induced is not a function of the initial pressure. There will be the same change in pressure for a given change in velocity, irrespective of the initial system pressure. For a given change in velocity occurring at low initial pressures there is perhaps increased risk of ingress (Fox *et al.*, 2015), while at high pressures there is perhaps increased risk of pipe failure, particularly if the pipe has already been fatigued by cycling loading due to transients (Brevis *et al.*, 2015). Most UK water companies utilise transient training facilities to ensure operational staff are aware of transient and limit the risk of inducing these through operations, such as turning a valve too quickly.

9.3.2 Quality

Drinking-water quality is an issue of concern for human health in both developing and developed countries. The risks arise from infectious agents, toxic chemicals and radiological hazards. The World Health Organization (WHO) produces, and regularly reviews, international standards on water quality and human health in the form of guidelines that are used as the basis for regulation and standard setting. These guidelines have been interpreted to form the EU Drinking Water Directive (98/83/EC). This Directive has formed the basis of strict standards for the quality of the public supply and laid down in national regulations by the DWI. DWI closely checks and regulates water quality in England and Wales, with the Drinking Water Quality Regulator for Scotland (DWQR) performing these functions there.

The drinking water quality criteria are set out in the Water Supply (Water Quality) Regulations 2000 that stipulate a maximum concentration or acceptable level of a large number of substances. In addition to these quality standards, the regulations stipulate the minimum frequency at which water samples should be collected and analysed and describe the methodology to create appropriate sampling zones. The regulations also provide advice on analysis, reporting, approval of materials for use in contact with drinking water, and the necessary actions if the standards are breached. Since the regulations came into force, they have been regularly subjected to amendment to take account of a number of EU Directives. The most important of these are the Water Supply (Water Quality) (Amendment) Regulations 1989 and 1999 that take account of the need to regulate cryptosporidium, nitrates and pesticides.

The standards defining the wholesomeness of the water supplied are also prescribed by the DWI. The object of these standards is to stimulate improvement in drinking-water quality and to encourage countries of advanced economic and technological capability, in Europe, to attain higher standards than the minimal ones specified in International Standards for Drinking-Water. The latter standards are considered to be necessary and attainable by every country.

At the same time, the industrial development and intensive agriculture of some European countries create hazards to water supplies not always encountered in other regions. Hence, stricter standards are demanded and justified (WHO, website). In the Water Act 2003, additional emphasis has been placed on consumer acceptability. Consideration needs to be given to any effect on the taste, odour and appearance of the water supplied to consumers. It is desirable, for example, that consumers continue to receive water of a hardness and mineral content that they are accustomed to receiving (DWI, 2004). The guidance also states that consideration be given to maintaining any existing quality agreements that the water company has with non-domestic customers. Some manufacturing processes require water with a fairly constant chemical composition. It might be possible for them to handle changes in composition, providing there is adequate consultation and advance warning of the changes. Frequent or unplanned changes need to be avoided (DWI, 2004).

UK water suppliers place the highest priority on assuring the quality of water provided to their customers. Figures for drinking water quality compliance in England and Wales to meet the strict UK and European standards are typically >99.95%. This impressive figure highlights the importance, as well as the effectiveness of the procedures used to ensure water quality. These procedures are typically based on risk assessment and risk management and are being further developed to work in a more holistic way, focusing on the whole of the water supply chain from source to tap (Water Science and Technology Board, 2005). This concept has given rise to the term 'water safety plans' (WSPs) and such plans are endorsed by the WHO. The WSP approach also requires that other stakeholders play their part in assuring water quality. The source to tap concept is a complex issue as all aspects of water quality from the management and the protection of water sources and through to delivery of acceptable water quality in buildings, including households, involves many stakeholders, some of whom are unaware of their roles and responsibilities.

The concentrations and levels of regulated substances are set out in tables of grouped parameters including aesthetic (such as colour, turbidity, taste and odour), microbiological (such as indicator organisms, pathogenic organisms and viruses), chemical (such as iron, manganese and lead) and disinfection residual (such as chlorine). These tables are subdivided into directive and national levels and concentrations.

Different standards are applied for the quality of water at different points in the network, primarily at the exit from the water treatment works, service reservoirs and at the point of use, usually the customers tap. At treatment works and service reservoirs, a combination of online and discrete sampling is used to obtain water quality data, while at the customer's tap data is derived from a random programme of discrete sampling. Hence samples are collected from customers' properties, selected at random from within the water supply zones defined by the water companies. A water supply zone is a region containing not more than 100 000 population (Chapter 5).

Sampling rates for different parameters vary and are generally defined with respect to the total population of the zone – for example, coliform bacteria and disinfection residual requires 12 annual samples per 5000 population. Full details are set out in the Water Supply (Water Quality) Regulations 2018. These discrete manually collected samples produce a small amount of spot data that is representative of the quality of the water at a single time at a given location. Such snapshots of system conditions provide no understanding of the system status prior to or after the data was recorded.

The analysis of discrete samples has to be completed by accredited laboratories that are recognised by the United Kingdom Accreditation Service (UKAS). This agency is the sole body recognised by the government to assess, against internationally agreed standards, organisations that provide certification, testing, inspection and calibration services. Once the samples have been analysed by an accredited laboratory the analytical results are returned to the water companies. Failure to meet the standards is judged by reference to the defined standards. Failure must be reported to the DWI. When a failure is recognised, or there are a large number of customer contacts, it is usual for companies to stimulate a programme of investigative sampling, over and above the routine samples. This is intended to capture the cause, extent and severity of the failure and to inform incident report to DWI, covered in Section 8.4.

Instruments continuously measuring water quality parameters generally require significant maintenance and the use of triple validation systems to ensure confidence and provide quality assurance. Application is generally limited to locations at treatment works and service reservoirs with measurement parameters including temperature, turbidity, colour, iron, manganese, free and total chlorine, pH, redox potential, dissolved oxygen and conductivity. Technologies for the continuous measurement of water quality parameters are improving, and recent advances take advantage of refinements in optical, thin and thick film printing on ceramic substrates and micro fluidics, to provide continuous reagent free probes. Some instruments even combine multiple parameters onto a single probe that can be inserted directly into mains or installed at customer boundary fittings.

Turbidity (the measure of light scattering due to particles in the water) time series data is increasingly being collected by water companies to provide insight and understanding of the processes resulting in discolouration. This is the most common reason customers contact water companies regarding water quality. The collection of chlorine time-series data is also increasing, while not strictly required by regulation the measurement of a residual implies limited risk of bacterial regrowth. Evidence around the relative costs and benefits of different deployment strategies of such instrumentation are emerging, and application is increasing.

It is envisioned that one day (in the very far future) quantity and quality instruments could be embedded into the pipes of water distribution systems at the time they are manufactured ('smart pipes'). An alternative vision is of autonomous robots living inside our pipe networks, providing a wealth of data (Mounce et al., 2021). Such innovations will enhance and move forward operational practice so that pipe replacement programs and other maintenance activities will be explicitly based on a network of sensors continuously relaying operational and regulatory data to point of use, enabling a shift from large reactive to truly proactive management.

9.4. Analysis

Much of the data analysis that is formally performed by water companies in England and Wales is used for regulatory reporting to Ofwat and DWI, including the quantification of leakage performance.

Ofwat regulates to make sure that water companies provide customers with a good quality, efficient service at a fair price. They monitor companies' performance and take action, including enforcement, to protect consumers' interests, and they set challenging efficiency targets for the companies. Each year Ofwat assesses the overall performance of water companies, ranking them in areas including leakage, customer satisfaction, pollution and expenditure. The information in the service delivery report, obtained through the annual performance reporting, provides transparency for customers, and wider stakeholders, to see exactly how water companies are performing on the issues that matter to them. The service delivery report monitors the performance of the 17 largest companies in England and Wales on key components of the outcomes framework.

For each performance commitment Ofwat assess whether a company has achieved or fallen short of its performance commitment target level. The service delivery report was designed to capture all aspects of service and to incentivise good overall service through comparative competition and transparent performance scores. Ofwat groups companies in three categories: sector leading, average and lagging. The categorisation is based on Ofwat's assessment of individual companies' performance across outcomes to an extent that this demonstrates that they are delivering value to customers – for example, companies that achieve a high proportion of key common outcomes in a cost-efficient manner are considered 'sector leading'.

As part of the price review, companies made performance commitments about the level of service they will deliver to customers. Delivery of most of these performance commitments is done under a mechanism of outcome delivery incentives. This largely financial incentive system could result in outperformance payments to companies if they exceed performance commitment targets, but conversely could result in penalty payments if targets are not met.

Drinking water quality compliance in the service delivery report is covered through the Compliance Risk Index (CRI). This index illustrates the risk arising from failures to meet drinking water standards throughout the supply system. The index is defined, calculated and reported by the DWI. As described later such failures invoke investigational sampling and analysis to determine the cause of failure.

Water companies in England and Wales are formally required to submit and gain approval for Distribution Operation and Maintenance Strategies (DOMS), as first set out in the DWI Information Letter 15/2002 (DWI, 2002). These plans are reviewed regularly and are considered as part of the water company five yearly Asset Management Plan (AMP) submissions to Ofwat. The purpose of a water company DOMS is 'to ensure customers are supplied with water of consistent or improving quality, through appropriate operation and maintenance of the distribution system'. They cover strategies for the proactive management of drinking water distribution systems, so that water companies meet and continue to meet drinking water quality standards. They typically include but are not limited to the following:

- arrangements for proactive, periodic, medium-term, system-by-system investigations
- monitoring of water quality at a local level leading to timely responsive maintenance
- control of operational activities related to risks to water quality
- regular inspection and maintenance for certain components of the distribution system related to risks to water quality.

9.4.1   Quantity

The most common instrumentation used to gather quantity data from water distribution networks are flow meters and pressure transducers. Such data is usually captured as a continuous time series that are analysed to identify performance shortfalls. Pressure data tend to be used primarily for monitoring pressure critical areas of DMAs, inlets and outlets of pumping stations and other critical assets. Pressure data informs when low pressure events are occurring and hence when pressure standards of service are met and that pumping stations are operating correctly, or vice versa. Analysis of pressure data is typically performed by examination of threshold levels, set with respect to regulatory standards and site-specific operational experience. Pressure is also an important data source used to inform network modelling (see chapter 6). Flow data is used primarily for leakage management but also in resource, statistical, hydraulic and water quality models.

In general, UK water companies are proficient at ensuring their customers receive a continuous supply. However, during transit through the distribution network, a significant proportion of the water entering the system is lost through leaks. In the UK, as much as 25% of distributed water is lost through leaks, with a national average of around 20%. Although leakage levels in England and Wales have dropped by 40% from a high of 5112 Ml/day in 1994–95 to 2954 Ml/day in 2019–20, they are still large – for example, the existing levels of leakage could still supply almost 20 million domestic users.

Leakage levels are estimated for company and customer assets. The customer supply pipes are owned by the property owners, however water companies must include leaks on customer pipes in the reported leakage and it is estimated that leakage on these assets contributed to 25% of leakage. Across Europe, reported urban leakage rates vary greatly between 3% and 50% (European Environment Agency, 2003). Hence, in places where large volumes of water are being lost there is considerable potential for significant efficiency gains through leakage management and reduction.

In the UK, leakage targets have previously been set to achieving an 'economic level of leakage' (ELL). ELL is defined as the point where the cost of identifying and repairing leaks is such that, to go beyond this level of expenditure, would incur unacceptably high costs per leak repaired. The best practice principals in the ELL calculation are set out in the Tripartite Study report (Ofwat, 2002). In 2009, the methodology of assessing the ELL was improved by including social, environmental and carbon cost in the assessment of the cost of leakage control and water loss. This led to the principle of sustainable economic level of leakage or SELL. Targets have been based on SELL principles within the UK water industry up to 2020 and incorporated within the water resource management plans in which the option of leakage reduction is considered against the cost of new water supply options.

In 2019 the English water companies made a Public Interest Commitment (PIC) to triple the rate of sector-wide leakage reduction by 2030. The National Infrastructure Commission's (NIC) report on 'Preparing for a drier future' (NIC 2018) recommended a long-term target to reduce leakage by 50% from 2018 levels by 2050. The water sector has also taken up this challenge of introducing significant reductions in leakage levels in the long term, drivers for this include recognition of the role that reducing wastage has in achieving net-zero targets.

Before leakage can be precisely estimated, accurate measurements of the volume of water entering the network and the volume used by all domestic and industrial users are required. However, leakage figures are built upon estimates of domestic and industry use components which themselves are subject to potentially large estimation errors – for example, accurate measurement of the domestic demand is not possible because only c.55% of household supplies in England and Wales are metered, while only major industrial demand is monitored, all other components are estimated.

Water balance is a commonly used leakage detection method. This involves the detailed accounting of water flow into, and out of, the distribution network or some part(s) of it. At the level of the whole system, this consists of a total water supply balance – that is, the summation of all water consumed (metered and un-metered) and not consumed (leakage, theft, exports etc.), which is compared to the total system input. Instead of monitoring the entire system, individual zones can be monitored such as in district flow metering. These balances help to identify areas of the network that exhibit excessive leakage. However, they do not provide information about the location of leaks. Local location techniques are commonly used for this (WRc, 1994).

The UK water industry initiated a significant research programme in the early 1990s to inform guidance on leakage management. The results of this work were published in Managing Leakage (WRc, 1994). One of the most important outputs of this research work was the concept of understanding the components of leakage and how these could be estimated. This provided practitioners with the capability of modelling leakage and the factors that affect it. The resulting Bursts And Background Estimates (BABE) methodology is now one of the commonly used methods for leakage estimation, and is incorporated within the IWA leakage methodology to determine unavoidable annual real losses and the infrastructure leakage index (Lambert, 2003; IWA, 2005). Other leakage that is not an identifiable burst is labelled as background leakage. It is clear from observations of the application of this method that background leakage represents a substantial proportion of the overall leakage estimate in many networks.

Flow data can also be used to directly evaluate leakage. A widely used operational approach is an analysis of measured minimum night flows – that is, the lowest flow supplied to a hydraulically isolated DMA. It is generally evaluated during the night hours, between midnight and five am, and is a measure of leakage as well as certain minimum night consumption. Estimating distribution losses from minimum night flow relies on the accurate estimation of the additional components that contribute to night flows. By monitoring night flows continuously, unusual changes in water volumes can be detected. The methodology of the industry until AMP6 was to rapidly respond to reactive burst reports as well as proactively find bursts and leakage within the DMA. The standard use of acoustic noise loggers and other emerging technology, such as using satellite leak detection and analytics, has seen a shift in the industry to develop effective solutions to provide points of interest for find activity rather than scanning whole zones for leakage. This change in strategy will aim to underpin a more cost effective find strategy across the industry.

Flow and pressure data analysis for leakage detection and reporting tends to be a semi-manual process with inherent inefficiencies that are prone to human error. A result of this is that leaks which are not obvious can potentially run undetected for extended periods (sometimes several months). Advances in instrumentation, telemetry systems and communications are leading to more data of a higher quality being gathered by water companies. Hence automated data analysis techniques are being developed to deal with these larger data volumes and to add consistency to, and remove human error from, the analysis. In addition, the constant flow of high-quality data is enabling water companies to develop online data analysis techniques using artificial intelligence or machine learning. Currently, these are being applied to analyse DMA data, in near real time, in order to automatically detect bursts and leakage (Mounce *et al.*, 2007, 2008, 2010).

As introduced in Chapter 6, hydraulic models of water distribution systems can be produced to provide a mathematical representation of the nonlinear dynamics of a water distribution network hydraulic performance. Such software is predominately used in an 'offline' form for activities such as strategic supply analysis, design of control strategies, network extensions, and maintenance planning. However, to be applied for day-to-day operational management, the models need to be populated with the most up-to-date data as is possible. A network model driven by the latest data can provide invaluable information of the system performance – for example, by extrapolating data from a relatively few measurement points the hydraulic performance at every point in the model may be estimated.

Preliminary efforts made to link simulation models with real time data from telemetry systems were reasonably successful (Skipworth *et al.*, 1999; Orr *et al.*, 1999). Improvements in data collection transfer and warehousing technologies have now enabled a new generation of online modelling techniques to be developed. A sufficiently accurate model can now be regularly updated with flow and pressure data sent directly from the field and can be used to detect many features of system performance to include

- hydraulic events such as low pressure
- illegal hydrant or valve operations
- automatically monitor standards and levels of service (Machell *et al.*, 2010).

It is anticipated such models will gain popularity and become widespread in their use, enabling almost real time optimisation of network operations in the future. Such data-driven hydraulic modelling are often a key component in emerging digital twins of water distribution networks. A digital twin uses the best available deterministic and artificial intelligence models driven from real-time sensor updates and historical performance data to provide insightful replication of the real world by way of effective visualisation.

9.4.2 Quality

The DWI has the power to prosecute companies who fail to meet the required water quality standards. With drinking water compliance in England and Wales being around 99.95%, prosecutions are rare. However, should such a prosecution be taken out against a water company for an offence under section 70 of the Water Industry Act 1991 (c. 56) – supplying water unfit for human consumption, a defence must be provided by the water company to show that it took all reasonable steps and exercised 'all due diligence' for securing that the water was fit for human consumption on leaving its pipes or that it was not intended for human consumption.

The major task in the use of the water quality data is to report whether the quality meets the regulatory requirements, primarily through the comparison of analytical results of discrete sampling with the defined standards. Results of this are then reported to the DWI for the purpose of demonstrating compliance. The information provided is compiled into drinking water quality data tables that indicate the extent to which the companies have, or have not, met each of the drinking water standards in force at the time of submission and are used for inter-company performance ranking.

Water quality data is available to the public. There is a statutory requirement that customers of water companies may enquire about their water supply through the Drinking Water Register, which provides access to drinking water quality information for every water supply zone in a company. Data analysis for this purpose requires that the records include name of the zone, population of the zone, water treatment works supplying water to the zone, details of any undertakings and results of analysis of water samples.

Water quality sample results are also fed back to operational teams to inform of deteriorating quality in order that remedial actions can be taken before standards of service are impacted. While the discrete sample data is sparse in both space and time relative to the scale and complexity of drinking water distribution systems, and the water quality interactions occurring within them extremely complex, data mining techniques are increasingly being shown able to extract understanding and actionable insight from such data. This is particularly true when the data is connected to other data, such as asset information, customer contacts, hydraulic modelling results (Mounce *et al.*, 2017; Speight *et al.*, 2019) and even from mining DWI reports (Mounce *et al.*, 2016). Data from continuous water quality instruments can also be used to supplement discrete sample data, since absolute values are no longer the primary concern, it can be changes in trends and patterns that become important yielding operational insight (Mounce *et al.*, 2015). Such data use might result in a quick remedial action or the bringing forward of planned maintenance such as mains flushing programmes. Water quality data is also used to generate warnings about impending problems, such as discolouration, and can initiate a program me of planned maintenance, such as mains flushing or service reservoir cleaning.

When water quality data analysis indicates an event, water companies have a responsibility to inform DWI as soon as possible. This is covered under the Water Undertakers (Information) Direction 2004. An event is defined as 'having affected, or likely to affect drinking water quality, or sufficiency of supplies and, where as a result, there may be a risk to consumers' health'. When notified of an event, inspectors of the DWI assess the information provided to establish if the event is an 'incident'. If deemed to be an incident, a detailed report is usually required from the company, including all possible data analysis. At this stage significant added value can be obtained from the analysis of continuous monitoring data to supplement the regulatory driven discrete sampling. The data analysis contained in the report should be such that the inspector can determine the cause, whether the incident was avoidable, the company's response, how the incident was handled, lessons that can be learned for the future, if there were any breaches of enforceable regulations, and whether the company supplied water that was unfit for human consumption. The outcome of such assessment can range from a simple letter, with or without recommendations, through to prosecution proceedings or a caution for a criminal offence.

In respect of the reporting of incidents to DWI, the response to incident management and contingency planning is covered under 'drinking water safety plans'. This is a 'source to tap' risk management approach that is key to the way in which water companies ensure a continuous supply of safe drinking water both now and for the future. Understanding risk enables effective controls to be implemented to safeguard water quality. Managing risk means the companies can anticipate problems to protect public health. When assessing risk to drinking water quality, the likelihood and consequence of hazardous events are ranked using a matrix to derive a risk score. This process is then repeated to take into account any control measures, to give an estimate of the residual risk. Summary reports are then submitted to the Drinking Water Inspectorate as required by the amended Water Supply (Water Quality) Regulations 2000. Information from the risk assessments is generally incorporated into water company business plans which are submitted to Ofwat and define the level of investment for maintaining and enhancing water supplies during the AMP planning periods. Identified risks are continually reviewed and the effectiveness of controls is monitored to ensure that the DWSP remains effective.

Network modelling, as it is with water quantity, can also be a valuable tool in the analysis and interpretation of water quality data. The requirement is again for an up-to-date, preferably online, solution. Applications can range from source tracking to identify sections of networks supplied from different sources and areas receiving mixed water, through the simulation of water age (hypothesised to be essential for the accurate simulation of all the various kinetic reactions and interactions that occur within the complex distribution system that acts as a highly variable high surface area reactor with high residence time (AWWA 2002; Machell et al., 2009; Machell and Boxall, 2014)), through to modelling of specific reactions. Modelling of disinfection residual and the formation of disinfection by products have received specific research attention (Rossman et al., 1994; Rossman and Boulos 1996; Kirmeyer et al., 2000; Mutoti et al., 2007).

Water quality modelling relies on the use of an accurate and calibrated hydraulic model that utilises the solution of transport and tracking algorithms as the basis for monitoring the quality changes to parcels of water as they travel through water distribution systems. The standard of calibration required for useful water quality modelling is generally higher than that required for quantity. Water quality models are gradually being adopted as investigative tools to better understand how water quality changes in distribution systems and to investigate the reasons for failed water quality samples, and as predictive tools to determine age of water, chlorine, trihalomethane and other chemical concentrations that are created or increase within the bulk water flow during transit in the pipe network.

Predictive modelling tools such as PODDS (Boxall and Saul, 2005; Boxall and Dewis, 2003) are used to assess the potential discolouration response to changes in hydraulic conditions. Because the mechanisms that determine the quality of the water emerging at a customer's tap are many and complex, current modelling approaches are known to have shortfalls. However, much research work is currently being undertaken, with longer term deliverables, to fully understand these mechanisms and to develop models that accurately reflect the complex processes and quality relationships that occur in water distribution systems. These modelling improvements are then written into the industry standard software for use by practising engineers and scientists.

9.5. Interventions

In this section the major pipeline intervention options that are commonly applied to water distribution systems are outlined.

9.5.1 Quantity

9.5.1.1 Mains renewal and replacement

Several techniques have been developed for mains renewal and replacement to minimise impacts to customers, society and the environment. This section outlines different options for mains replacement and the circumstances in which these may be applied. Relining techniques are covered under the quality heading as this is the major driver for such techniques, although they can also offer structural performance improvement.

9.5.1.2 Open cut

The most obvious approach to mains replacement is the excavation of a trench to lay the new water main, Figure 9.4. This approach may be selected as the most effective option depending on the location and the method of reinstatement. Common factors to consider are the route, the location and size of the multi utility buried infrastructure, the impacts to traffic, shops and businesses and pedestrian access, and the size diameter of the main to be replaced – that is, the new main is likely to be of a larger diameter than that of the existing pipe.

9.5.1.3 Directional drilling

In certain situations, it may be more appropriate to use directional drilling rather than replacing the pipe in a trench (Figure 9.5). This involves drilling a pilot tunnel with a precision-guided drilling rig then pulling the new pipe back through the tunnel. This option minimises the impact on the highway and is often applied in situations where mains need to traverse a strategic crossing such as a watercourse or other such significant obstacle.

Figure 9.4 Open cut (Images reproduced with permission from Thames Water)

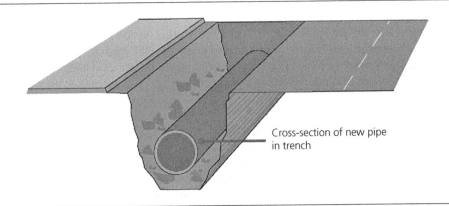

Cross-section of new pipe in trench

Figure 9.5 Directional drilling (Images reproduced with permission from Thames Water)

Rod drills through underground

Hooks on pipe and pulls back through tunnel

New pipe

Rodding system

Distance between pits approximately 100 m

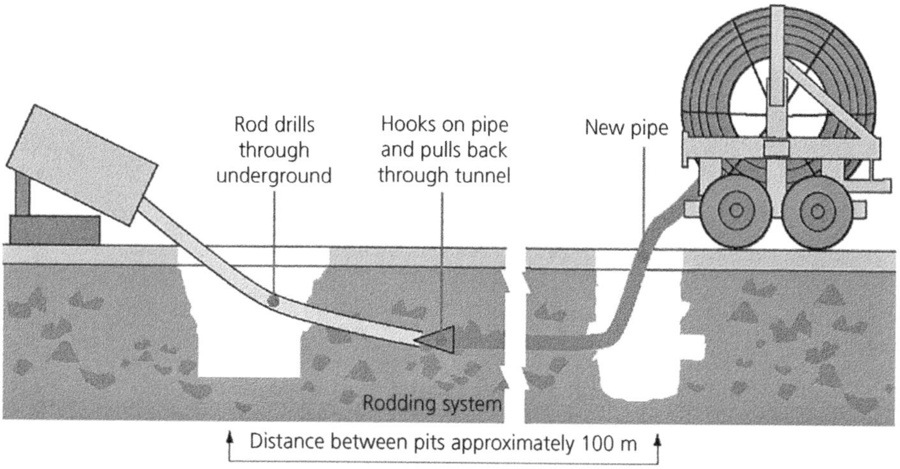

9.5.1.4   Pipe bursting

Pipe bursting can be applied in most circumstances where the new main is of a similar diameter to the existing main. Work is done underground, with only two small pits dug at each end of the section of pipe that is being replaced (Figure 9.6). In simple terms a steel rod is pushed into the existing, old main, which is then used to pull a cutting tool back through the pipe. This breaks up the old pipe as the rod returns and at the same time the new pipe is pulled into the space left by the old pipe. This approach offers the advantage of minimising the above ground impact and access to the highway.

9.5.1.5   Inserting a new pipe/sliplining

The further option for mains renewal is to insert or 'sleeve' a new plastic pipe inside that of the current main (Figure 9.7). The obvious constraint on this approach is whether the reduction in diameter will compromise the hydraulics, thus preventing delivery of an adequate level of service.

Figure 9.6 Pipe bursting (Images reproduced with permission from Thames Water)

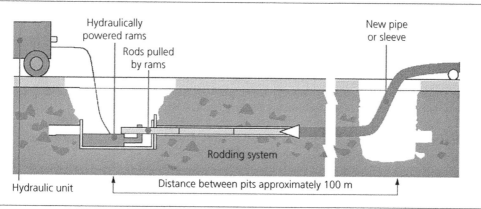

Hydraulically powered rams

Rods pulled by rams

New pipe or sleeve

Rodding system

Hydraulic unit

Distance between pits approximately 100 m

Figure 9.7 Inserting a new pipe (Images reproduced with permission from Thames Water)

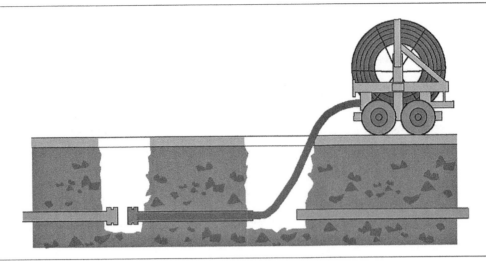

Network analysis would be used to model this impact and test the applicability of the approach. However, in most cases, although the pipe would be of a reduced diameter, the new smoother pipe often results in little or no impact or reduction in level of service.

9.5.1.6 Pump scheduling, valve regulation and system optimisation

The rising cost of energy is one of the greatest challenges facing the water industry, and investment in the assets which offer the opportunity for increased operational efficiency is one way to reduce such costs. However, it should be recognised that energy savings may also be realised through optimising the operation of existing assets and infrastructure and, hence, it is important to consider both approaches.

Energy cost reduction can be realised in several ways by

- moving energy consumption to cheaper tariff bands
- reducing peak demand charges
- running pumps more efficiently
- choosing the shortest path from source to destination
- choosing the lowest cost source of water (raw and treated).

Operators need to choose the lowest cost solution without sacrificing reliability or levels of service. Production of treated water needs to be directed to the cheapest sources of water, while pumping should be moved to the lowest cost tariff periods. Where possible, the storage capacity of reservoirs should be optimised, using greater volumes of turnover, to satisfy demand during expensive electricity periods. Such an approach will realise benefits in terms of

- energy cost savings
- efficiency gains and, therefore, carbon emissions reductions
- improved water quality through lower water age and improved turnover in reservoirs
- planned rather than reactive operating culture
- a structured and measured approach enables risk assessment of operating decisions.

9.5.1.7 Moving energy consumption in time

Energy companies typically charge more for energy consumption when demand is high. Scheduling pumps and flows into lower cost energy periods, while still satisfying storage and pressure requirements, will reduce energy expenditure aligned to these tariff structures.

9.5.1.8 Reducing peak demand charges

A peak demand charge is a penalty fee imposed by energy companies for the highest electrical load recorded in a month. A proactive and measured approach to pump scheduling helps companies avoid peak demand charges by choosing the minimum number of pumps required to run concurrently.

9.5.1.9 Improving pumping efficiency

Pumps are designed to run most efficiently at a particular pressure and flow rate. Understanding how operating individual pumps and combinations of pumps at the most efficient point on their efficiency curves will lead to optimised scheduling and more efficient use, requiring less electricity to pump the same volume of water.

9.5.1.10 Selecting the lowest cost source or path

As water transmission and distribution systems have evolved over many years there are often a number of routes by which water can be delivered to a customer. Often more than one treatment plant can service the same area, providing back-up for planned and unplanned operational failures. Network Models can assess the use of different sources to reliably meet demand, and can be used to optimise and select the lowest cost source to supply an individual customer within a water supply zone. Similarly, a combination of modelling and near-real-time information taken from the network can be used to optimise variations in the demand pattern during the day to ensure the lowest cost path to customers at all times of the day, for example by optimising the use of storage tanks.

There are a number of technical challenges to overcome when attempts are made to optimise transmission through valve and pump operations. These include

- optimising against multiple sources of water, storage or demand points
- optimising fixed speed drive (FSD) and variable speed drive (VSD) pumps and parallel pump sets
- maximising the use of lowest cost water source within a number of constraints, e.g. abstraction limits or raw water quality
- maximising the use of off-peak electricity periods within any constraints
- forward predicting daily pattern of water consumption in near-real-time, and optimising pumping schedules immediately
- interfacing to existing SCADA and telemetry systems to collect near-real-time data and implementing changes remotely
- reacting to unanticipated changes to production capacities, as well as setting availability of pumps and storage for planned operations.

A number of tools and solutions exist to undertake this mode of operation, which include all or various combinations of the following.

- Technology that can take real-time or near-real-time data from telemetry systems to understand the boundary conditions or dynamics of the water distribution system.

231

- A network model, or models, linked with an optimiser, to arrive at optimised solutions for the operation of the network, and modelling changes to pump combinations and required valve changes. Predictions need to schedule forward-looking changes towards an appropriate horizon (24 to 48 hours ahead).
- Network models also offer the opportunity to run offline scenarios for strategic planning, particularly in relation to planning maintenance and refurbishment programmes to infra and non-infrastructure assets, and risk assessment for undertaking such work.
- Technology that can control assets remotely through interfaces driven by the optimised pump and valve schedules, with back up functionality for manual or closed loop operation – that is, safe mode, interventions.

9.5.1.11 Pressure and demand management

Pressure management or the use of pressure reduction as a method of reducing leakage is an accepted approach that is widely used across the industry. Water losses are reduced as a consequence of the following.

- Leakage that is pressure dependant; as defined by the standard orifice equation, discharge through an orifice is a function of the size of the opening and the square root of the pressure head, thus reducing pressure will reduce the volume of leakage.
- Maintaining a stable network; reducing variations in pressure, there is lower propensity for mains, fittings and connections to rupture, reducing the likelihood of the main sources of leakage.

A practical guide to pressure management for leak reduction is provided by Thornton (2003), including extension of the standard orifice equation to consider fixed and variable area leaks and the impact of pressure variations as well as absolute values.

Other benefits that pressure management include are as follows.

- Un-regulated demand or 'open-tap use' will be reduced by reductions in pressure.
- Reduced water volume, as a result of leakage reduction will reduce required distribution input, hence will give subsequent reductions in power and chemical costs for water production.
- Environmental savings – that is, reduced water volume – will reduce wastewater treatment, translating into a reduced environmental impact.

Pressure management or reduction is generally implemented through the installation of a pressure reducing valve (PRV). Typically installed on a bypass, and of a reduced diameter to the host main, water is constricted within the valve body and directed through an inner chamber controlled by an adjustable spring-loaded diaphragm and disc. As the inlet pressure fluctuates, the PRV ensures a constant flow of water at a functional pressure, as long as the supply pressure does not drop below the valves pre-set outlet pressure. There are a number of types of PRV that may be used for pressure management and, hence, in any pressure management scheme, the performance of the valve should match the specific operational requirements and conditions of the network. The continual development of control equipment, coupled with enhancement in optimisation tools, data insight and real-time modelling capabilities, means that pressure management continues to be one of the key opportunities to reduce leakage and bursts.

When implementing pressure management schemes, it is important to consider the following.

- Customer perception. Where pressures are reduced, some customers may perceive a degradation of service, even though adequate levels of service are maintained. When the PRV scheme is commissioned, it is recommended that pressures are reduced over a number of steps. The magnitude and number of steps should be governed by the overall magnitude of pressure to be reduced.
- Sensitive or vulnerable customers that may be significantly affected by pressure reduction. These may include commercial users with fire suppression systems or dialysis patients requiring a specific minimum pressure for systems to operate. Consideration should particularly be given to ensuring the provision of adequate firefighting flows in the network following the implementation of a PRV scheme.
- Maintenance. The PRV should be inspected and maintained throughout its asset life. The costs of maintaining a PRV may be equal to, and often exceed, the costs of new valve. The installation's design should facilitate ease of maintenance and eventual replacement. Not maintaining the PRV will lead to an unstable network and reduction in overall leakage benefits.

The scale of a pressure management scheme is obviously an important consideration. Not simply the magnitude of pressure reduction, but also the number of properties encompassed in the scheme. Generally, the larger the number of properties, the greater the resulting benefits and cost benefit of the scheme. Ideally the number of pressure management schemes should be optimised or rationalised, minimising the number of PRVs to be monitored and maintained. However, there is often a reluctance of treating pressure management in a more holistic sense, due to levels of historical investment in pressure management and quantified risks of level of service failures to large numbers of customers.

Other opportunities for demand management are realised from the following.

- Promotion of water efficiency, through structured and targeted campaigns for user awareness of water use and how it can be minimised, both through changes in behaviour, the use of grey or untreated sources for activities, such as irrigation, and the application of various forms of water saving devices fitted to points of consumption, such as taps, toilet cisterns and shower heads. Water companies have now been tasked with the need to drive water efficiency with their customers to achieve regulatory targets.
- Water metering, with customers being charged for the volume of water used and disposed of by way of the sewerage network. In general, awareness of charging for actual water used drives customer behaviour to reduce consumption.
- Rationalisation and optimised pumping, using reduced pumping delivery heads to reduce pressures across the water distribution network. This can be applied where customers' available pressure can be reduced, similar to application of a PRV for pressure management, or where greater turnovers of storage volumes at services reservoirs can be achieved, by, for example, reducing pumping durations. The reduction in pressure results in reduced volumes of water lost through leakage.

9.5.2   Quality

This section focuses on the approaches and techniques associated with undertaking mains rehabilitation, driven, primarily, by water quality considerations.

9.5.2.1   Mains re-lining, swabbing and scouring

The approaches discussed here are seen as aggressive techniques to return the internal condition of mains to varying degrees approaching their original 'as-installed' condition. These are

- pigging or swabbing
- air-scouring
- re-lining.

9.5.2.1.1   PIGGING OR SWABBING

Pigging and swabbing are probably the most effective and proven approaches to cleaning pipelines to improving serviceability and reduce or defer capital investment programmes, capable of removing biofilms, sediments, tuberculation and other scales. A bullet shaped pig or swab is inserted into the main and is driven by the hydraulic force of mains pressure. As the pig or swab is pushed through the main, its abrasive force on the pipe wall cleans the pipe with the residual debris pushed along the pipe in front of the device.

A main may be pigged or swabbed several times (progressive pigging / swabbing), where the 'aggressiveness' of the pig or swab may be gradually increased, based on the number of repeats and type or size of the water main being cleaned. Careful selection is required as a too aggressive clean can exacerbate water quality problems by removing tuberculation from unlined iron pipes and causing accelerated corrosion, bleeding and deposition of material into supply.

Deploying pigs or swabs requires launch and receiving pits. For distribution size mains (i.e. 4 to 6 inch / 100 to 150 mm) hydrants often provide ready-made access. For mains of larger diameter, it is often necessary to install a 'Y' or 'T' entry facilities, and these have to be specially installed as part of preparatory works.

A technique that uses a crushed ice slurry to replace the pig or swab (University of Bristol, 2010) has been developed. A slug of ice is inserted into the main through a fire hydrant, driven by mains pressure and is then flushed out with the resulting debris. The consistency or density of the slug is varied dependant on the required abrasiveness. Uptake and reports of the efficiency of this technique vary (Sunny *et al.*, 2017).

9.5.2.1.2   AIR-SCOURING

Air-scouring is another effective technique for mains cleaning. By isolating a section of water main between an entry and exit point, typically between two hydrants, the section is purged of standing water with a high volume of low-pressure air at a high velocity. By opening the valve upstream of the entry point, slugs of water are then produced allowing the passage of controlled volumes of water into the section. The compressed air, mixing with the water slugs, creates a highly turbulent disturbance that travels through the main, removing sediment and bio-film material. Not as abrasive as pigging or swabbing, air-scouring does have the distinct advantage of, typically, being a third of the cost of pigging or swabbing.

9.5.2.1.3   RE-LINING

Following the abrasive cleaning of any iron main, it is recommended that the cleaned main is relined to provide an inert surface to reduce any further potential for internal corrosion. See Figure 9.8.

Historically, mains were lined with a cement mortar lining (typically 4 mm thickness), sprayed on the cleaned pipe wall. Water quality and water chemistry can have an impact on this form of liner,

Figure 9.8 Relining (Reproduced with the permission from © WRc)

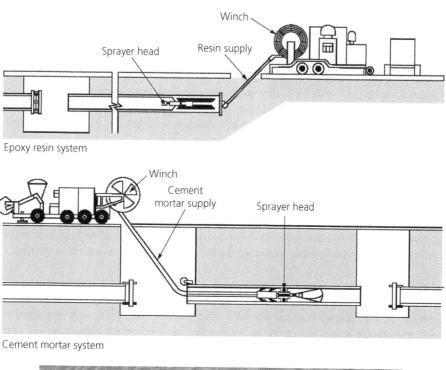

Epoxy resin system

Cement mortar system

leading in some cases to the lime in the mortar leaching away, destroying the liner and leaving a sand residue. Curing time for this form of lining can take one day or more before successfully re-commissioning the main.

More recently, epoxy or polyurethane (PU) spray lining systems have been developed. A thin lining of resin (typically 1 mm thickness) is sprayed onto the cleaned surface of the main. Semi-structural liners with a greater thickness (typically 2 to 3.5 mm) are being developed to reinforce the structural capabilities of the host main. A particular benefit of the epoxy liners is their much-reduced curing time before the main can be re-commissioned, typically epoxy or PU lined mains can be commissioned within 2 to 8 hours of relining.

Attention needs to be paid to maintaining service connections when applying liners. Commonly these are reinstated by the deployment of special drilling pigs, however some of the latest epoxy and PU spays employ a vaporising technique such that service connections are not covered by the lining.

9.5.2.2   Mains flushing

Historically, water companies have flushed water mains in response to customer complaints or following maintenance work. In recent times, companies are implementing regular, structured flushing programmes as part of their strategy to maintain water quality in the distribution system. Flushing can be a remedy to aged water in the network and/or removal of accumulated material by scouring clean internal pipe surfaces to reduce the risk of discolouration events occurring.

Analysis has shown that the particles responsible for discoloration in UK networks are small and light (Boxall et al., 2001). Their behaviour is rarely dominated by gravity driven self-weight and invert deposits of sedimentary material are rare, other than in dead-ends, very oversized pipes or tidal points on network loops. Research has shown that material accumulates over the entire sur-face of distribution mains in cohesive layers. The daily demand patterns and the resultant shear stress is used to describe how the material accumulates in layers, and these layers are remobilised in the main by increasing the shear stress above that found in the daily pattern (Boxall and Saul, 2005; Vreeburg and Boxall, 2007; Husband et al., 2008). Discolouration events often occur due to an increase in shear stress such as caused by a burst or valve operation that may be some distance upstream of where the contact occurs.

Mains are flushed over specific lengths, typically by the opening of a fire hydrant with the attach-ment of a standpipe. A flow meter may also be attached to monitor and regulate the aggressiveness of the flushing operation and turbidity instrumentation added to measure the material mobilised and removed. Water quality samples may also be taken during the course of the flush for further analysis. Flushing operations are typically conducted during the night to minimise disruption to customers. Figure 9.9 shows a monitored night-time flushing operation being conducted by the University of Sheffield in association with a UK water company.

Ideally, the hydraulic forces required and imposed during flushing should be estimated for each pipe in a network (Friedman et al., 2003). However, the effort involved, consequence (customers potentially exposed per pipe) and onsite practicalities often mean full DMA unidirectional is not justifiable. Effective flushing is usually achieving with simple universally applied target velocities, such as 1.0 to 1.5 m/s, but with care and attention to flow routes and to avoid unexpected mobilisa-tion of upstream material. Such flushing is practically limited to smaller diameter pipes due to the volumes of water involved; it is rarely a viable option for large diameter trunk mains.

As a highly effective and efficient alternative UK water companies have been applying the idea of 'flow conditioning' to trunk mains (Husband and Boxall, 2016). This involves the carefully

Figure 9.9 Monitored flushing operation being undertaken by the University of Sheffield

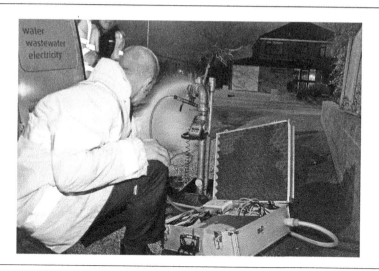

managed increase of flow at normal peak hour to mobilise a small amount of material, well below regulatory limits, with sequential repeats slowly increasing the imposed force, resulting in the cohesive layers most at risk of mobilisation being gradually removed. Confidence in simulation results and real-time monitoring are essential for the success of this method. However, a concern with this approach is that the mobilised material is retained within the system and passes on to the downstream network. Research has shown that it is the long-term concentration of material rather than short-term peak loading that controls material accumulation, with benefits being observed in downstream networks where trunk main flow conditioning is imposed regularly (Sunny *et al.*, 2020).

Further information on mains rehabilitation techniques can be found in WRc (1989).

9.6. Decisions

The decision to operate and maintain the distribution network is made in accordance with the needs of the consumers it supplies and is governed by the regulatory standards set out to protect defined levels of service.

In the first instance the design of the network should be sufficient to provide a wholesome supply of water at a sufficient quantity on demand. The decision-making process is instigated either proactively or reactively. Proactive operation and maintenance are triggered where the operator takes action to meet expected changes – for example, a change in demand due to seasonal tourism. The operator will take action in accordance with either a locally or globally prescribed operating regime, defined within an operation and maintenance schedule. Reactive operation and maintenance is the more common approach. The network is operated and maintained in response to it failing to meet the defined criteria or when deviation from performance standards is observed through monitoring.

Despite encouraging developments in smart network networks, which increasingly enables early detection of sub-optimal network performance, the consumer is often the first to observe that the level of service is not being achieved. Such customer observation may be clearly identified where the level of service has been completely lost, such as in the case of a burst main, or it may be a more subjective assessment of the acceptability of the product's aesthetic appearance.

In practice, a typical water company will react both proactively and reactively to operate and maintain the network to achieve the required level of service. The key levels of service around which a typical operation and maintenance schedule is designed are, in England and Wales, collectively known as the performance standards and are measured through the performance commitment and outcome framework. How a water company takes decisions to maintain or improve system performance against the standards will vary from one organisation to another. However, because quantity and quality performance, and hence problems, are closely related, the decision process does not strictly differentiate between quantity and quality aspects and investigations to determine the cause and identify a solution, and therefore, tend to include both parameters.

Broadly speaking, the following functions within an organisation will be involved in the decision-making process.

9.6.1 Customer contacts

The majority of decisions will be initiated following contact from the consumer. Most companies provide an operational contact centre to deal with questions and observations from consumers and resolve them. In some cases, problem resolution may be possible on first contact where known historical performance issues are held on corporate databases, but, for the majority of contacts, the customer observations, such as discoloration, visible leaks and so on, are recorded and forwarded to appropriate company sections to be dealt with.

9.6.2 Field management

Where the performance issue is identified before the customer becomes aware of it, the problem will be passed directly to a 'field management function' for further investigation. A field manager will often have direct access to resources and business processes which can be used to identify the issue through desktop evaluation or, if further information is required (e.g. where there is a loss of supply) by deployment of a field-based technician.

Advances in technology and business process now mean that this process is mostly automated using work management and scheduling tools that select the most appropriate resource to respond. The field-based technicians attend site to try to understand further the nature of the issue faced by the consumer. Where this issue can be easily resolved (e.g. by re-configuring the network or undertaking minor maintenance such as replacement of a stop tap), the query is closed immediately, and feedback is provided directly to the consumer or relayed to them through the operational contact centre.

9.6.3 Network asset management

Where resolution of the problem is more complicated (e.g. where the consumer is experiencing intermittent pressure variations), the issue is usually referred to a network or asset management function. Most water companies have a performance or asset management function that is sub-dived into sections that focus on specific issues such as leakage, water quality or structural integrity. These functions provide a more proactive approach to network operation and maintenance. Within this function, the distribution engineer will often have an overarching strategic plan such as the Asset Management Plan (AMP), and defined within this plan will be the specific performance standards and levels of service that the water company intends to maintain or exceed.

The AMP is supported by more detailed operation and maintenance schedules such as Distribution Operations and Maintenance Strategies (DOMS). These strategies will set out the timeframe over which the network will be monitored, evaluated and maintained. For simple network configurations, a distribution engineer may utilise corporate performance figures such as flow and pressure data and asset records to determine the nature of the issue. Where the issue is more complicated and it is necessary to undertake additional flow and pressure monitoring in conjunction with hydraulic analysis to identify the problem, support may be required from specialist analysts. Once the cause has been identified the distribution engineer will promote a solution to the problem.

The solution, and those responsible for its implementation, will depend on the complexity of the solution procedure. For minor maintenance, such as rezoning, the solution can be implemented by the field-based technicians or partner organisations who undertake repair and maintenance. However, where more complex solutions are required involving capital maintenance on the network, the problem will often be passed to investment functions that will prioritise the work according to cost and benefit analysis and implement the solution through an investment and resource management programme.

9.6.4   Additional monitoring and analysis

Where the distribution engineer identifies a divergence from the defined performance standards or levels of service it will be required to undertake additional analysis. In order to make the monitoring and evaluation of the network more manageable, they will typically be subdivided into water supply systems. These are notional hydraulically defined areas with one or more source waters and which may contain one or more water quality zones. In turn these are made up of DMAs. The DMA concept, in its current form, was introduced during the 1990s to improve the control, management and reporting of water losses in the network but, today, it is often the foundation of all proactive monitoring and operation.

The keys pieces of information used in the decision-making process are flow, pressure, consumer contacts, water quality sample analysis, and network data such as asset records, consumption records, connections, burst records and pipe samples. All of this information will be at the disposal of the distribution engineer when investigating variance from performance standards. The quality of this information varies from one water company to the next. Where good quality data is available, accessible and well maintained, the ability of the distribution engineer to monitor, evaluate and make good decisions with regard to the operation and maintenance of the network is greatly enhanced. The distribution engineer will access and evaluate data according to the needs of the required operational and maintenance activity and the performance standards which are being managed.

In the case of leakage, this is an ongoing process with flow, pressure and demand information constantly being collected and processed to identify differences between the observed and target leakage level. The most commonly utilised approach is to monitor the nightline (Section 8.3.1). Much of the analysis and leakage detection is still undertaken manually, but, as technology advances, there is the potential for a greater degree of automation and on-line decision making.

For water quality, the processes associated with the operation and maintenance of the system are applied over a longer timeframe, often assessing water quality data and consumer contacts over a period of months and years, to identify trends in network deterioration. Similarly, pressure variation and structural integrity will be assessed over a longer timeframe to identify trends. The distribution engineer will then decide whether to take small corrective interventions or develop longer term investment programmes to address a range of performance issues, thereby making the investment much more cost effective.

The challenges facing distribution engineers, operations staff and managers over the next 10 years relate as to how they can more effectively make decisions across large and complex water networks as a whole network and without the need to sub-dive the network to the extent as it is today. In addition, the expectation of the consumer is increasing all the time and changes to the regulatory framework, such as the introduction of the Service Incentive Mechanism in England and Wales, will undoubtedly lead to a need for new tools and techniques. The ability, in future, to locate, diagnose and repair faults on the network without interruption to supply and disruption to the public will become increasingly necessary.

9.7. Evaluation

The water industry in the UK, including regulators and water companies, have clearly positioned themselves as a service industry, where the customer experience is a primary consideration. A key factor in achieving and improving this service provision is the need to move from a reactive to a proactive strategy for the effective operation and maintenance of distribution systems.

At the present time, it is usually the customer who is the first to know of a service problem and customer contacts alert the water companies who, subsequently, react to the event. This approach has recognised shortfalls, and this has moved the industry to invest in research that is directed at the introduction of a better understanding of system performance in near real time and to stimulate a proactive operation and maintenance strategy. There is, therefore, a move towards near real time monitoring and modelling where systems are appropriately monitored and data are analysed to provide the best available knowledge of the system status such that failures are anticipated and decisions made to intervene prior to service interruption or failure. Such an approach requires a paradigm shift and a move away from data collection and analysis, driven by regulatory requirements, to a system where near real time data collection and analysis is used to best inform water companies of the condition and performance of their significant infrastructure asset base.

Reacting to water supply system events, such as low water pressure or poor water quality, as and when they occur, with little to no anticipation of the events, describes the current operational and maintenance strategy in most UK Water Companies. The usual outcome of such a reactive strategy is to undertake a repair, initiate maintenance or change operational practice in order to return the supply system to again meet the regulatory requirements. It is recognised that there are major problems with this approach, in that, by waiting for an event to occur, the company is not fully in control of the performance of the system and standards of service are often compromised. The customer often experiences poor or failing service for long periods before the company is aware or able to react. Once aware, it is then necessary to carry out an investigation and subsequently propose, plan and implement a solution. All this takes time, during which the effects of the impact of the event may become widespread and the solution procedure more complex and costly.

Once the solution has been implemented, further observations and analysis are completed to ensure that the system status again meets compliance standards. As a consequence, the reactive management approach is one where it is usual to quickly get the system resources back into production, whether it be supply, infrastructure or people. What is also clear is that, over the years, water companies have developed significant expertise in reactive management and now have the ability to

- rapidly identify the root cause of events
- devise different creative possible solutions that are proven and innovative
- select the best remedial option and to quickly implement the chosen solution, calmly and in control in the midst of a crisis, to resolve the problem.

Some argue 'why change' but current thinking and advances in technology and communications highlight that there is considerable potential for the industry to move to more proactive strategies for the maintenance and management of system performance and operation.

Proactive planning and management involves anticipating system events such as operational problems, water quality failures and customer demands prior to their occurrence. This requires forward planning to agree the desired future operational state of the system and subsequently to implement a strategy and schemes that will deliver and create the preferred future state. This approach allows a company to actively control the outcome of planning and investment and to actively shape the future. To move to such a proactive approach also provides the opportunity to reduce the potential of future threats by managing risk, to capitalise on a potential future opportunity and subsequently to optimise efficiency gain and cost benefit.

Proactive planning and management involves

- planning for the short and long term to meet company, regulatory, operational and customer service requirements
- taking calculated risks
- encouraging innovation
- working closely with technical and customer service staff and regulators
- undertaking customer satisfaction surveys.

Clearly, the proactive approach is the one that provides most opportunity for operational stability and control, and tools and methods are constantly being improved and some are now being implemented (Savic *et al.*, 2008; Bicik *et al.*, 2009).

In order to manage proactively, it is essential to understand all aspects of the supply system, including the current system status and of the way in which operation and maintenance strategies influence this performance. The various regulatory requirements provide a base framework for the performance requirements of the system, and developments in instrumentation, communications technologies, computing power and data processing methods are enabling companies to better understand the hydraulic and water quality characteristics of systems operation in near real time. Hence, these systems provide companies with a timely feed of data that is transformed into information on which future operation and maintenance decisions may be based, and such that the proactive management of water supply systems may become a reality. In time, these systems will be the normal mode of system operation and will vastly improve over the coming years.

The vision for the future is real time operational control of all aspects of water supply in order to continually optimise storage, flow, pressure, water quality, energy use, leakage and carbon emissions. This will be achieved by introducing even smaller, cheaper, robust instrumentation at key locations that will have intelligent data analysis geared to record and report only data that is outside normal operational bounds. Continued technological development will drive the introduction of smart assets that have local measuring capabilities built in at the manufacturing stage and the embedding of data processing functionality with self-diagnostic algorithms.

Transmission of data may, by this time, be through the fabric of the assets or even the body of the water within the system itself. Data from the field will be passed directly to hydraulic, water quality and performance models which in turn will feed into operational models and decision support systems. The models and systems will be capable of accepting operational feedback and will automatically control key assets to continually optimise system operation. The future is bright but will require significant investment to maintain the current provision of safe and wholesome water throughout the UK water distribution systems.

REFERENCES

AWWA (American Water Works Association) with assistance from Economic Engineering Services (2002) *Effects of water age on distribution system water quality.* Total Coliform Rule, Distribution System White Papers. http://www.epa.gov/safewater/disinfection/tcr/pdfs/whitepaper_tcr_waterdistribution.pdf, [Accessed: April 2nd, 2007].

AWWA (2007) *Evaluating Water Loss and Planning Loss Reduction Strategies.* AWWA RF Report 91163.

Bicik J, Kapelan Z and Savic DA (2009) Operational perspective of the impact of failures in water distribution systems. *Proceedings of the World Environmental and Water Resources Congress.*

Brevis W, Susmel L and Boxall JB (2015) Investigating in-service failures of water pipes from a multiaxial notch fatigue point of view: A conceptual study. *Proceedings of the Institution of Mechanical Engineers, Part C: Journal of Mechanical Engineering Science* **229(7)**: 1240–1259.

Boxall JB, Skipworth PJ and Saul AJ (2001) A novel approach to describing sediment movement in distribution mains, based on measured particle characteristics. *Proceedings of the International CCWI (Computing and Control for the Water Industry) Conference.* De Montfort University, UK.

Boxall JB and Dewis N (2003) Identification of discolouration risk through simplified modelling. *Journal Environmental Engineering, ASCE (American Society of Civil Engineers).* EWRI 2005.

Boxall JB and Saul AJ (2005) Modelling discolouration in potable water distribution systems. *Journal Environmental Engineering, ASCE* **131(5)**: 716–725.

Collins RP, Boxall JB, Karney BW, Brunone B and Meniconi S (2012) How severe can transients be due to a sudden depressurization? *Journal of the American Water Works Association* **104(4)**: E243–E25.

European Environment Agency (2003) *Indicator Fact Sheet (WQ06) Water use efficiency (in cities): leakage.*

DWI Information Letter 15/2002

DWI (2004) Guidance on the Water Quality Aspects of Common Carriage.

Farley B, Boxall JB and Mounce SR (2008) Optimal location of pressure meters for burst detection. *10th Annual International Symposium on Water Distribution Systems Analysis, 17–20th August,* Kruger National Park, South Africa.

Friedman MJ, Martel K, Hill A *et al.* (2003) *Establishing site-specific flushing velocities.* AWWA Research Foundation, Denver, CO, USA.

Fox S, Shepherd WJ, Collins RP and Boxall JB (2015) Experimental quantification of contaminant ingress into a buried leaking pipe during transient events. *ASCE Journal of Hydraulic Engineering* **142(1)**: 1–10.

House of Lords (2006). Water Management, Volume I: Report, House of Lords Science and Technology Committee, HL Paper 191-I, 6th June 2006.

Husband PS, Boxall JB and Saul AJ (2008) Laboratory studies investigating the processes leading to discolouration in water distribution networks. *Water Research* **42(16)**: 4309–4318

Husband PS and Boxall J (2016) Understanding and managing discolouration risk in trunk mains. *Water Research* **107**: 127–140.

IWA (2005) *Water Loss Group: IWA Task Force: Best Practice Performance Indicators for Non Revenue Water and Water Loss Components: A Practical Approach.* IWA, Halifax, UK.

Kirmeyer *et al.* (2000) *Guidance manual for maintaining distribution system water quality.* AWWA and AWWA Research Foundation, Denver, CO, USA.

Lambert A (2003) Assessing non-revenue water and its components: a practical approach. The IWA Water Loss Task Force, Water21, Article No. 2, pp.50–51.

Machell JM, Boxall JB, Saul AJ and Bramely D (2009) Improved representation of water age in distribution networks to inform water quality. *Journal of Water Resources Planning and Management, ASCE* **135(5)**: 382–381.

Machell J, Mounce SR and Boxall JB (2010) Online modelling of water distribution systems: a UK case study. *Drinking Water Engineering and Science* 3: 21–27.

Machell JM and Boxall JB (2014) Modelling and field work to investigate the relationship between the age and the quality of drinking water at customer's taps. *Journal of Water Resources – Planning and Management* **140(9)**: 1943–5452.

Morrison J (IWA Water Loss Task Force) (2004) Managing leakage by District Metered Areas: a practical approach. *Water 21* **5**: 44–47.

Mounce SR, Boxall JB and Machell, J. (2007) An Artificial Neural Network/Fuzzy Logic system for DMA flow meter data analysis providing burst identification and size estimation. *Water Management Challenges in Global Change.* Ulanicki *et al.* (eds), Taylor and Francis, pp 313-320, ISBN 978-0-415-45415-5.

Mounce SR, Boxall JB and Machell J (2008) Online application of ANN and Fuzzy Logic system for burst detection. *Proceedings of the 10th Water Distribution System Analysis Symposium, Kruger National Park.*

Mounce SR, Boxall JB and Machell J (2010) Development and verification of an online artificial intelligence system for detection of bursts and other abnormal flows. *Journal of Water Resources Planning and Management* **136(3)**: 309–318

Mounce SR, Mounce RB and Boxall JB (2012) Identifying sampling interval for event detection in water distribution networks. *Journal of Water Resources Planning and Management, ASCE* **138(2)**: 187–191.

Mounce SR, Gaffney JW, Boult S and Boxall JB (2015) Automated data-driven approaches to evaluating and interpreting water quality time series data from water distribution systems. *Journal of Water Resources Planning and Management, ASCE* **141(11)**: 1–11.

Mounce SR, Mounce RB and Boxall JB (2016) Case-based reasoning to support decision making for managing drinking water quality events in distribution systems. *Urban Water Journal* **13(7)**: 727–738.

Mounce SR, Ellis K, Edwards JM *et al.* (2017) Ensemble decision tree models using RUSBoost for estimating risk of iron failure in drinking water distribution systems. *Water Resources Management* **31(5)**: 1575–1589.

Mounce SR, Shepherd WJ, Boxall JB, Horoshenkov KV and Boyle JH (2021) *Autonomous robotics for water and sewer networks*. HydroLink 2021-2 Artificial Intelligence. https://www.iahr.org/library/infor?pid=10799 [accessed 7 April 2022]

Mutoti G, Dietz JD, Imran S, Taylor J and Cooper CD (2007) Development of a novel iron release flux model for distribution systems. *Journal of the AWWA* **99(1)**: 102–111.

NIC (2018) Preparing for a drier future: England's water infrastructure needs. https://nic.org.uk/app/uploads/NIC-Preparing-for-a-Drier-Future-26-April-2018.pdf [accessed 4 April 2022]

Ofwat (2002) *Future approaches to leakage targets for water companies in England and Wales – Tripartite Study, Best practice principles in the economic level of leakage calculation.* Ofwat, London, UK.

Orr C, Bouulos P, Stern C and Liu P (1999) Developing real-time models of water distribution systems. *Modelling and optimisation applications* (Water engineering and management series, 18 3-4) by D. Savic and G. Walters (eds.). vol. 1, Research Studies Pr.

Prasad TD and Walters GA (2006) Minimizing residence times by rerouting flows to improve water quality in distribution networks *Engineering Optimization* **38(8)**: 923–939.

Rossman LA, Clark RM and Grayman WM (1994) Modeling chlorine residuals in drinking water distribution systems. *Journal of Environmental Engineering, ASCE* **120(4)**: 803–820.

Rossman LA and Boulos PF (1996) Numerical methods for modeling water quality in distribution systems: a comparison. *Journal of Water Resources Planning and Management* **122(2)**: 137–146.

Savic, DA, Boxall JB, Ulanicki B *et al.* (2008) Project Neptune: Improved Operation of Water Distribution Networks. *Proceedings of the 10th Annual Water Distribution Systems Analysis Conference (WDSA2008)*, Kruger National Park, South Africa.

Shucksmith JD, Boxall JB, Staszewski WJ, Seth A and Beck SBM (2012) On site leak location in a pipe network by cepstrum analysis of pressure transients. *Journal of the AWWA* **104(8)**: E457–E465.

Skipworth PJ, Saul AJ and Machell J (1999) Predicting water quality in distribution systems using artificial neural networks Source. *Proceedings of the Institution of Civil Engineers, Water Maritime and Energy* **136(1)**: 1–8.

Speight V, Mounce SR and Boxall JB (2019) Identification of the causes of drinking water discoloration from machine learning analysis of historical datasets. *Environmental Science: Water Research and Technology* **5**: 747–755.

Sunny I, Husband PS, Drake N, Mckenzie K and Boxall JB (2017) Quantity and quality benefits of in-service invasive cleaning of trunk mains. *Drinking Water Engineering and Science* **10**: 45–52.

Sunny I, Husband PS and Boxall JB (2020) Impact of hydraulic interventions on chronic and acute material loading and discolouration risk in drinking water distribution systems. *Water Research* **169** : 115–224.

Thornton J (2003) Managing leakage by managing pressure: a practical approach. *The IWA Water Loss Task Force, Water 21* **3**: 43–44.

University of Bristol (2010) *Ice pigging*. http://www.bristol.ac.uk/red/techtransfer/scilicopps/ice-pig.html [Accessed 20 May 2010]

Vreeburg J and Boxall JB (2007) Discoloration in potable water distribution systems: a review. *Water Research* **41(3)**: 519–529.

Water Science and Technology Board, Committee on Public Water Supply Distribution Systems: Assessing and Reducing Risks, National Research Council (2005) *Public water supply distribution systems: assessing and reducing risks, First Report.* National Academy Press, DC, USA.

WRc (Water Research Centre) (1989) *Planning the rehabilitation of Water Distribution Systems.* WRc, Swindon, UK.

WRc (1994) *Managing Leakage, Report A.* U.K. WRc, Swindon, UK.

REFERENCED LEGISLATION

The EU Drinking Water Directive (1998) European Council Directive 98/83/EC of 3 November 1998 on the quality of water intended for human consumption, Official Journal of the European Communities L 330, 5th Dec 1998.

The Water Supply (Water Quality) Regulations 2018, Stationery Office, London.

Water Industry Act 1999, HMSO, London, UK. (supersedes Water Industry Acts 1945, 1973, 1989, 1991)

Water Undertakers (Information) Directive 2004. TSO, London, UK

Dragan A. Savić and John K. Banyard
ISBN 978-1-83549-847-7
https://doi.org/10.1108/978-1-83549-846-020242012

Chapter 10
Asset planning and management

Zoran Kapelan
Delft University of Technology, The Netherlands

John K. Banyard
Independent consultant, UK

Mark Randall-Smith
RS Analytical Solutions Ltd, Bournemouth, UK

Dragan A. Savić
KWR Water Research Institute, The Netherlands and University of Exeter, UK

10.1. Introduction

The concept of asset management as a specific discipline in the (UK) water industry goes back to the mid 1980s. The privatisation of utilities (gas, electricity, telecoms and so on, a group to which water companies are also normally considered to belong) brought conventional financial disciplines to previously 'nationalised' industries. Not only did the newly privatised utilities have to meet strict performance targets, they also had to do so within clearly defined financial constraints, monitored by their investors and other providers of capital.

One of the disciplines introduced by a more rigorous financial regime was the need to consider carefully the question of depreciation. All assets wear out, and conventional accounting practice requires this to be reflected in the value of the assets shown on the balance sheet. This is achieved by predicting the life of each asset and then reducing the value in the balance sheet to zero over the chosen life of the asset. It is essential that this book life is achieved otherwise the whole residual value of the asset has to be written out of the accounts when the asset is taken out of service. Since financially this sum is subtracted from what would otherwise be profit, it can be extremely damaging to a company's performance if premature write off of assets is common place, however that does not mean there can be unlimited spending to preserve assets, since that would also damage the profitability.

Because the utilities are distinguished from other companies by having extremely large asset bases, this resulted in considerable attention being applied to the management of the asset base, and the discipline of asset management has emerged as a result of the importance of being able to control and forecast the serviceability of the asset base. Furthermore, it has been adopted by other organisations to assist in management of large asset bases such as railways, and the concepts are widely adopted for utilities whether they are state owned or private companies.

The objective of this chapter is to outline some of the main issues and potential solutions related to the planning and management of water supply and distribution system assets. This includes

- relevant background information (Section 10.2)
- regulatory aspects (Section 10.3)
- main asset management drivers and issues (Section 10.4)
- asset performance indicators (Section 10.5)
- asset assessment techniques (Section 10.6)
- interventions used to improve assets' condition/performance (Section 10.7)
- asset ageing and deterioration issues (Section 10.8)
- asset failure consequence modelling (10.9)
- an integrated, whole-life costing based framework for making more informed asset management decisions (Section 10.10).

Finally, a summary is made and future challenges identified in the last section of this chapter.

10.2. Background

The replacement value of fixed tangible assets in the possession of all water (and sewerage) companies in England and Wales is well over £250 billion with an annual turnover of about £12 billion (from Water UK website as of 2018). According to the same source, the total annual capital investment was £4.9 billion, of which £2.5 billion was spent on maintaining the networks alone. Average Totex (total expenditure) has been at around £10 billion a year since 2000, and average Capex (capital expenditure – money spent on assets) has been between £5 billion and £6 billion a year, reaching the highest point in that range (£6 billion) between 2015–2020 (Ofwat, 2022). The quantity of assets that require careful planning and management is often vast, especially in the larger water utilities and companies – for example, the total length of water mains in the UK is approximately 350 000 km.

As understanding of the asset base has increased, so has the complexity of what is meant by asset management. The widely accepted definition of asset management can be found in the ISO 55000 standard, which states that it is *'the coordinated activity of an organization to realize value from assets'*. The standard defines an asset as an *'item, thing or entity that has potential or actual value to an organisation'*, while *'realisation of value will normally involve balancing of costs, risks, opportunities and performance benefits'*. Therefore, asset management is more than doing things to assets – it is about using assets to deliver value and achieve the organisation's business objectives (IAM, 2015). An alternative definition is adopted by the MWH's Asset Group (TAG) (Box 9.1)

Box 9.1   Asset management definition by MWH TAG

'Asset management is a business discipline for managing the life cycle of assets to achieve a desired service level while mitigating risks and it encompasses management, financial, customer, engineering and other business processes.'

These definitions cover a number of important aspects of asset management.

- Firstly, asset management is a *business* discipline and as such, the success is measured using business indicators (e.g. profit).
- Secondly, assets need to be managed through the whole *life cycle* of an asset – that is, from planning and purchasing to installing, maintaining and eventually disposal.

- Thirdly, one of the main asset management goals is to achieve the *target service level* at the lowest reasonable overall cost, while mitigating various *risks*.
- Finally, it is a complex activity encompassing a large number of *diverse* processes, not all of which are of engineering nature (a fact often overlooked by engineers).

Because it is so complex, it is necessary to tease apart some of the different aspects to better understand them and how they inter-relate with other elements. It is important, although, to always keep the above explanations in mind; it can be very misleading to base decisions on any single element.

For the purpose of asset management, all water supply and distribution system assets can be broadly classified into *infrastructure* assets (assets that are part of the water supply and distribution network) and *non-infrastructure* assets (e.g. treatment works, HQ buildings, vehicles etc.). Accountants would normally differentiate between fixed assets and mobile plant, and so it can be immediately seen that reference to the normal accounting practices will not meet the need, and that a new system of classification will need to be established, but one which uses data from the conventional accounting systems.

Another helpful classification is the one which separates the below and above ground assets. This important distinction is made as the location of assets relative to the ground level has a major effect on the way asset condition and performance are monitored and also the way assets are maintained. The below ground assets include primarily pipes – that is, water trunk and distribution mains and service connection pipes. The above ground assets include water sources (wells/boreholes, open reservoirs with dams, river captures etc.), treatment works, various pumps/pumping stations, tanks/reservoirs and so on – see Chapters 5 and 6 for a more detailed description of these elements. The big difference is that 'above ground assets' can be relatively easily inspected and accessed for maintenance. With buried pipes, the problems are very different; while closed circuit TV inspection may be possible to inspect particularly the inside of sewers, the ground/pipe interface is always hidden. However, the pipe network is made up of what may be interpreted as a number of asset classes, each comprising an assumed set of homogenous assets. Therefore, statistical techniques can be applied to help understand the overall serviceability of the network.

10.3. The regulatory framework

The issue of water supply and distribution has economic but also political, social and environmental dimensions. Also, the nature of the business is often referred to as a 'natural monopoly'; one cannot choose the water supplier, it is pre-determined on a geographic basis. As a consequence, the planning and management of water supply and distribution system assets is often regulated by the government.

The water industry in the England and Wales is regulated by the following main bodies:

- the Water Services Regulation Authority (Ofwat)
- the Environment Agency (EA)
- the Drinking Water Inspectorate (DWI).

All three regulators are established as statutory bodies to empower them for practicable implementation of the legal requirements of Parliament and policy aims of the government.

Ofwat is the economic regulator created under the 1989 Water Act (privatisation) as a non-ministerial government department directly responsible to Parliament. Ofwat was, therefore, created when the water industry in the UK was privatised, to protect the customers from the monopolistic nature of the business – that is, effectively the role of the economic regulator is to make up for the lack of market competition. Ofwat currently regulates 11 water and sewerage companies and six water only companies in England and Wales.

The Ofwat duties were laid down in Section 2 of the Water Industry Act 1991 as updated by Section 39 of the Water Act 2003 and subsequent legislation. The main duties are as follows (Ofwat, 2009):

■ To protect the interests of consumers, wherever appropriate by promoting effective competition.
■ To secure that the functions of all water companies are properly carried out and that they are able to finance their functions, in particular by securing a reasonable rate of return on their capital.
■ To secure that companies with water supply licences (i.e. those selling water to large business customers, known as licensees) properly carry out their functions.
■ To secure the long-term resilience of water companies' water supply (and wastewater) systems and to secure that they take steps to enable them, in the long term, to meet the need for water supplies (and wastewater services).

Ofwat regulates companies' performance by a number of different instruments, including the following (Ofwat, 2009):

■ Setting price limits that reflect what each company needs to charge to finance the provision of services to its customers. These are reviewed every five years based on business plans submitted by the companies. The business plans are based, among other things, on companies' asset planning and management plans which specify in detail what each company will deliver during the analysed five year period in order to maintain or improve the levels of different services.
■ Monitoring the activities of the companies. Until 2010–2011, each company provided detailed information to Ofwat about their performance each year. This annual data submission (or 'June return') was published to allow customers and stakeholders to understand each company's performance and provide Ofwat with details on a wide variety of activities including levels of customer service, new additions to the network and leakage information. This information also allowed Ofwat to compare performance levels between companies. Annual performance reporting has continued from 2011 onward but in a slightly different form.
■ Enforcing companies' licences. Companies operate under licences granted by the Secretary of State for Environment, Food and Rural Affairs and by the Welsh Assembly Government to provide water and sewerage services in England and Wales, respectively. The licences impose conditions on the companies which Ofwat enforces.

The EA and Natural Resources Wales (NRW) are the environmental regulators for England and Wales, respectively. Their principal aims are to *(a)* protect the environment and *(b)* improve the environment and promote sustainable development. They play a central role in delivering the environmental priorities of the central government and the Welsh Assembly Government through different functions and roles, including pollution control. They advise their respective Governments on environmental programmes and improvements required to meet the needs of national legislation but also the relevant EU Directives.

DWI is the potable water quality regulator. It is responsible for assessing the quality of drinking water in England and Wales, taking enforcement action if standards are not being met and appropriate action when water is unfit for human consumption. Legal standards are set out in the Water Quality Regulations, most of which are derived from the EU Drinking Water Directive.

Effectively, the DWI and EA/NRW set the quality standards for both potable water and sewage discharges, against a background of statutory obligations, and are responsible for enforcing those standards. Ofwat is responsible for setting water charges that allow those standards to be achieved by an efficient water company that uses to the full the physical and financial facilities available to it. It is important to stress, yet again, that Ofwat takes an economic view and will not be obliged to support expensive solutions simply because they appeal to technologists. Equally, Ofwat will make assumptions about the available sources of finance that would be available to a well run company; it has no obligation to ensure the continuance of any company that has squandered its financial resources.

Although Ofwat has been referred to here as the English and Welsh situation, the arrangements in both Scotland and Northern Ireland are similar but with their own independent regulatory bodies. This concept of independent regulators has become more common around the world, even though there are differences in structure and responsibility from country to country.

10.4.   Asset management drivers

Asset management in the UK is subject to a number of different pressures coming from various sources.

- First, when acquiring assets, companies must realise that they will only be given the finance once through customer charges. If they select inappropriate assets then the cost of replacement will fall on their owners (in England and Wales the investors).
- Like all physical assets, water supply and distribution systems are ageing and, as such, their condition and consequently performance is deteriorating with time, even though this can be slowed down or accelerated depending on the maintenance regime.
- The performance of these systems may also deteriorate due to population growth/increased urbanisation (e.g. increased water demand, changing spatial and temporal patterns of demand etc.) and climate change (reducing water available at source while increasing demand).
- There is an urgent need to reduce greenhouse gas emissions associated with water supply and distribution so that the water sector can contribute to meeting respective countries' targets and make these systems more sustainable in the long term.
- At the same time, customer expectations are increasing in terms of better service expected – for example, fewer interruptions to supply, better water quality, reduced negative impact on the environment, improved sustainability of various solutions, to name a few.

The above pressures affect all water utilities (both public and private), but the effect can be increased by the regulators' own agendas – for example, an economic regulator may wish to emphasise customer service, or a quality regulator may wish to emphasise the need to control provision of discoloured water to customers. This introduces a problem that will be identified in more detail later in this chapter, but it is useful to introduce the idea now. The drivers highlighted by regulators do not necessarily identify all of the issues that need to be taken into account when seeking to optimise the management of the asset base.

Regulators emphasise the issues that are currently of national importance, such as 'sustainability'; in doing so they assume that the more mundane issues such as maintaining borehole pumps are being properly addressed. If they are not, then the utility may well be disappointed when it seeks to increase charges in order to replace the pumps in a shorter timescale than its competitors. The consequence of this is that, for successful asset management, there is a need for more data than simply that which is required to respond to the regulators' demands.

To illustrate the complexity of this, a number of different issues can be considered that have been or are being addressed to improve the quality of the water service in the UK:

- Water *supply/demand balance* management including *security of water supply* and *demand efficiency* issues.
- *Leakage management.*
- *Water quality* management including aesthetic, bacteriological and chemical aspects of the problem.
- *Hydraulic capacity* including pressure related issues.
- *Energy consumption and cost* related issues.
- *Overall resilience* of the water service provided.
- *Customer service*, including issues related to consumer bills, complaints, provision of information to consumers etc.
- *Sustainability* including climate change, renewable energy, carbon footprint and other related issues.

Some of these are inter-related such as water supply/demand balance, and leakage. Others like Customer Service include issues that have no real relationship to asset management as defined here – for example, the speed of addressing customer complaints may well depend on successful outsourcing of the service.

However, this does provide further illustration of the dangers of relying solely on the regulators' measures when managing an asset base. For call centre operatives to respond properly to operational problems there needs to be good communication between the operational staff and the customer service staff. This will require IT systems such as geographical information systems (GIS). Regulators do not wish to micromanage utilities so they do not specify the need for GIS, nor the type of GIS that should be adopted. However, without a good GIS, it is most unlikely that customer service targets can be met; it is for management to make those decisions, not regulators.

10.5. Asset performance indicators

The planning and management of water utility assets should be undertaken with the objective of achieving or maintaining a desired and defined level of service to the customers and the environment at the lowest reasonable cost. This, in turn, is achieved by actively controlling the performance of relevant assets and systems, which may involve improving the assets or simply operating them in the appropriate manner.

A wide range of performance indicators (PIs) may be used to measure serviceability and performance of assets including those found in water supply and distribution systems. These PIs are used to measure the current performance but also to identify shortfalls against desired targets which are then used in turn to drive future asset management plans.

Performance indicators are also used for benchmarking – that is, when comparing performances of different water companies, either within the UK or internationally. However, this can be misleading unless care is taken to ensure that all parties are measuring the same indicator in exactly the same way.

Different definitions of performance indicators exist highlighting various aspects of their use. Some of the examples include

- standard and manageable way of measuring performance
- an indication as to how to progress towards a target
- a representation, numeric or otherwise, of the state of, or outcomes from, an organisation or any of its parts or processes.

Ofwat has introduced high level 'customer service' indicators to set target standards for the water companies which are shown below in Table 10.1 (original Director General (DG) indicators relevant for the clean water service) and Table 10.2 (additional indicators, introduced later on).

Table 10.1 Ofwat's original DG Water and Customer Service Indicators

Indicator	Title	Brief Description
DG2	Risk of low pressure	The number of connected properties that have received, and are likely to continue to receive, pressure below the reference level when demand for water is at a normal level. The reference level of service is defined as 10 m head of pressure at a boundary stop tap with a flow of 9 l/min.
DG3	Supply interruptions	This indicator shows the number of properties experiencing (unplanned) interruptions to their water supply for 3–6, 6–12, 12–24 and more than 24 h. Further guidance on assessing this indicator has been provided over the years.
DG4	Restrictions on water use	This indicator shows the percentage of a company's population that has experienced water usage restrictions. Water usage restrictions can be divided into voluntary reductions (encouraged by a publicity campaign), hosepipe restrictions, Drought Orders restricting non-essential water use and Drought Orders imposing standpipes or rota cuts.
DG6	Billing contacts	This indicator shows the total number of billing contacts that consumers made during the report year and the time each company took to respond to them. The time is measured in two bands: within five working days and in more than ten working days.
DG7	Written complaints	This indicator identifies the total number of written complaints received during the report year and the time taken to respond to them. The time is measured in two bands: within ten working days and in more than 20 working days.
DG8	Bills for metered customers	This indicator shows the percentage of metered consumers who receive at least one bill during the year based on a meter reading taken by either the water company (or its representative) or the consumer.
DG9	Ease of telephone contact	The aim of this indicator is to identify the ease with which consumers can make telephone contact with their local water company and their satisfaction with the way the company handled their call.

Table 10.2 Ofwat's other water service indicators

Drinking water quality: Operational Performance Index	DWI's Operational Performance Index (OPI), based on the following six parameters: iron, manganese, aluminium, turbidity, faecal coliforms and trihalomethanes.
Drinking water quality: percentage mean zonal compliance	The method that the DWI uses to assess and report on compliance with the drinking water standards is based on zones and is known as percentage mean zonal compliance. This is calculated for the eight parameters: E. coli, odour, taste, nitrate, aluminium, iron, lead and pesticides.
Security of Supply Index (SoSI)	Used to assess each company's compliance regarding its duty to ensure the security of its water supplies. At a company level, index scores reflect the size of any deficit against the company's estimate of target headroom in each of its resource zones and the proportion of consumers in each resource zone that is exposed to headroom deficits. For calculation details see Ofwat (2008c). Once calculated the SoSI values are banded as follows: A – no deficit against target headroom in any resource zone (SoSI = 100) B – marginal deficit against target headroom (SoSI = 90-99) C – significant deficit against target headroom (SoSI = 50-89) D – large deficit against target headroom (SoSI < 50).
Leakage performance	The amount of water leaked. A distinction is made between distribution, customer supply and service reservoir leakage. Reported in megalitres per day (Ml/d), litres per property per day (l/prop/day) and cubic metres per kilometre of main per day (m^3/km/d).
Water efficiency	Measured by customer supply pipe repairs and replacements, number of cistern devices distributed to households, number of household and non-household water audits and total savings/costs.
GHG emissions	Measured by total operational Green House Gas (GHG) emissions – that is, without GHG emissions released in the construction of assets or materials.

The method is called the 'overall performance assessment (OPA)' method and is used to measure and incentivise performance across the broad range of services provided to consumers and the environment. It allows Ofwat to compare the quality of the overall service, and tells consumers and other interested parties about how their local water company has performed relative to other companies.

Ofwat measures the serviceability of the water mains networks (i.e. water infrastructure) by using the following indicators (Ofwat 2008c):

- the extent of low pressure problems (DG2)
- the scale of interruptions of supplies to consumers – unplanned interruptions to supplies greater than 12 h (DG3)
- compliance in respect of the level of iron in water (mean zonal compliance)
- the number of burst water mains
- distribution losses (component of overall leakage).

Ofwat measures the serviceability of water treatment works, service reservoirs and pumping stations (water non-infrastructure assets) by using the following indicators (Ofwat 2008c).

■ The percentage of the total number of samples taken at water treatment works containing coliforms.
■ The number of water treatment works where enforcement action was considered because of contravention of the coliforms standard.
■ The number of service reservoirs with coliforms detected in more than five per cent of samples, which is the current standard.
■ The number of water treatment works where turbidity (water clarity) exceeds a threshold value within the permitted range. This data can be affected by raw water quality at some types of treatment works, and thus does not provide a guide to serviceability in some cases.
■ Unplanned maintenance (company specific).

When reporting the above indicator values, all companies are asked to provide the confidence grades for the reliability and accuracy of information submitted. The four Ofwat's reliability bands are as follows:

■ A – sound textual records, procedures, investigations or analysis properly documented and recognised as the best method of assessment
■ B – as A, but with minor shortcomings – for example, include old assessment, some missing documentation, some reliance on unconfirmed reports, or some use of extrapolation
■ C – extrapolation from limited samples for which grade A or B data is available
■ D – unconfirmed verbal reports, cursory inspections or analysis.

Ofwat's seven accuracy bands are as follows: ± 1%, ± 5%, ± 10%, ± 25%, ± 50%, ± 100% and 'X' or very small numbers where accuracy cannot be calculated or the error could be more than ±100%.

The International Water Association (IWA) has developed an alternative PI system. The latest, third edition of this PI system (Alegre *et al.* 2016) provided a library of more than 160 PIs covering both water and sewerage services classified into the following main categories:

■ water resources indicators
■ personnel indicators
■ physical assets indicators
■ operational indicators
■ quality of service indicators
■ economic and financial indicators.

Each IWA performance indicator has a number of its own properties (e.g. ID, description, units etc.), the set of associated variables (used to calculate its value) and the contextual information (info relevant to the PI analysed). Similarly to Ofwat, IWA has confidence grades for the accuracy and reliability of PI values obtained. For technical use, four categories exist for data accuracy (0–5%, 5–20%, 20–50% and >50%) and three for data reliability (*, ** and ***). Accuracy and reliability are assessed at the variable level and then propagated to the relevant performance indicator value. For the purpose of public communication, all PI values are simply classified as 'good', 'average' or 'bad'. The IWA system is of course an option for water utilities and their regulators around the world, or they may prefer to develop their own approaches.

If details of two of the Ofwat indicators are analysed, it can be seen clearly why companies cannot simply concentrate on these high-level indicators if they are to run their businesses successfully.

Let us start with the drinking water OPI in Table 10.2. This is taken from the DWI annual report, which actually contains performance against over 50 parameters defined by the EU, failure against which may result in regulatory action by the DWI against the company. In the case of cryptosporidium, which does not even feature in the EU list, there is a strong likelihood of prosecution of the company if this limit is breached. So simply managing the distribution system against the eight parameters is an almost certain route to failure.

A second example is slightly more complex but just as damaging to a company's long term prospects. Let us consider supply interruptions in Table 10.1. Imagine that there is a high lift pumping station supplying a small town by way of a terminal reservoir. The station contains only two pumps: one duty, one standby. Initially, all works well, when the duty pump fails the standby starts automatically and there is no noticeable interruption to supply. The regulatory standard is being met, so if that is the only indicator being used, when the duty pump reaches the end of its life, there is no history of failure in customer service to justify replacing it. Instead, the station continues to operate without any standby. As time passes the customer service will deteriorate depending on the ability of the maintenance team to carry out repairs before the terminal reservoir empties. In real life, the operators will soon bring the problem to the attention of management and, hopefully, the issue would be addressed proactively before customers suffer what could be a major disruption in supply.

These examples do not highlight failures in the regulatory indicators – they simply demonstrate that the regulator takes a very high-level overview of the state of each company. However, if the company is to meet the economic regulator's targets, it must have a far better awareness of the state of its assets than is provided by regulatory targets and indicators. In the language of total quality management, it needs to be able to identify the root causes of problems, not just the symptoms, even though the presence of the symptoms may be sufficient for regulatory assessments.

10.6.   Asset assessment techniques

10.6.1   Asset inventory

Banyard and Bostock (1998) describe the development of a suite of computer programmes that allows them to define a long-term capital programme in order to optimise the delivery of all regulatory outputs. To do this the starting point is the development of an asset register, which is a database containing all of the company's assets.

At this point, it is necessary to consider what level of detail is required – for example, it is of very limited benefit to record that there is a water treatment works at a certain location. What is required is to consider what assets collectively make up the waterworks. Most water treatment works will have been constructed in phases, and hence some assets will be older than others. Additionally, it needs to be recognised that while concrete tanks may have a life of 60–70 years, mechanical and electrical equipment such as pumps and compressors will only have a life expectancy of 20–25 years. Telemetry equipment will have an even shorter life. It becomes clear, therefore, that an asset register has to be established that will distinguish between these very different assets.

The easiest way to tackle this problem is to decide at what level assets might be abandoned and replaced by newer assets once their useful life has expired. So each pump would be separately listed including details of its manufacturer, date of installation, output and power consumption.

But there would be no separate listing of pump casing or pump impeller. If the pump impeller needed to be replaced it would be a maintenance item paid from revenue, not capital. This issue will be revisited later in the chapter.

It is also helpful to record details of tank dimensions so that meaningful cost models can be developed. It is stressed here that all data must be capable of audit, so there can be no question over the integrity of the data, which after all is a foundation block of asset management. It is almost impossible to deliver effective asset management if there is uncertainty about which assets are actually being managed.

For pipe systems, the introduction of GIS systems has been a huge step forward. The ability to display the pipe network against an up to date map background has been invaluable and has also lead to significant reassessment of the amount of pipework actually in the ground. As with the above ground assets, this needs to be supplemented with other information such as the pipe material, pipe internal diameter, date of laying and so on. A word of warning is appropriate at this point: it is essential that the data cannot only be displayed, but can also be exported to other databases for manipulation with other information. Many of the early GIS systems were very restricted in allowing such export of data, and frequently required purpose-written software before manipulation was possible. This, at times, led to separate databases being developed which, frequently, were out of step with one another. The modern GIS systems have avoided this problem.

The next step is to supplement the fundamental asset data with an assessment of the 'serviceability' of the assets.

10.6.2 Assessing serviceability

Skipworth *et al.* (2002) distinguish between three major types of performance indicators:

- Condition PIs: used to measure the physical condition or state of an asset
- Performance PIs: used to measure the behaviour of the asset(s) in terms of its performance
- Serviceability PIs: used to measure the quality of service received by customers and the ability of the assets (and their operator) to maintain the quality of service.

The Water Research Centre (WRc) had produced Condition Grades for various water industry assets and these were adopted by many companies. They complied with the preferred Ofwat grading of 1 – 5 with 1 being excellent and 5 being awful. However there were no similar Performance Grades and it was left to individual companies to develop these, resulting in a lack of consistency between companies.

An example of condition PIs are the Ofwat's pipe condition grades:

- Condition grade 1: No failures, fully complies with modern standards.
- Condition grade 2: No significant failures (minimal impact on service performance), not quite consistent with modern standards.
- Condition grade 3: Deterioration beginning to be reflected in service levels or increased operating costs.
- Condition grade 4: Considerable corrosion affecting service performance, nearing the end of useful life, frequent bursts.
- Condition grade 5: Substantially derelict and source of service problems, no residual life.

257

As can be seen from the above, the condition grades are of a descriptive nature and involve implicit performance information which was often criticised in the past. Ofwat have moved away from this approach as far as their Periodic Reviews are concerned, but it still has value within companies for their day to day management, and is worth further exploration here.

Firstly, there needs to be much greater clarity between Condition and Performance Grades, and a more numerate approach is also helpful. Condition Grades describe the physical condition of the asset: whether it is structurally sound, whether it is safe to operate (particularly electrical equipment) etc. For pipes, the number of bursts per km per year would be one possible numerate assessment. Performance Grades indicate whether or not the asset is capable of meeting the required output. So the output from a pump measured in volume and head would be graded in terms of how far it fell short of requirements.

A simple example that illustrates how this approach operates is as follows: Consider a pipeline which can deliver the desired flow, but fails 8 times per km per year. Such a pipe would be rated 5 for Condition and 1 for Performance. In contrast to this, a new pipeline that had been badly designed and was incapable of delivering the flow required would be rated 1 for Condition but 5 for Performance.

It is now possible to examine a more complex problem to see how these Condition and Performance Grades can assist a company in managing its asset base. Consider a situation where an existing pipeline is no longer capable of delivering the necessary flow because a new housing development is being constructed. The options for dealing with this problem might be:

(a) replace with a new pipe
(b) duplicate with a new pipe that would be sized to make up the deficit in capacity
(c) renovate the existing pipe and add a second pipe only if necessary.

The problem facing any large company is that these options will be evaluated on the basis of individual subjective judgements unless company-wide standards have been defined. However, the individual preferences of different members of staff are not a satisfactory way of controlling the investment in a huge asset base. Thus standardised Condition and Performance Grades are a major step toward objective decision making. Further, some of the decision making can be standardised across the company. The decision on whether or not relining is economic is driven by a net present value (NPV) calculation comparing the NPV of a new main today with the NPV of relining now and replacing when the relined pipe is worn out. Once this calculation has been made, it can be simplified into a rule that says no pipe will be relined unless it has a certain life after relining, which can be linked with a Condition Grade. So for example, it could be said that only pipes with a Condition Grade 3 or better are to be relined.

Finally, there is a common misunderstanding that Condition and Performance Grades in some way lead to a position whereby all assets are replaced when they reach the end of their book lives. This is incorrect. Assets should generally only be replaced when they are life expired – that is, when they reach Condition Grade 5. They may be retired before that if upgrading performance is best achieved by early retirement, but this should be determined through discounted cashflow (DCF) analysis (see Chapter 11.8.2 for details of DCF analysis), not subjective judgement. In any system there will be a mix of asset condition and performance and this will constantly change as assets age.

10.6.3   Risk of asset failure

The use of Condition and Performance Grades has been criticised because it fails to take account of the risk of asset condition/performance failure, and to some extent, this is a fair criticism. In the mid-eighties, Germanopolous et al. (1986) considered the case of a Service Reservoir serving a large town that had only 4 hours nominal storage, when the water service provider's standard was 18 hours. However, the town had never run out of water. Either the company standard was too high (nominal retentions are usually between 14–18 hours for design purposes) or there were other factors at work. Germanopolous et al. (1986) showed by use of statistical analysis that the town was extremely well protected because there were numerous independent feeds serving the reservoir, and it was highly unlikely that they would all fail at the same time. This example showed the dangers of simply taking company standards and failing to understand the precise situation. Mathematically because of the number of feeds into the reservoir, the nominal capacity understated the resilience of the actual situation, or put another way the effective capacity of the reservoir was well above the WSP's standard.

Unfortunately, this idea of incorporating 'Risk' may lead to potentially damaging conclusions and great care is needed. In the general case, the risk of some failure is defined as a function of the Likelihood and the Consequence of that failure. In the context of asset management, this function is usually defined as a simple product of the likelihood and the consequence components of risk (despite the obvious drawback that small likelihood and large consequence can be represented by the same risk number as in the case of a large likelihood and the small consequence). In the case of water distribution systems, the consequence is usually equated with 'population affected' (or some similar measure). From here it is only a very small step to concluding that money should be applied to protecting the large centres of population at the expense of small village communities because that is the route to minimise risk. The fallacy in this approach is that the statutory requirements are normally for all customers to receive a defined minimum level of service. There is no defence to regulatory failure to argue that only a few hundred people were supplied with unwholesome water because the money had been applied to giving extra protection to a large city. However if it is understood that there must be a minimum level of service for all customers in most aspects of a water utilities work, then risk analysis does provide valuable tools in deciding where any further funds can be best applied to obtain best value for money.

It is certainly possible to trawl operational staff for their views of which schemes should be pursued to deliver risk reduction. These schemes can then be evaluated to see what actual risk reduction they would bring, and then risk quantification methods applied to help distinguish between competing schemes and groups of schemes. The output would be similar to the idealised plot below, and the Utility is left where it wishes to draw the cut off point, either in terms of funds available or residual risk (see Figure 10.1). Thus the concept of risk is a valuable contribution to asset management but its use must be fully understood and controlled accordingly.

10.6.4   The Common Framework

Although the use of Condition and Performance Grades had proved very useful, the regulator for England and Wales was concerned that it did not properly meet his needs, and sought an alternative approach. This resulted in UKWIR the industry common interest research body commissioning Tynemarch to produce an alternative. The report was published in May 2002, and can be obtained from UKWIR (2002).

Figure 10.1 An idealised trade-off curve between the net present cost of asset management interventions and the service performance benefit (risk)

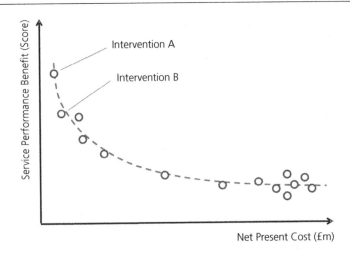

It sought to build on the regulator's preferred four-stage process for assessing Capital Maintenance needs (capital monies used to replace worn out assets). Not to be confused with the maintenance of assets):

- Historical Analysis to identify levels of expenditure in previous years and levels of Customer serviceability
- Forward looking Analysis to determine Future Needs
- Identify any changes that make continuing at historical levels unreasonable.
- Make the case for a revised level of Capital Maintenance.

The Capital Maintenance Common Framework (UKWIR, 2002) was then developed incorporating the following:

- Capital Maintenance should normally be justified on the basis of current and forecast risk analysis of asset failure – that is, the likelihood and consequence of asset failure should be assessed. Consequences are to be expressed in terms of both service levels and costs.
- Service includes both service to customers and to the environment to capture statutory obligations that are not represented by standard customer service indicators.
- A least-cost approach must be demonstrated for the capital expenditure/operating expenditure (Capex/Opex) trade-offs.

The Framework deliberately did not specify detailed methodology as the regulator wished to encourage innovation and diversity of approach between companies. It has turned out to be a very comprehensive way of assessing Capital Maintenance needs, but is very data hungry. Unfortunately, much of the data required was not routinely collected before the publication of the Framework and this has hampered progress. It is possible to synthesise missing data using expertly facilitated teams of practitioners, but such approaches are not always fully accepted by regulators.

Following on from the introduction of the Common Framework, the UK water industry has contributed to the updated expenditure planning framework (UKWIR, 2015). The new framework provides approaches and tools for service forecasting tools.

10.7. Asset interventions

In the asset management context, the term 'intervention' is typically used in a very broad sense denoting any activity of capital or operational nature that is used to maintain and/or improve the condition, performance and/or serviceability of a single or group of assets (including water supply and distribution system as a whole).

A wide range of interventions is usually available. Within water distribution system asset management, interventions are often classified into infrastructure (pipe related) and non-infrastructure interventions (applied to other, typically above ground assets).

One of the generic intervention classifications often used is as follows (interventions shown in the order of increasing costs and impacts):

- Routine asset maintenance (e.g. regular pipe cleaning using jetting, scouring or some other method)
- Asset repair (e.g. a local pipe patch - a pipe repair to take care of a recent burst)
- Asset refurbishment or rehabilitation (e.g. pipe relining)
- Asset renewal (e.g. pipe replacement).

Another common classification of water distribution system asset interventions is into *capital* or Capex and *operational* or Opex type interventions. The accounting conventions allow expenditure to be charged to Capex only if it creates an asset that will exist for more than one year. Creating an asset can mean extending the original book life of an existing asset, but not simply allowing it to meet its intended life.

In the above list, Routine Maintenance and Asset Repair would normally be Opex and the other two Capex type interventions. The decision of the source of funding will vary in detail based on the accounting policies of individual companies. However, the fundamental issue is not based on the magnitude of the expenditure but whether or not it extends the life of the asset beyond the originally assumed book life. So if the expenditure is simply to allow the asset to achieve its book life it should be charged to revenue and if it extends the book life it can be charged to capital and the residual value of the asset in the balance sheet will be increased accordingly together with its new book life. The scope for variation arises because, while most companies will have a minimum value below which all expenditure will be charged to revenue, it should be noted that the converse (i.e. a maximum value that can be charged to revenue or Opex) never occurs although many engineers seem to believe that it does.

Interventions can be applied to a single asset (e.g. pipe replacement) or groups of assets (e.g. flushing of a group of pipes to mitigate the risk of discoloured water). Sometimes the intervention covers large parts of a water distribution network (e.g. active leakage control by means of regular sounding, or systematic flushing of pipes to prevent discolouration events).

Each intervention comes with a cost attached to it. Developing accurate intervention cost models, especially the parametric ones where the cost of an intervention (e.g. building of a new service reservoir) is a function of several parameters (e.g. reservoir volume, material etc.) is not an easy task and it often requires data that is not readily available (e.g. data on past engineering works and related costs, data on various technological details and related unit costs etc.).

Different interventions have different levels of impact on asset condition and therefore also generally on performance. This is true not just for interventions of different types (e.g. pipe cleaning will change the pipe's roughness coefficient – that is, hydraulic capacity only, while pipe replacement may also change the pipe diameter and material thus changing also its structural performance), but also for the interventions of the same type (e.g. cement-mortar and epoxy are two pipe relining techniques that will result in different hydraulic and structural characteristics of the relined pipe).

A single intervention is likely to have a simultaneous impact on future values of a number of performance and/or serviceability indicators, e.g. pipe replacement will affect hydraulic capacity, burst rate, leakage etc. Quantifying the impact of interventions on one or more asset/system performance and especially serviceability indicators is a complex issue often requiring development of specialised mathematical models. Some of these models may require running various physically based and computationally expensive system simulation models to obtain more accurate predictions. For example, to quantify accurately the effect of replacing several pipes in a distribution system on the pressure at a critical point in the system, a hydraulic simulation model of the water distribution system should be run with and without modified pipe characteristics – that is, before and after interventions are applied. Even worse, in some cases, it is very difficult (if not impossible) to quantify the impact of some interventions on some assets and related indicators simply because accurate impact models do not exist (e.g. how to accurately quantify the impact of active leakage control on leakage reduction?). Some examples of consequence modelling approaches and tools are presented in Section 9.8.

Asset interventions, especially the ones of capital type, are often costly and as such need to be timed well. This is especially important for the proactive type of asset management where assets should be replaced just before they come to an end of their useful life. This, of course, is easier said than done. To achieve an optimal mix of asset interventions over some future planning horizon, especially at the system level, some sort of asset management decision support tool should be used (see Section 9.9). The reason for this comes from a large number of possible solutions (i.e. combinations of different intervention types and their timing) which makes it a very difficult, if not impossible, job for a human brain.

10.8. Asset deterioration
10.8.1 General
Asset deterioration is one of the key drivers behind large investments required to manage water company assets. The UK water and infrastructure systems are particularly affected by this as many pipes were buried under the ground a long time ago. The average age of water mains in the UK in 2008 was approximately 45 years (Ofwat, 2008c), with some pipes being older than 100 years.

Deterioration in asset performance is usually a consequence of ageing but can also result from other external factors/processes. For example, depending on the type of source water the hydraulic performance of unlined ferrous pipes may reduce with age due to the internal corrosion process, leading to an increase in a pipe's absolute roughness or, in some cases, reduction in its effective internal diameter. However, even where a pipe is not subject to the above corrosion, its hydraulic performance may reduce in the future owing to a need for it to convey increased flow rates as a result of, for example, urbanisation and/or climate change. Therefore, asset deterioration can be expressed in terms of its condition deterioration but also in terms of its performance deterioration. Note that the two are not necessarily always fully correlated (Hall *et al.*, 2006).

Physical asset (i.e. condition) deterioration is a consequence of various mechanisms that may take place over time. These processes are often asset specific and usually involve complex physical, chemical and/or bacteriological processes. An example of asset deterioration processes is the corrosion of metallic (i.e. ferrous) water pipes. Corrosion of these pipes is a process of ion release from the pipe into the water (internal corrosion) and/or the surrounding soil (external corrosion).

Quantifying and predicting asset deterioration, especially in terms of performance, is critical for successful asset planning and management. This was less important in the past when reactive asset management was used – that is, where assets were replaced following a major failure. The proactive approach aims to replace the asset just before it comes to the end of its service life (e.g. just before the pipe starts bursting frequently in short periods of time). However, predicting asset deterioration is a difficult task. To illustrate this, examples of pipe deterioration models are presented in the next section.

10.8.2   Pipe deterioration modelling

Two general types of pipe deterioration models exist: physically-based models and statistically-based models. The former aim to describe the physical mechanisms underlying the pipe deterioration processes, while the latter aim to develop predictive models by applying various statistical analyses to the data available.

Bearing in mind the complexity of pipe deterioration processes (see the previous section), it is no surprise that there are very few physically based models developed so far. Most of these models are limited in scope – that is, they typically apply to very specific types of pipes and conditions. An example of such a model is that developed by Rajani and Tesfamariam (2007) which aims to estimate the time to failure of cast iron water mains.

Opposite to physically-based models, quite a few statistically-based pipe deterioration models have been developed so far (Kleiner and Rajani, 2001). These models can be further classified into models aiming to predict condition deterioration and those aiming to predict performance deterioration. In this context, a pipe's deterioration performance is usually defined in terms of its burst rate (equal to the number of pipe bursts per unit time) or its burst frequency (burst rate divided by the pipe's length). Regardless of the type of predictions made, almost all statistical models developed aim to link burst rate/frequency to various pipe attributes and other explanatory factors. Examples of these factors (also known as the covariates in the statistical literature) include pipe material, age, diameter, external loading (traffic on the road above), quality of workmanship during pipe installation, surrounding soil conditions and other factors (Boxall *et al.*, 2007).

According to Kleiner and Rajani (2001), all statistical pipe deterioration models can be broadly classified into deterministic, probabilistic multi-variate and probabilistic single-variate group-processing models.

The deterministic models are regression type models that predict pipe break rates/frequencies using two or three parameters (typically time/age, pipe material and diameter). The models can be further classified into time-exponential (Shamir and Howard, 1979; Walski and Pellicia, 1982) and time-linear models (McMullen, 1982; Kettler and Goulter, 1985; Jacobs and Karney, 1994), depending on the shape of the curve used to approximate the deterioration period on the Bath Tub curve (Figure 10.2).

This curve is often used in the reliability theory to depict an asset's (or system's) deterioration expressed as the asset's failure rate λ with time t. Depending on the statistical technique used, the probabilistic multi-variate models can be further classified into proportional hazards models (Marks *et al.*, 1985; Andreou *et al.*, 1987), accelerated lifetime models (Lei and Saegrov, 1998) and time-dependent Poisson models (Constantine and Darroch, 1993). Probabilistic single-variate group-processing models can be classified into cohort survival models (Herz 1996), Bayesian diagnostic models (Kulkarni *et al.*, 1986), semi-Markov process-based models (Gustafson and Clancy, 1999) and break clustering models (Goulter and Kazemi, 1988). For details of these models and their respective pros and cons, please refer to Kleiner and Rajani (2001).

Several authors identified the need to aggregate pipes into homogeneous groups in order to conduct a more effective analysis (Shamir and Howard, 1979; Lei and Saegrov, 1998). Shamir and Howard (1979) were the first to suggest that data groups ought to be considered homogeneous with respect to the causes of failure. Pipe material, diameter and age, with or without additional factors such as soil types and/or land use above the pipes, have been widely adopted as grouping criteria to emphasise their influence on failure (Lei and Saegrov, 1998; Herz, 1996). While easier to develop and more accurate than the models aiming to make burst rate predictions at the single pipe level, these models are less useful for generating detailed asset management plans. As a result, pipe level models have started to appear recently (Le Gat and Esibeis, 2000; Economou *et al.*, 2008).

Figure 10.2 The Bath Tub curve

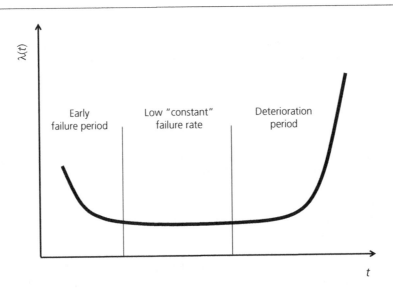

264

The data used for the development of statistically-based pipe deterioration models typically come from two sources.

- The asset database, containing basic information about pipes (material, diameter, age and so on).
- The incident database, containing information about observed pipe burst events (i.e. incidents).

Three major problems associated with these data sets in the UK are as follows.

- Data sets have been recorded for relatively short periods of time (most of the UK water companies started collecting this data in the early 2000s).
- Both databases often have a lot of missing and/or incorrect information.
- Data collection is also affected by network configuration changes (e.g. the recent widespread use of pressure management/network calming interventions) which are likely to have had a significant effect on masking true asset condition by suppressing burst rate during much of the period since burst data started being collected routinely.
- In many cases, linking the two databases is not straightforward as locations of bursts in the incident database tend to be recorded based on the nearby property rather than the pipe itself.

In the remainder of this section, two pipe deterioration models are presented in more detail. The first model is based on the evolutionary polynomial regression (EPR) technique (Berardi., 2008; Giustolisi and Savic, 2006). The second model is the Bayesian-based zero-inflated non-homogenous Poisson process (NHPP) model by Economou *et al.* (2008).

10.8.2.1 Example 1: the EPR model

The EPR model is a deterministic model aiming at predicting the burst rate/frequency for a single or a group of pipes as a function of a number of potential explanatory factors. The EPR method is essentially a smart regression approach. Unlike the conventional regression approach where the exact mathematical model structure (e.g. a function linking the pipe burst rate and explanatory variables) has to be assumed before the regression, in EPR only the following, very general, polynomial form of this function (or similar) has to be assumed.

$$Y = a_0 + \sum_{j=1}^{m} a_j \cdot \left(X_1\right)^{ES(j,1)} \cdot .... \cdot \left(X_k\right)^{ES(j,k)} \cdot f\left[\left(X_1\right)^{ES(j,k+1)}\right] \cdot .... \cdot f\left[\left(X_k\right)^{ES(j,2k)}\right] \qquad (10.1)$$

where
Y is the predicted burst number/rate/frequency
X_k is the *k*-th explanatory variable
ES is the matrix of unknown exponents
f and *g* are functions selected by the user
a_j are unknown polynomial regression coefficients (e.g. model parameters)
m is the number of polynomial terms (in addition to the bias term a_0).

The EPR model works in two main loops (Giustolisi and Savic, 2006).

(a) In the external loop, EPR uses the multi-objective genetic algorithm (MOGA) to determine the unknown ES values.
(b) In the internal loop, given the selected ES values, EPR uses the singular value decomposition-based least-squares method to determine the values of unknown polynomial regression coefficients *aj*.

The EPR model starts its search with all possible explanatory factors and automatically determines the significant ones, which are, typically, few in number. This is achieved by the means of **ES** values: every time a value of the exponent **ES**(j,k) becomes equal to zero this means that the value of k-th input variable $\mathbf{X}_k$ is insignificant – that is, effectively deselected from the model shown in Equation 10.1. This way a lot of time is saved as the user needs to run the EPR model only once, as opposed to the conventional regression approach.

In addition to the above, because the EPR model uses the MOGA method it generates a whole set of Pareto optimal solutions ranging in different complexity and accuracy of the burst prediction models identified. More importantly, all the models generated are fully transparent as they are presented in the equation format, which makes it easier to inspect them for any potential illogical relationships and so on. An example of such an equation is

$$BR = 0.084904 \cdot \frac{A}{D^{1.5}} \tag{10.2}$$

where
BR is the burst rate (number of bursts per year)
A is the pipe age
D is the pipe diameter.

The EPR method has already been successfully used to develop real-life pipe burst prediction models in the UK (Hall *et al.*, 2006), including the recent application to develop all-pipe predictive models for the entire networks of two water companies in the UK.

10.8.2.2 Example 2: the zero-inflated NHPP model

The zero-inflated NHPP model is a stochastic, Bayesian-based pipe burst prediction model (Economou *et al.*, 2008). It is based on the non-homogenous Poisson process which is flexible enough to capture the non-linear relationship of the failure rate (here pipe burst rate) with time, at the same time allowing for the inclusion of suitable pipe explanation factors (Loganathan *et al.*, 2002). Furthermore, the model was adjusted to account for possible zero-inflation that may well exist in pipe failure data due to the fact that many pipes never experience a break during the observation period. Finally, the model was calibrated using the Bayesian-based approach which provides a natural framework for updating the model parameters – for example, every five years when new business plans have to be submitted to the regulator (Section 10.3). The Bayesian approach also provides an effective way of dealing with short observed data sets by using the relevant prior estimates of model parameters which are based on engineering judgment rather than data. These priors can then be updated once the newly observed data becomes available.

The Zero Inflated NHPP model by Economou *et al.* (2008) uses the following function to approximate the deterioration period on the Bath Tub curve (Figure 10.2)

$$\lambda(t,\mathbf{x}) = \theta t^{\theta-1} e^{\beta \mathbf{x}} \tag{10.3}$$

where
λ is the pipe burst rate
$\theta > 0$ is the deterioration curve shape parameter
$\mathbf{x} = \left(1, x_1, \ldots, x_q\right)$ is a vector of related explanatory variables

Figure 10.3 Example of the zero-inflated NHPP model output

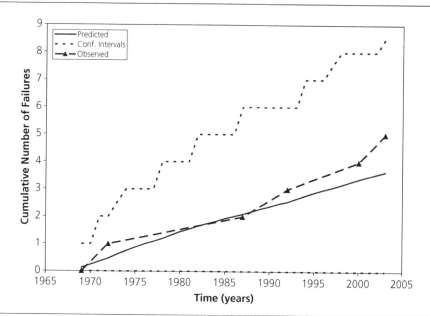

$\beta=\left(\beta_0,\beta_1,\ldots,\beta_q\right)$ is a vector of model parameters

t is time.

The values of θ and β are determined by using the aforementioned Bayesian based calibration approach.

The zero-inflated Bayesian model makes pipe burst predictions at the single level, which is a significant advantage over many other models available nowadays. Being a stochastic model, it enables prediction of confidence intervals in addition to mean values. This is demonstrated in Figure 10.3 which depicts the zero-inflated NHPP model 'predicted' and 'observed' cumulative number of bursts over time for a single pipe in a large North American town. The calibration period is 1969–1998 and the validation period is 1999–2003.

10.9. Failure consequence evaluation
10.9.1 General

Section 9.7 addresses the process of asset deterioration and some of the modelling approaches that have been applied to predict the frequency of failure. In compiling the information needed to identify the actual 'risk' presented by assets (see Section 9.5.3), and hence to carry out the cost-benefit calculations which drive asset intervention decisions (as explained further in Section 9.9), it is useful to apply models which assess the consequence of failure.

The principal service impacts relevant to water infrastructure intervention decisions include the hydraulic service measures monitored by Ofwat (Table 10.1) – that is, interruptions to supply and low pressure, together with discolouration. Leakage is also a desirable addition to this list, although the complex relationships affecting deterioration and the difficulties in consistently modelling the impact of active leakage control on leakage levels hinder this. Secondary impacts may also be included where their costs may be determined – for example, the disruption arising to traffic caused by an unplanned burst repair, or the biological water quality risk created by supply interruption of long duration.

As referred to elsewhere in this chapter, the impact of inadequate pipe capacity on system pressures is best assessed either by use of a hydraulic model or system monitoring, or a combination of both. In most cases, this type of deterioration (caused either by physical asset deterioration through corrosion or increasing demand) is gradual or is at least planned, and may be investigated selectively on an 'as needed' basis linked to planned development or low pressure monitoring data. Such capacity inadequacies are, of course, relevant to integrated asset management decisions. However, the less predictable service impacts are those which arise from a sudden asset failure, predominantly a water main burst, and it is on this aspect that the rest of this section focuses.

10.9.2   Supply interruptions (DG3) and low pressure (DG2)

Regardless of the frequency with which a pipe is predicted to burst, its impact on customer service in terms of loss of supply depends on its position in the local network – that is, connectivity, number and relative position of valves and so on.

Assessing the hydraulic impact that the failure of a pipe would have is ideally carried out using hydraulic models. Some hydraulic simulation packages offer a batch run facility which allows the model to be set up to run a suite of scenarios sequentially. By setting up these runs to represent the closure of each pipe in the system in turn, then running a simulation and logging the results (specifically identifying the nodes at which there is either no supply, or where pressure is inadequate relative to a target minimum), it is possible to assess the number of customers affected through the demand information used to build up the model. This allows a pipe-specific consequence to be established for each pipe, which, when factored by the predicted failure frequency for the cohort to which the pipe belongs, provides a risk number for each asset in its current position (or at some future horizon, to reflect ongoing deterioration). If this is compared with the risk number associated with, for example, a replacement pipe with a much lower predicted failure frequency, the risk mitigation impact and its value may be determined as part of the cost benefit evaluation.

When carrying out a large-scale selection of asset interventions as would typically be the case for preparation of water companies' business plans at the price review (and which companies are increasingly moving towards as part of their 'business as usual' processes), the practicability of using hydraulic models to cover all areas is, however, limited. Many companies (at least in the UK) do not have a comprehensive stock of hydraulic models, particularly not models which have been calibrated to the high standard necessary to determine the impact results with sufficient confidence; and even if they do, the time taken to run the models may be prohibitive. Alternative pragmatic approaches have therefore also been applied using GIS tracing, allowing these pipe-specific consequences to be determined in a relatively crude but nevertheless effective way.

GIS scripts have been written and applied by several water companies to determine the number of houses that would be affected as a result of pipe failure. There are three pre-requisites for this approach to be feasible.

(a) The GIS must contain connectivity information to allow tracing along the appropriate pipes only.
(b) Information must be present regarding the number of properties connected to each pipe.

(c) Information regarding valve locations and settings must be comprehensive and accurate.

Two separate logical approaches have been taken.

- Logic that tests whether each individual pipe is a 'sole feed' to an area of demand and, if it is, counts the number of properties connected downstream of that point. The sole feed test includes an assessment of the likely adequacy of alternative paths based on pipe size. This approach provides an indication of the initial impacts in the event of a catastrophic pipe failure. Its disadvantage is that, to determine what is 'downstream', there is a fourth pre-requisite: hydraulic direction of flow must be known which, in looped systems particularly, can be difficult to determine consistently without a hydraulic model.
- The second tracing analysis returns the number of properties that would be isolated, assuming that valves are closed to allow the repair of each pipe in the system in turn. The number of properties is the sum of properties contained within the isolated boundary together with any that lie outside of that area but are cut off from their 'source' as a result of the isolation. This does not require a knowledge of hydraulic flow direction and is arguably a more authentic representation of the real problem.

Figure 10.4 shows an example of DG3 consequence tracing from a pipe using the second approach.

The ability to determine impacts of failure on low pressure is clearly limited without a hydraulic model. However, there is scope for refinement of the GIS type approaches by developing more sophisticated rules which, perhaps analysed against hydraulic model results using a technique such

Figure 10.4 Example of DG3 consequence tracing

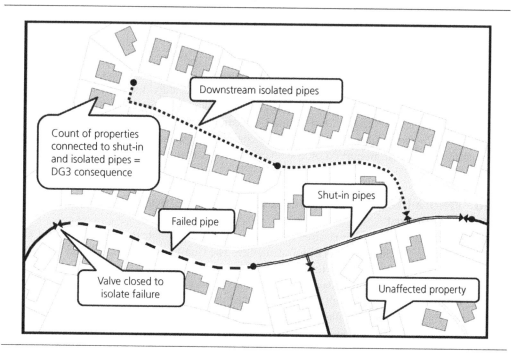

as neural networks, assess the likely extent of inadequate pressure service when particular combinations of failed and surviving pipe capacities apply.

10.9.3 Discolouration modelling

Models have also been developed which provide an indication of the risk that each pipe in a system poses to discolouration. Although not itself a hydraulic measure, discolouration is closely associated with hydraulic changes in a system when there is a burst or operational action, and therefore requires a hydraulic model.

A modelling tool known as the Discolouration Risk Model (DRM) was developed in the early 2000s (Dewis *et al*, 2005) which, similar to the DG3 batch run approach described above, simulates the sequential failure of each pipe in a system in turn and its effects on the hydraulic velocities in other pipes in the system. Each failure is assumed to comprise two phases: first, the application of an additional demand at the pipe to represent the water lost from the burst, and second the closure of the pipe for repair. A matrix is established summarising the impacts in terms, with the risks being greatest in circumstances where a pipe which has excess capacity under normal conditions (and, therefore, has a low conditioning shear stress, i.e. self-cleaning condition, acting at the pipe wall) is forced to work hard to an extent which could cause discolouration (i.e. a mobilising shear stress is introduced resulting in transportation of material accumulated at the pipe wall). Other factors reflect further considerations that could influence the magnitude of the risk, such as the source water type or length of ferrous trunk mains upstream of the system.

The DRM concept has since been updated in the form of the Discolouration Propensity Model (DPM) (McClymont *et al.*, 2010); DPM applies the same concepts but, unlike DRM which was based on water velocity, calculates discolouration risk more accurately by applying the cohesive layer shear stress theory first identified by Sheffield University through its PODDS (Prediction Of Discolouration in Distribution Systems) programme (Boxall *et al.*, 2001).

Discolouration is a complex phenomenon and the industry still has much to learn before being able to predict precisely its occurrence and scale of impact. However, the tools currently available provide an initial means of establishing a consequence value at individual pipe level. The points made above about the impracticability of applying hydraulic models on a wholesale basis are, of course, equally applicable here, so some initial screening of water systems by readily available measures, such as the number of discolouration complaints that have been received, is usually applied to filter the small proportion of systems with the greatest apparent risk for more in-depth hydraulic analysis.

10.10. Whole-life costing based asset management

10.10.1 The concept

Whole-life costing is a term that is often used in an engineering context, particularly civil engineering with respect to the provision of infrastructure. The Construction Research and Innovation Strategy Panel (CRISP, 2009) define the whole-life costing (WLC) as 'the systematic consideration of all relevant costs and revenues associated with the acquisition and ownership of an asset'. Therefore, the WLC concept considers all the costs incurred throughout the asset's life – that is, all the costs arising from the asset's installation, provision, operation, maintenance, servicing and decommissioning. It is important to understand that these costs will be used as input data to economic analysis – that is, they will be used as part of a DCF analysis as explained in Section 11.8.2. They in no way replace such analysis.

Skipworth *et al.* (2002) developed a new water infrastructure asset management framework that makes use of the WLC concept (Figure 10.5). The costing framework takes into account both activity-based costs (e.g. due to interventions) and aforementioned life-cycle costs associated with assets and system as a whole. In this way, costs (e.g. leakage costs) are linked to system performance (leakage) through the quantities that drive costs – the cost drivers (volume of leakage). Therefore, as a system's performance changes with time as a result of deterioration and/or interventions, so the changes in costs are tracked.

In the above framework, the optimal set of interventions to be applied to the analysed water distribution system (or part of it) over some long-term planning horizon (typically 20–50 years, although shorter time periods can be considered as well) is determined by choosing the set of interventions that will lead to the lowest NPV of total whole-life costs over this planning horizon subject to target levels of service and a number of regulatory, environmental and other constraints (all defined over the same planning horizon). The whole-life costs include the cost of interventions (both Capex and Opex) but also any other costs arising in the asset management process (e.g. maintenance cost, regulatory penalties etc.). Based on ownership, all the costs are classified into private (i.e. water company) and social (i.e. public/society and environmental) costs (Skipworth *et al.*, 2002). All the costs can also be classified into fixed or variable costs.

Every time an intervention occurs somewhere on the planning horizon (e.g. a pipe is replaced at some point in time), the associated intervention cost is accounted for and the impact of this intervention on the relevant asset(s) and the overall system performance (e.g. minimum pressure at critical point) is quantified by using relevant impact models (e.g. a system hydraulic model). The estimated PI values are then compared to the corresponding targets and, if target constraints are not met, penalty type costs are added to the overall sum of costs. At the same time, if the intervention

Figure 10.5 WLC based asset management framework (Based on Skipworth *et al.* (2002))

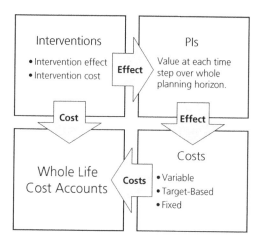

271

applied has an impact on some other PI(s)/cost(s) (e.g. future pipe burst rate, the associated cost(s) such as maintenance costs related to the number of pipe bursts) are evaluated too and added to the total whole-life cost of the analysed system.

In addition to the above, deterioration of various system elements is modelled to take into account the deteriorating asset/system performance in periods of time without any interventions.

10.10.2 A demo decision support tool

The above concept is encapsulated in a demo software tool. This decision support type tool has the following main modules

- WLC Accounting module
- Network Performance module
- Decision Tool module.

The WLC Accounting module stores and calculates all whole life-costs mentioned in the previous section. It is an accounting type module that attempts to identify, store and report all costs.

The Network Performance module is used to store the information about the analysed water distribution system and the associated performance models. This module uses GIS to store the physical asset data (e.g. pipe material, diameter etc., pumps' installed capacity, head-flow and other curves etc.) but also the information on connectivity of network elements. In addition to this, the Network Performance module stores all the performance (e.g. a pipe burst model) and other models (e.g. a hydraulic model) that are used to simulate different aspects of network's performance with time under changing conditions (e.g. deterioration of assets, changes in demand and changes due to interventions). This way, the impacts of various interventions on the analysed system's performance can be quantified.

The Decision Tool module performs the optimisation of interventions by minimising the whole-life asset management costs over the given planning horizon subject to target system performance. It uses the optimisation technique called Genetic Algorithms (Goldberg, 1989) which is capable of solving large, complex, non-linear optimisation problems like the asset management problem. The Decision Tool makes use of both WLC Accounting and Network Performance modules and generates as an outcome a list of interventions to be applied to the system assets (together with associated timing). In addition to this, the decision tool generates optimal whole-life cost (Figure 10.6) and associated performance indicators' profiles over the planning horizon (Figure 10.7). The Decision Tool can also solve alternative optimisation problems – for example, the problem where the optimal intervention selection is driven by the maximisation of water distribution system's performance subject to limited budgets available.

The demo tool and the WLC based methodology behind it have provided the basis for the development of a well known commercial asset management tool, WiLCO (Engelhardt *et al.*, 2005) and then as Enterprise Decision Analytics (EDA) to allow the user to visualise 'what if' operation and maintenance scenarios, providing decision support by way of WiLCO whole-life costing accounts and long-term costing profiles. This tool is one example of several that have now been used by major UK water companies and other infrastructure industries (transport, gas) to help manage their assets.

Figure 10.6 WLC profiles generated by a demo tool (Skipworth *et al.*, 2002)

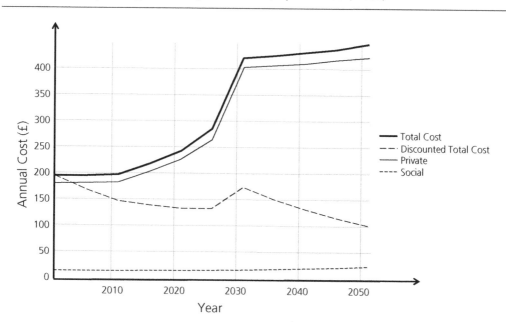

Figure 10.7 Leakage profile generated by the Demo Tool (Skipworth *et al.* 2002) was 9.7 in 1E

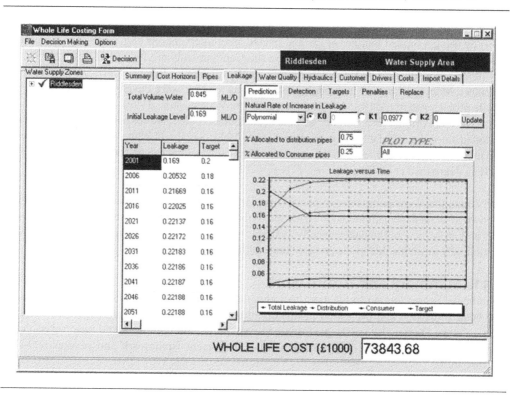

10.10.3 Other asset management modelling approaches and tools

Many other asset management approaches and tools exist nowadays. One of the first asset management methodologies developed for optimal strategic investment planning in the water industry was the GAasset model and software (Miller *et al.*, 2001). Developed in the mid-1990s, this was probably the first asset management approach that suggested the use of an integrated and optimised approach for selecting the best future asset investment planning strategy, incorporating a WLC assessment over a pre-determined planning period.

Another important asset management approach is the methodology developed as part of the large European Union Fifth Framework Programme research project entitled 'Computer Aided REhabilitation of Water Networks' or CARE-W (Saegrov, 2005). The overall aim of the project was to establish a framework for water network rehabilitation decision making. The work on the CARE-W project resulted in a computer-based system that comprises a suite of tools, providing the cost-efficient system of maintenance and repair of water distribution networks, with the aim of guaranteeing the security of water supply that meets social, health, economic and environmental requirements. The CARE-W system consists of software dealing with fundamental instruments for estimating the current and future condition of water networks – that is, PIs, prediction of network failures, calculation of water supply reliability, routines for estimating long-term investment needs, as well as selection and ranking of rehabilitation projects. These tools are integrated and are operated jointly in the GIS-based prototype software. The principal outputs obtained are a prioritised list of rehabilitation asset candidates and the associated short- and long-term strategic plans with project costs.

Many other tools exist that can be broadly classified as asset management tools (Beuken *et al.*, 2016). The main difference between these and the aforementioned tools is that these tools tend to address specific aspects of the asset management problem (e.g. water distribution system rehabilitation) by using a limited number of PIs and related asset interventions (e.g. Capex type interventions only). Examples of commercial tools are the CapPlan software by MWH Soft (2009), the Darwin Designer software by Bentley Systems (2009) and the OptiRenewal software by Optimatics (2009). Many customised methodologies and software that were developed to solve specific real-life problems exist too (Randall-Smith *et al.*, 2006; Walters *et al.*, 1999). All these commercial tools have their origins in a number of research methodologies and associated prototype tools developed since the mid-1990s (e.g. Simpson *et al.*, 1994; Savic and Walters, 1997). A review of these methodologies, together with the relevant historical timeline, can be found in Lansey (2006) and Mala-Jetmarova *et al.* (2018).

Finally, note that some companies, especially the larger water utilities in the UK, have developed their own asset management methodologies and tools, which are often customised to suit their own needs and, in particular, the data currently available to them. However, details of these models are not publically available due to the private nature of business in the UK water sector.

10.11. Summary and future work

A number of issues related to the asset management of water supply and distribution systems have been addressed and presented in this chapter, including key asset management drivers, the regulatory framework, asset performance indicators, interventions and associated deterioration models, the whole-life costing asset management framework and several real-life software tools.

Based on the material presented, it can be concluded that a lot of good work has already been done in the area of asset management of water supply and distribution systems. Having said this, a number of challenges remain to be addressed in the future. Some of the key challenges remaining are as follows (shown in no particular order).

- Development of new technologies and methods resulting in assets and related water supply and distribution systems that are more cost effective yet reliable, easier and safer to install and use, more energy efficient and sustainable. Examples include new water treatment equipment and related structures, improved flow and pressure regulating devices and so on.
- Development of new technologies and methods for more cost-effective and energy-efficient observation of condition and, especially, performance of water supply and distribution assets and systems. Examples include autonomous inspection robots, more energy-efficient equipment for real-time pressure and flow monitoring, new sensors for real-time water quality monitoring, equipment for automatic detection of various anomalous events including pipe bursts/leaks, various equipment failures, discolouration events and so on.
- Development of new technologies for more cost-effective and reliable asset interventions including cleaning, repair, rehabilitation and renewal of pipes and other assets. An example would be the development of new burst/leakage repair methods which can be applied more quickly and efficiently than the existing ones.
- Development of improved asset management models, including
  - (a) more accurate asset performance models, in particular better asset deterioration models that can make accurate long-term predictions at the single asset level;
  - (b) improved impact type models that can more accurately predict the effects of complex interventions on single/grouped asset(s) and overall system performance, including more realistic hydraulic network models with modelled pressure-dependent demands and leakage, water quality models able to more accurately predict discolouration events and various other water quality changes in the network (e.g. concentration of disinfection by-products), models to more accurately estimate green house gas emissions (i.e. carbon footprints) and so on;
  - (c) new asset management modelling tools that require less parameter tuning, and generally less specialist expertise which, in turn, would make these tools usable by a large number of smaller, typically publically owned water utilities run by municipalities worldwide.
- Development of approaches for assessment of intangible benefits and costs, focusing on, among others, customer expectation and stakeholder involvement.

As it can be seen from the above, there are plenty of challenges remaining in the future.

REFERENCES

Alegre H, Baptista JM, Cabrera E *et al.* (2016) *Performance Indicators for Water Supply Services*, 3rd edn. IWA Publishing, London, UK.

Andreou SA, Marks DH and Clark RM (1987) A new methodology for modelling break failure patterns in deteriorating water distribution systems: Theory. *Advances in Water Resources* **10**: 2–10.

Banyard and Bostock (1998) Asset management – investment planning for utilities. *Proceedings of the Institution of Civil Engineers – Civil Engineering 1998* **May**: 65–72.

Bentley Systems (2009) DarwinDesigner software overview, available at http://www.bentley.com/en-US/Products/WaterCAD/Darwin-Designer.htm, last accessed 15 Sep 2009.

Berardi L, Kapelan Z, Giustolisi O and Savic DA (2008) Development of pipe deterioration models for water distribution systems using EPR. *Journal of Hydroinformatics* **10(2)**: 113–126.

Beuken R, van Vossen J, Trietsch E *et al.* (2016) Comparing results of four decision support software tools on mains replacement. *Proceedings of the Computation and Control for the Water Industry (CCWI)*, Amsterdam, Netherlands, 7th to 9th November 2016.

Boxall JB, Skipworth PJ and Saul AJ (2001) A novel approach to modelling sediment movement in distribution mains based on particle characteristics. In *Water Software Systems, vol. 1: Theory and Applications (Water Engineering & Management)*. B Ulanicki, B Coulbeck and J P Rance, Research Studies Press, Hertfordshire, UK, pp. 263–273

Boxall JB, O'Hagan A, Pooladsaz S, Saul AJ and Unwin DM (2007) Estimation of burst rates in water distribution mains. *Proceedings of the (ICE) Institution of Civil Engineers – Water Management* **160(2)**: 73–82.

Constantine, A.G. and Darroch, J.N. (1993). In S. Osaki S and Murthy DNP (eds.), *Pipeline reliability: Stochastic models in engineering technology and management*. World Scientific, Singapore.

CRISP – Construction Research and Innovation Strategy Panel, (2009), Whole Life Cost Forum website, http://www.wlcf.org.uk, last accessed 11 September 2009.

Defra (2008) *Future water – The Government's water strategy for England*. Stationery Office, London, UK, p.98.

Dewis N and Randall-Smith M (2005) Discolouration Risk Modelling. *Proceedings on the 8th International Conference on Computing and Control for the Water Industry* **2**: 223–228.

Giustolisi O and Savic DA (2006) A symbolic data-driven technique based on evolutionary polynomial regression. *Journal of Hydroinformatics* **8(3)**: 207–222.

Goldberg DE (1989) *Genetic Algorithms in Search, Optimization, and Machine Learning*. Addison-Wesley, Reading, MA, USA.

Goulter IC and Kazemi A (1988) Spatial and temporal groupings of water main pipe breakage in Winnipeg. *Canadian Journal of Civil Engineering* **15(1)**: 91–97.

Gustafson JM and Clancy DV (1999) Modelling the occurrence of breaks in cast iron water mains using methods of survival analysis. *Proceedings of the AWWA Annual Conference*. Chicago, IL, USA.

Economou T, Kapelan Z and Bailey T (2008) A zero-inflated Bayesian model for the prediction of water pipe bursts. *Proceedings of the 10th International Water Distribution System Analysis Conference*, Kruger National Park, South Africa.

Engelhardt MO and Skipworth PJ (2005) WiLCO – State of the art decision support. *Proceedings of the International Conference on Computing and Control in the Water Industry*, Exeter, UK.

Germanopolous, Jowitt and Lumbers (1986) Assesing the reliability of supply and level of service for water distribution systems. *Proceedings of the ICE* **80**: 413–428.

Hall M, Kapelan Z, Long R and Savic D (2006) Deterioration Rates of Sewers. *Ewan Group Report No. PP/05/051*, published as *UKWIR Report No. 06/RG/05/15*, p. 97.

Herz, R. K. (1996), 'Ageing processes and rehabilitation needs of drinking water distribution networks', *Journal of Water SRT - Aqua* **45(5)**, 221–231.

IAM (Institute of Asset Management) (2015) Asset Management – An Anatomy. https://theiam.org/media/1486/iam_anatomy_ver3_web-3.pdf, last accessed 17 June 2022.

Jacobs P and Karney B (1994) GIS development with application to cast iron water main breakage rate. *Proceedings of the 2nd International Conference on Water Pipeline Systems*. BHR Group, Edinburgh, Scotland.

Kettler AJ and Goulter IC (1985) An analysis of pipe breakage in urban water distribution networks. *Canadian Journal of Civil Engineering* **12(2)**: 286–293.

Kleiner Y and Rajani BB, (2001) Comprehensive review of structural deterioration of water mains: statistical models. *Urban Water* **3(3)**: 131–150.

Kulkarni RB, Golabi K and Chuang J (1986) *Analytical techniques for selection of repair-or-replace options for cast iron gas piping systems – Phase I.* PB87-114112, Research Institute, Chicago, IL, USA.

Lansey KE (2006) The Evolution of Optimizing Water Distribution System Applications. *Proceedings of the 8th International Water Distribution Systems Analysis Symposium*, Cincinnati, OH, USA.

Le Gat Y and Eisenbeis P (2000) Using maintenance records for forecast failures in water networks. *Urban Water* **3**: 173–181.

Lei J and Saegrov S (1998) Statistical approach for describing lifetimes of water mains – case Trondheim municipality, Norway. *Proceedings of the IAWQ (International Association of Water Quality) 19th Biennal International Conference on Water Quality*, Vancouver, Canada, pp. 21–26.

Loganathan GV , Park S and Sherali HD (2002) Threshold break rate for pipeline replacement in water distribution systems. *Journal of Water Resources Planning and Management, ASCE* **128(4)**: 271–279.

Mala-Jetmarova H, Sultanova N and Savic D (2018) Lost in optimisation of water distribution systems? A literature review of system design. *Water* **10(3)**: p.307

Marks HD *et al.* (1985) *Predicting urban water distribution maintenance strategies: A case study of New Haven Connecticut.* US Environmental Protection Agency (Co-operative Agreement R8 1 0558-01-0).

McClymont K, Keedwell E, Savic D and Randall-Smith M (2010) Mitigating discolouration risk with optimised network design. *9th International Conference on Hydroinformatics*, Tianjin, China.

McMullen LD (1982) Advanced concepts in soil evaluation for exterior pipeline corrosion. *Proceedings of the AWWA (American Water Works Association) Annual Conference*, Miami, FL, USA.

Miller I, Kapelan Z and Savic DA (2001) GAasset: fast optimisation tool for strategic investment planning in the water industry. *4th International Conference on Water Pipeline Systems*, York, UK.

MWH Soft, (2009) CapPlan Software Overview available at http://www.mwhsoft.com/page/p_product/ CapPlanwater/capplanwater_overview.htm, last accessed 15 Sep 2009.

Ofwat (2008a) *Financial performance and expenditure of the water companies in England and Wales 2007-08.* Annual Report, ISBN: 1-904655-45-9, p.39.

Ofwat (2008b) *Service and delivery – performance of the water companies in England and Wales 2007-08.* Annual Report, ISBN: 1-904655-46-7, p.52.

Ofwat (2008c) *Service and delivery – performance of the water companies in England and Wales 2007-08 – Supporting Information.* Annual report, p.76.

Ofwat (2008d) *Preparing for the future – Ofwat's climate change policy statement.* Ofwat, London, UK, p.28.

Ofwat (2009) http://www.ofwat.gov.uk/legacy/aptrix/ofwat/publish.nsf/Content/Regulating Companies.html, last accessed on 25 Aug 2009.

Ofwat (2022) *Investment in the water industry.* https://www.ofwat.gov.uk/investment-in-the-water-industry, last accessed on 17 June 2022.

Optimatics (2009) OptiRenewal software overview available at http://www.optimatics.com/go/software/optirenewal/optirenewal, last accessed at 15 Sep 2009.

Rajani B and Tesfamariam S (2007) Estimating time to failure of cast-iron water mains. *Proceedings of the ICE – Water Management* **160(2)**: 83–88.

Randall-Smith M, Rogers C, Keedwell E, Diduch R and Kapelan Z (2006) Optimized design of the City of Ottawa water network: A genetic algorithm case study. *Proceedings of the 8th Annual Water Distribution System Analysis Symposium*, Cincinnati, Ohio, USA.

Savic DA and Walters GA (1997) Genetic algorithms for least-cost design of water distribution networks. *Journal of Water Resources Planning and Management, ASCE* **123(2)**: 67–77.

Shamir U and Howard CDD (1979) An analytic approach to scheduling pipe replacement. *Journal of the AWWA* **71(5)**: 248–258.

Simpson A, Dandy G and Murphy L (1994) Genetic algorithms compared to other techniques for pipe optimization. *Journal of Water Resources Planning and Management, ASCE* **120(4)**: 423–443.

Skipworth PJ, Engelhardt MO, Cashman A *et al.* (2002) *Whole Life Costing for Water Distribution Network Management*. Thomas Telford, London, UK.

Saegrov S (2005) *CARE-W – Computer Aided Rehabilitation for Water Networks*. IWA Publishing, London, UK.

UKWIR (2002) *Capital Maintenance Planning: A Common Framework*. Report No. 02/RG/05/3, UKWIR, London, UK.

UKWIR (2015) *Framework for Expenditure Decision Making: development of Service Forecasting Approaches*. Report 15/RG/05/43, UKWIR, London, UK.

Walski TM and Pelliccia A (1982) Economic analysis of water main breaks. *Journal of the AWWA* **74(3)**: 140–147.

Walters GA, Savic DA, Thurley RWF *et al.* (1999) Optimal design of water systems using genetic algorithms: some recent developments. *Proceedings of Computing and Control for the Water Industry*, Powell R and Hindi KS (eds.), Research Studies Press, Baldock, UK, pp.337–344.

Dragan A. Savić and John K. Banyard
ISBN 978-1-83549-847-7
https://doi.org/10.1108/978-1-83549-846-020242013

Chapter 11
Finance, regulation and risk in project appraisal

Iain McGuffog
South West Water, Exeter, UK

Adrian Cashman
Independent Water Management Consultant and University of the West Indies, Barbados

John Banyard
Independent Consultant, UK

11.1. Introduction

No two projects are the same, even for engineering projects that may have many similarities and commonalities. They are unique in their composition, location, size, time, environment, people and purpose, and, as such, there will be uncertainty over their performance and effectiveness. Therefore, all projects to some extent or another require appraisal. How much time and effort (the degree of detail and expertise) put into an appraisal varies with the nature of the project under consideration. In recent years there has been a significant increase in focus on environmental and social impacts of projects, moving beyond commercial and financial criteria. Some of this is down to government policy and regulatory expectations, but the reputation of an organisation and the risk and opportunity that are reflected in environmental, social and governance factors for organisations and their investors is arguably even more significant as a driver for this change in focus.

The purchase of a table for an office will not be subjected to the same degree of scrutiny as, say, the acquisition of a new corporate headquarters, but even the office table could have ethical considerations, even though you wouldn't expect to conduct an environmental impact decision before purchasing it, compared to analysis that a new headquarters might require in order to meet planning requirements and corporate, social and financial objectives. But behind these two examples there are some basic facts that are needed before proceeding. First, who or what body is making the decisions, and what is their interest? The answer to this can have a significant impact on the type and scope of an appraisal as well as on the level of detail required. It has a direct influence on the information required and to be presented. For example, the answer will influence whether a financial analysis is required or an economic one. It could also determine the level of information required to make the decision, and the governance that surrounds it. Understanding the environment in which a project appraisal is to be conducted is fundamental to understanding its purpose, and the shaping and presentation of the results.

Knowing the audience and purpose will assist the engineer in identifying the processes to be followed and the requirements that are to be met. In almost all cases there will be requirements that the process of decision-making should be formalised and follow appropriate methodologies. The formal process and methodologies adopted should demonstrate that a rational and documented basis on which a decision was made has been followed – the governance process for the decision. This is an important part of risk and opportunity for any project. It is difficult to conduct a financial or economic project appraisal successfully unless there is a clear understanding and description of uncertainties and the value chain involved in the project – the inputs and outputs involved and the influences on them in the future that affect the outcome for investments, the environment, stakeholders and wider society.

This chapter on project appraisal techniques first seeks to address why projects should be examined from an economic and financial standpoint. Having been satisfied that there are good reasons, the basic concepts and factors influencing project appraisal and its techniques are introduced to the reader. Next, the question of where the necessary funding for projects comes from is considered, before discussing the various alternative techniques that are available and the circumstances under which they might be used. We focus on the governance and management of risk involved. Discounting is a fundamental part of many techniques, and so the principles that are behind it are introduced, followed by the presentation of evaluation techniques, along with examples of how they can be applied. Finally we consider the topic of regulation and how this is shaping project appraisal.

11.2. What affects project appraisal?

Why should there be concerns about how projects are appraised and decisions made? One very basic reason is that projects require the input of scarce resources in order to be realised. A range of different types of resources can be scarce, often referred to as capitals. Such concepts are increasingly being used in corporate financial reporting. The six capitals that affect project appraisal are broadly.

(a) **Financial capital:** funding and financial resources available to the organisation.
(b) **Manufactured capital:** tools, plant, equipment and infrastructure with a productive capacity.
(c) **Intellectual capital:** knowledge, information, human resources and relationships that help to define the organisation and achievement of its purpose.
(d) **Human capital:** the capabilities, skills and experience of the company's employees.
(e) **Social capital:** relationships and ability of people to work together for common purposes.
(f) **Natural capital:** stocks of natural resources, environmental assets and the ecosystems that link them to affect productive capacity and services.

These same capitals are also useful categories to describe the outcome of a project, recognising there is often an overlap between them, particularly between human and social capital.

A legitimate question is to ask whether the proposed project constitutes the best use of those scarce resources. The question is even more pertinent when there is more than one course of action available to achieve the desired result. Under these circumstances, choices have to be made, and the grounds on which they are made explicit.

It is expected that projects and options should be assessed in a structured way, making use of comparisons based on a decision analysis technique. This would include some form of economic appraisal. Economic appraisal can be thought of as the counterpart of and complement to a technical appraisal: both consider performance relative to a benchmark requirement and often relative to other alternative options or courses of action.

An important step is to identify the boundary of the project appraisal – its context. This can often be through engaging with the stakeholders involved in the project to be clear about their priorities and purpose for the project, and the factors that are key to the decision. The link to risk and opportunity governance is important – for example, if the risks identified in a construction project include an environmental impact, then you would expect to have a greater focus on natural capital as part of the project appraisal, and social capital if external stakeholder relations are likely to be impacted by the outcome. Similarly, if skills in the supply chain to deliver the project are identified as a key factor in successful delivery, then intellectual, social and human capital factors are likely to feature. Governance matters, as the expectations of a range of decision makers, whether governments, regulators, clients, a board and, for utilities, even procurement rules will normally be significant in determining what project appraisal and factors to be considered in them are appropriate.

Drummond *et al.* (2006) identify three reasons for adopting a systematic approach as to whether or not to commit resources. Firstly, without a systematic approach, it is difficult to clearly recognise what are the relevant alternatives available to achieve a desired objective. Secondly, the viewpoint assumed in an analysis is important. In other words, from whose perspective should a project be considered and appraised? Lastly, a systematic approach is an important means of managing uncertainty, although not an infallible one.

One of the functions of an appraisal process is to identify areas of uncertainty or lack of information and, by doing so, gradually improve the level of knowledge and detail of a project proposal. A systematic approach identifies what data are required to refine the appraisal and, in doing so, improve the performance of the project by managing uncertainties.

In the past, the development of options was seen as being very much a technical activity, to be undertaken by specialists in the area. This was on the presumption that with their experience and expertise they would 'know best' what would work and what would not. This view has been increasingly questioned. Requirements for stakeholder involvement and degrees of public participation have in some cases broadened the pool of knowledge involved in the generation of options. And this has been seen as a good thing, although it is by no means widespread. The idea behind this is that broadening the range of participants brings different types of knowledge to the process.

In doing so, more potential options are generated – all of which should be treated seriously and equally. This, it is argued, leads to better projects and decision-making. Appraisal implies the exercising of judgement, usually against some norm – in some cases, this may be a 'do nothing' or 'business as usual' option. So, proposed projects or alternatives are measured against each other and/or against whatever is the appropriate baseline. This point is illustrated in Figure 11.1, where the 'with project' benefits would be compared with the 'without project' alternative – business as usual. Appraisal as a process is also a way of assessing the degree of confidence in the preferred option that emerges from the assessment. It can serve to highlight key uncertainties and the potential risks that might be associated with the project.

Project appraisal is, therefore, not just something that informs decision-makers: it is a method of communication that, at best, provides a way of sharing information and understanding between diverse parties. In other words, it is not an arcane tool for a small clique of people who can understand the intricacies of the process but, rather, it should be your tool for clear communication.

In particular, trying to avoid 'blind spots' in decision making will help to deliver better long-term outcomes. A blind-spot is where data on an uncertainty that affects a project decision exists, but the organisation lacks the understanding and perspectives on how it may affect the outcome. The timing of a decision can also be important – there can be an option value of waiting for better

Figure 11.1 Comparison of projects

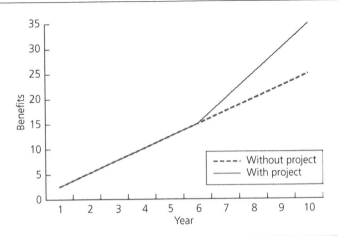

information to reduce uncertainty, or a decision to pilot a project rather than a decision to fully implement at scale for the ultimate objective. This has to be balanced against an opportunity cost of indecision in terms of the cost of analysis and appraisal, or even just reserving the limited pool of financial resources for the potential project can be resolved. All of this requires an understanding of the circumstances and governance around a decision, rather than just focusing on the mechanics of project appraisal.

11.3. Financial or economic appraisal

We describe above the types of resources that may need to feature in project appraisal and how the purpose of this appraisal should be to identify the best use of these scarce resources. We also describe the importance of identifying the boundaries of the decision, and the need for stakeholder engagement to understand these. This is clearly linked to the governance surrounding the decision, and how uncertainties that affect project delivery are identified and managed.

Most of this description has come from the perspective that an economic appraisal of a project is required, one that directly considers scarce resources and the value from their alternative uses. There is an important distinction to be made before proceeding further. Economic appraisal is not the same as financial appraisal, and so it is necessary to be clear as to which one of the two is to be conducted. Financial appraisal considers the expenditures and revenues associated with a project from the point of view of the entity or project participants, and it looks at the money needed to finance a project.

Economic analysis considers the various effects of a project from the point of view of its impact on the wider economy, and by extension wider society and the environment. In other words, economic analysis takes a broader view than a financial analysis. The commonality between the two is that both are conducted in monetary terms. Furthermore, the two are complementary to each other: for a project to be economically sound it must also be financially sound. If not, then the economic benefits associated with the project will not materialise.

Note that although both types of analysis are conducted in monetary terms, the overall project appraisal may still include non-monetary criteria. A financial analysis that screens out options because of unacceptable social or environmental impact is still a financial analysis. A qualitative ranking of environmental impacts that results in an option that may not have the best financial outcome in the financial analysis is still a financial analysis. It is the decision making that has considered other criteria and the overall appraisal of a project can be informed by more than one type of analysis.

Economic analysis takes on board societal aspects and includes them in the evaluation by comparing the financial value of resources used and the financial value of social, environmental and economic impacts with the extra benefits that would be generated for the economy as a whole. The measurement of positive and negative impacts is included through the assessment of a financial value for units of increased consumption and for foregone units of consumption of goods and services.

There can be a number of sources of the values for these impacts. If there is a market price available that is directly an outcome of the project (e.g. if the value of carbon emissions is reflected in energy prices), then the economic impacts may be included in an appraisal. In other cases we need a proxy for an actual market price where it cannot be observed, or because the impact on society would be unacceptable if left to market forces (a 'market failure' that occurs without perfect competition, which has a constrained set of economic assumptions that rarely or arguably never apply fully in practice). Proxy values attempt to identify a willingness to pay for the beneficial outcome being delivered or willingness to accept compensation for the loss of the scarce resource as a result of the project. The judgements on what is valued and how willingness to pay values are considered is controversial, particular for elements such as biodiversity where there is a question about whether our understanding of the world and society is sufficient to allow meaningful economic values to be identified. However, if we are to conduct an economic appraisal we have to quantify all the relevant factors so they can be incorporated on top of a financial analysis, irrespective of the controversy and challenges in ascribing financial values to social and environmental factors.

This is particularly important to water distribution systems which are rarely implemented as competitive markets and, where competitive market pressures are used, both the project outputs and prices are usually subject to government control of both quality and economic regulation. Even for the regulated water company where the government has set out policies towards what investments are carried out, there may often be good reasons for not then limiting the project options appraisal to a financial analysis, given that the economic factors may have been reserved for policy makers and regulators. Particularly where there is economic regulation which has some independence from government policies, an economic assessment will be required. The private investor in a regulated water company may want to see more than just a financial analysis.

Water service providers (WSPs) undertake a wide range of projects, many of which require some form of evaluation in order to assist decision-makers. They will need to be prioritised, as not all improvements can ever be made at the same time, because of financial constraints such as end customer affordability or efficient financing by investors. Some projects will be of such a nature that they have a significant 'public good' element to them. For these projects, which have the potential to impact on society, the economy or the environment, it is often the case that the government believes that it should have a say in how such projects are evaluated, even where it delegates detailed decisions to private companies and their economic and quality regulators.

Over time, the boundary between whether a project should be evaluated from a financial or from an economic standpoint has become blurred. Infrastructure projects are subject to a range of assessment requirements by authorities, requirements that have accumulated over a number of years. One of the reasons for the imposition of requirements is that infrastructure projects have impacts on society and the environment which might not otherwise be taken into consideration due to 'deficiencies in the market'. In addition, investors are increasingly interested in environment, social and governance credentials as financial markets recognise the economic value from better, more sustainable management of the other capitals than just financial. Better uncertainty management reduces long-term financial risk to investors, which is building a market for environment, social and governance (ESG)-focused investors and green financing, where the cost of finance is linked to meeting environmental targets as part of the agreed outcome.

For this reason, directives, regulations and policies have been put in place to ensure that consideration is given to some of the wider impacts that water sector projects, among others, might have on a wider environment. Governance and rules however will still matter to what factors should be included in the project appraisal, and what techniques used. As we show as an example in Box 11.1, moving investment decisions beyond a financial analysis has become less and less controversial over time. There are two key reasons for this. Firstly, a financial or limited economic appraisal may be proved to have too short-term or too limited a focus, from the perspective of long-term sustainability or acceptability to customers and stakeholders. Secondly, even where there appears to be an optimal and economically efficient level of service and investment where the benefits equal the costs, this is often a static balance which may not have accounted for future uncertainties. At its simplest, a sudden shift in Government policies, such as the risk of water supply restrictions or water quality requirements, can make project appraisal very complex when faced with all the interacting factors that contribute to a water supply system.

The point is that in the majority of cases a purely financial evaluation will not be adequate. But on the other hand, it will not be the case that a full economic appraisal will be required either. In other words, any project appraiser must recognise what the various regulatory and governance requirements are that have to be complied with.

Box 11.1 A journey through the 'economic level of leakage'

Ofwat defined the economic level of leakage as 'the level of leakage at which it would cost more to make further reductions in leakage than to produce water from another source'. In 2005 the House of Lords' Science and Technology Committee asked if this was the most appropriate concept to apply. In their opinion, it focused exclusively on the relative costs, paying insufficient attention to the environmental impact that the development of additional resources might have.

In other words, they were questioning whether the application of a financial analysis that excluded costs associated with environmental impacts was the proper approach to employ. Their recommendation was that something more than a financial evaluation was required.

By 2009 the concept of the 'Sustainable economic level of leakage was developed', which included a limited range of social and environmental factors, such as traffic disruption from mains bursts and the environmental impacts of abstracting water. However, this did not go as far as including customer willingness to pay for leakage reduction in the valuation. Customer direct valuation for leakage tended to focus on dislike of leakage as a sign of water companies being poorly run for the long term, and also risks such as flooding and ice in winter from visible leaks, rather than purely economic or social factors.

In advance of the 2019 price review, Ofwat observed that leakage reduction had stagnated in the water industry since leakage targets based on an economic level were first introduced in 1997 (with a few notable exceptions). Ofwat observed that leakage targets varied more significantly between companies than could be explained by differences in either economic cost or social and environmental values. Ofwat were also concerned that the industry was not innovating sufficiently in how leakage was tackled, given the digital and data technology and smart monitoring technologies that should be emerging – the cost assumptions in the economic level of leakage could well be too high if based on history. The sustainability landscape had also changed, with Government policy shifting with climate change towards more drought resilience, larger strategic water resource schemes mixed with longer term demand reductions from both leakage and customer consumption. The cost of this challenge is uncertain, particularly the degree to which the cost of identifying and fixing leaks may increase at lower levels of leakage, and whether technology could offset these increasing marginal costs.

Rather than targets based on an economic level, Ofwat instead set a leakage target at the 2019 price review that was based on a minimum 15% reduction for each company, with additional marginal costs only allowed for companies with lower relative levels of leakage. The water sector responded by including leakage as one of a set of Public Interest Commitments that reflected areas of wider social and public value. In the case of leakage there was a target for a 50% reduction in the 2017 industry level of leakage by 2050, with a 25% reduction by 2030. The Government adopted a variant of this target based on recommendations from the National Infrastructure Commission, as well as setting an ambition for consumption reductions.

The role for economic appraisal therefore has become nuanced, but no less important, given the range of factors companies will now need to consider when putting forward their short term leakage reduction plans to meet. Economic appraisal is required for the timing of leakage reduction alongside other supply and demand options, as well as other regulated targets. Even though leakage has arguable become a policy decision rather than being set at an economic level, economic analysis is necessary to demonstrate that the financial costs reflect the best value (not necessarily the least financial cost) of meeting both leakage ambitions and water resource resilience more generally.

11.3.1 Factors to consider in economic appraisal

The overall objective of economic appraisal is to look at value for money by considering the associated costs and consequences of a course of action. It tries to answer the question 'Is this the most effective and beneficial use for the scarce resources, compared with the other options or opportunities available?' In order to do this, it is necessary to have ways of being able to compare alternatives on an equal basis using the same measure for all the alternatives, including 'do nothing'. By far the most common means of doing this is to denominate all costs, expenditures, benefits and impacts in monetary terms.

That said, it also needs to be remembered that decisions are very often not made purely on economic grounds. Economic appraisal of a project must be seen as an input into a decision-making process, a fact that tends to be overlooked. It can at times lead to a degree of dismay over the decisions that are eventually made. The economic appraisal of a project is not an end in itself, in the same way that a technical appraisal is not an end in itself. Both are part of a larger picture. There are other questions that need to be considered – for example, what are the various categories of unknown factors and risk associated with a proposal? How these might impact on a project proposal should be evaluated, especially if there are associated cost consequences. These must be taken into consideration if the affordability of a project is to be properly determined, especially if there will be costs associated with mitigation measures. All things considered, the objective with the appraisal of a project is to be satisfied that a course of action is chosen that has a (required) degree of certainty that it will realise the largest set of net benefits as compared with other alternatives.

In addition to this, there are other considerations, such as distributional issues: who bears the costs and the benefits and what is their relationship to the project proposed? An example of this in the water sector arises with the use of catchment management schemes and natural capital solutions, particularly those that involve payment of grants to farmers to reduce nutrient run off into water courses, or to housing developers to build more water efficient houses or avoid surface water run-off. Regulation, historically, used to object to water companies investing in such schemes if the costs were recovered through water customer bills, on the grounds that it was consistent with the 'polluter pays principle'. However, with a longer timeframe and focus on the least cost way of the water sector achieving its objectives, the objections have reduced, with consideration that to minimise costs and protect the environment for the long term a 'beneficiary pays' approach can sometimes be justified. At this point the water sector becomes an agent for wider society beyond provision of a product or direct service.

It is often the case that costs are borne by one set of stakeholders while the benefits are enjoyed by others. This gives rise to equity considerations, which project appraisal techniques such as cost–benefit analysis (CBA) are not that well equipped to handle without modification. To some extent, the reasons why or by whom an appraisal is required will go some way to determining such considerations and, therefore, the completeness of the information that will be required to be generated.

An appraisal is the means of communicating ideas and describing how far a set of proposals meets required objectives. This involves a series of sequential steps: the identification of the alternatives and a determination of their impacts; measurement of the resources required and quantification of the impacts; determination of the value of the inputs, the outcomes and the impacts; and comparison of the alternatives in terms of their costs and consequences. It is good practice to include sensitivity analysis and test the robustness of your assumptions (Green, 2003).

It is clear from much of what has been said that the appraisal of a proposal requires the gathering of data and the generation of relevant information that has to be brought together as a first step. Preparation is a key element, as there must be a reasonably good idea of what is to be appraised before it can be undertaken. This is not to say that no appraisal can take place before all the facts are known. Appraisal in one form or another is an ongoing part of the development of a proposal; it is an iterative process. But when significant decisions have to be taken, then a more rigorous approach is required, and for this the information requirements are more demanding. Stakeholder involvement and the governance process associated with the appraisal is usually informative in understanding uncertainty.

In summary, the key factors to consider in defining a project appraisal, whether a financial or economic appraisal, is the primary focus, and the types of appraisal techniques to use to support decisions are

- understanding the baseline: the commercial environment and market, the regulatory context (both economic and quality regulation)
- time: both the timeframe for the project and for the project decision/project appraisal

- is this a one off, standalone, discrete project, or is the project more complex, with multiple linked decisions and options that require optimisation
- a clear understanding of what trade-offs between competing factors are involved
- the governance involved in the decision, including the risks and opportunities associated with the project; if the governance process has already specified the project as an output (build this), rather than as an outcome (achieve this), then the project appraiser will need to want to appraise options targeted at the output – possibly a more constrained exercise than for outcomes
- a communication process, having considered governance and including appropriate stakeholder engagement, to identify what factors are relevant and sources of information to inform them
  - understanding social, environmental, community and cultural factors
  - distributional impacts – such as the impact on the wellbeing of particular customer groups
  - the scope and boundaries of the project – for example, whether the scope of carbon emissions is limited to operational impacts or embedded carbon and supply chain consequences
- how will decision making be informed by the project appraisal e.g. for a financial appraisal is the criteria a positive net present value (NPV) of future cash flows, or a ranking of internal rates of return (IRR) from an investment? Or is it an economic appraisal that requires a cost benefit analysis, and are there other criteria or constraints to be applied alongside this
- how is the time value of money to be reflected and what discount rate is used to achieve this?

We explore a selection of these issues in the rest of this chapter.

11.3.2   Concepts of value in analysis
Net present value (NPV)
NPV is the present value of the cash flows that arise from the project, compared to the value of the initial investment. The net reflects that it is the net value of cash inflows and cash outflows. Because it involves cash flows, it does not include indirect benefits of a project in the appraisal, even if financial value can be attributed to them. The NPV of a project includes the return that investors require to make the initial investment, including elements of risk perceived by investors. It is a financial appraisal as it considers financial cash flows only.

Cost–benefit analysis (CBA)
CBA operates from the exact same principles as NPV. The difference from NPV is that it extends the costs and benefits considered from financial cash flows to any other factors that are considered relevant to the objectives of the project. This also highlights the difference between costs and value – costs are the amount incurred to produce, service or maintain an asset – whether it's physical, environmental, financial or intellectual. Value is the use (utility) from that asset to consumers, the environment or society. A cost should be an objective and measurable thing, but a value is a perception of worth, which in economic terms is assumed to be revealed through purchase decisions in a market. When there is no market price, for instance when consumer surveys are used to establish a willingness to pay for environmental improvements, what we are measuring is perceptions of the worth of consumers of the value of the environment. These perceptions of value in themselves may not be robust for the long term, because perceptions of risk from environmental changes are poorly understood. This is important to understand how time is taken into account in project appraisal.

Costs, benefits and time

For both NPV and CBA calculations discounting can be used to reflect the time value of money. This is different to inflation, because generally both NPV and CBA calculations will be calculated at a constant price base (often prices at the start of the project), particularly in utility industries where price controls are often set relative to consumer inflation indices.

A project appraisal is assuming that costs are sunk into a project at the start. When considering different timing options for a project, there is the potential to calculate the *real option* value of a project by including uncertainty into the analysis. For instance, does investment in a pilot project that would result in more information that could reduce uncertainty (about costs of benefits) have a higher CBA/NPV than sinking the investment up front? Real options also consider the principle of an opportunity cost – a potential benefit that could have been delivered with investment but is forgone in its absence. Effectively the opportunity cost is the net benefit of a project that is not delivered over the time period if delay in making a decision.

This links to another project appraisal question – what is a cost and what is a benefit? Costs in project appraisal should directly link to the project in question – the investment, operation, maintenance and financing of the assets involved. Everything else is, therefore, a benefit – this includes costs that are saved indirectly as a result of a project (e.g. lower production costs if leaky pipes are replaced).

Time within an NPV calculation is dealt with through *discounting*. This is the process of converting a value in a future time period to a the current period, usually the point at which an investment is sunk at the start of a project. For a financial appraisal, a discount rate is needed that reflects the value of alternative uses of this potentially sunk investment. This is normally the cost of capital for the investor who carries the risk associated with the investment – an investor whose rate of return reflects the alternative investment that they could make and the returns available from this portfolio of investments.

11.4. Specifying objectives – the value chain

It is important in any project appraisal to understand the objectives. This starts with the core needs of the project, and this depends on the governance and risk perspectives for the recipients of the appraisal (the decision makers). The wider the governance perspectives, the wider the project appraisal that is needed. For instance, if the project appraisal decision is as simple as whether to invest to deliver a particular output (e.g. 10 km of new water mains to serve a new industrial estate) then a financial appraisal may be sufficient. But if the same investment is looked at from an outcome perspective (e.g. the impact on the rest of the network if the new industrial estate is supplied from the existing network compared to the network upgrade) then social and environmental factors and a CBA from an economic perspective could be more appropriate.

The purpose of an organisation is often helpful in understanding how it will balance financial resources with other contributors to its value chain. How it describes its relationship to employees and society through its purpose statement and organisation values can provide useful information as to what outcomes should be considered in project appraisal. Where an organisation is not clear on this, then the appraiser may need to find ways of teasing this out through engaging with the project sponsor. This is framing the project appraisal in terms of the outcomes that are relevant to the project. While the six capitals are often useful to infrastructure firms, mapping outcomes to the 17 United Nations Sustainable Development Goals can help to tease out what factors should be considered within the appraisal. These tools can be helpful if you have questions from a project specification that are not clear as to what the scope of your assessment should be – they can be considered as a communication tool. For instance, Goal 6 (Clean Water and Sanitation) is directly relevant to water distribution projects. Social goals of coverage of water supply systems and affordability apply. Goal 3 (Good Health and Well-being) links health valuations into water quality aspirations. Goal 4 (Quality Education) links to staff skills and community education benefits. Generally for the water sector all 17 goals will have relevance and metrics that could feature in a project appraisal.

11.4.1   Communication

Having specified the objectives, a key issue is communication. Whatever the outcome of a project appraisal, the results have to be communicated to an audience. This should be done in a way that is clear and comprehensible to all parties and not just those familiar with the intricacies of economic or technical analysis. Even something as simple as a net present value or a benefit–cost ratio should not be assumed to be obvious to the intended audience.

Regardless of the intended audience, a good practice is to assume that whoever is reading the appraisal knows very little about appraisal techniques. This can be done visually, such as describing the value chain of the project, or through the use of decision trees that explain the criteria (governance steps) involved in the decision and what the outcome of each element of the project appraisal that informs each criterion is. The presentation of the appraisal should be set out in clear and simple steps.

11.4.2   Externalities and valuation

Project appraisal that goes beyond estimating the direct market observable costs and benefits of the project will need to consider externalities (Box 11.2) – a cost or benefit caused by the project that is not financially received as a cash flow to the project. Those that are expected to be incurred or received will be valued as a market price – whether this is a construction cost that arises from a tender process, or the future energy price to move water around a network.

For externalities, the values often cannot be observed through a market and, therefore, need an approach to value them. Some of the alternative valuation approaches that can be used are

- observing prices in equivalent markets (e.g. a carbon price from carbon sequestration or carbon credit markets)
- revealed preferences through looking at other markets (e.g. how house prices vary by location because of proximity to local ecosystems)
- from government policy (e.g. the social cost of carbon set out in HM Treasury Green Book)
- benefit values transferred from a specific study (e.g. survey of visitors and users of a river) to another project (e.g. value of leaving water in the river environment rather than abstracting)
- general social impacts (e.g. assumptions on the value of education on productivity or the value of work experience in terms of future job prospects)
- general economic values (e.g. the gross value added by an industry that can only exist with a water supply)
- stated preference studies – establishing economic values through consumer market research, in particular exploring trade-offs and preferences between different outcomes through market research – for instance, asking customers their value for different levels of river quality improvements compared to the consequential impact on future water bills from that investment.

Box 11.2 Externalities

An externality occurs when the action of one party directly or indirectly changes the options available to a third party and the generating party does not have to account for that change. Therefore, any associated cost is not reflected in the prices charged for any goods or services provided.

Can you think of the externalities associated with a wastewater treatment works? What are the similarities and differences from the externalities associated with leakage?

Repeating a point made earlier, the important point to illustrate is that having understood the governance process covering the project appraisal, the output of the analysis needs to be presented in a way that communicates back to the objectives, particularly where there may be competing objectives involved. One challenge that can be made to CBA is that there needs to be confidence that the benefits will actually arise in practice. The more that these relate to an externality (particularly habitat and environmental benefits), the greater the challenge involved. Generally, these are also harder benefits to quantify, particularly for factors that fall outside of customers' experience. For any project, it is important that you are clear about how delivery of the project will be monitored and how the benefits will be measured in practice, including through the construction or delivery process. The challenge to engineers that what is delivered should be in line with a client's requirements can be assisted by the governance process on what benefits are included in a project appraisal.

There is a wide range of literature and practical examples of how customer research can be used to obtain willingness to pay values that can be used for social and environmental values in CBA. There remains controversy about any values that are used from customer research, particularly through stated preference surveys which try to understand customer trade-offs between different aspects of water services that are rarely experienced by individual customers. Ofwat noted (following their 2019 price review) the very wide range that were obtained by different water companies for the same service aspects, including through seemingly similar surveys. More recent analysis suggests that both the way that service risks are presented and the scope of the service improvements affects the valuations obtained. Before using willingness to pay values in a CBA it is important to understand the context in which the valuation was obtained. It is good practice to triangulate between a range of willingness to pay values, by assessing the relevance of the context of the source of the values and then weighting them. The location, age and service descriptions of the survey can be compared to the project situation to judge how relevant the social and environmental consequences (such as reflected within a customer willingness to pay to reduce leakage or supply interruptions) may be. If a high, central and low value is used for social and environmental benefits in a CBA, then the sensitivity testing will help to explore the robustness of the benefit–cost ratio and the decisions that have been made.

11.5. Costs

In any project there is a need to account for expenditures, to identify them and categorise the nature of the expenditure. Traditionally, the water industry, like others where infrastructure is a significant component, distinguishes between capital and operational expenditures, especially for accounting purposes. Although Ofwat moved towards a totex (capital and operating cost) basis for cost assessment to reduce the perception of bias towards capital investment in a world where natural capital and catchment management were becoming increasingly important rather than 'end of pipe' solutions being the first choice; in practice, accounting (and project appraisal) requires capital costs and operating costs to be considered separately. A better distinction (also used by Ofwat) is to consider the enhancement investment (the one that delivers the change being appraised) from the change in operating and future capital maintenance costs that result from that investment.

Other distinctions that are made include direct and indirect costs and internal and external costs – which have been touched on above. Within any one of these there will be a number of subdivisions that categorise costs according to a schema. Different industries and organisations will often have their own schema, developed to meet their particular reporting and cost control needs. Also, regulatory bodies (e.g. Ofwat) have their own reporting requirements.

In project appraisal, there are two questions that should be asked that can inform the identification and quantification of expenditure. First, what would be the most appropriate categorisation? Secondly, what level of detail is appropriate for the appraisal? The answers depend, to a certain extent, on what stage in a project cycle the appraisal is being carried out. In the initial concept stage, a few broad cost categories would suffice as the level of cost information available and the relative accuracy of the figures will not be of the highest. As a project progresses and more information becomes available, more details concerning the proposals are tied down, and then the granularity of expenditure information becomes much finer. A word of caution: information on expenditures is to be used to inform decisions and not to be an end in itself.

System costs

If the project being considered is part of a larger system, the question might arise as to whether to include or exclude expenditures on the larger system. Here the test would be that if the expected benefits of a project cannot be realised without expenditures on the larger system, then one should include those expenditures. The cost boundary must include all expenditures required to achieve the expected benefits.

Sunk costs

In contrast to the systems costs described above, where the project uses existing facilities but does not require any 'upgrading' to realise the benefits, then the prior expenditures should not be included. They are regarded as 'sunk' costs, and have no bearing on the project being appraised. The only proviso that there is to this would be if there were opportunity costs associated with the prior project. In other words, using the prior project would foreclose on other potentially more beneficial opportunities. Under these circumstances it would be the opportunity costs that would need to be considered: the benefits that would be foregone.

Capital costs

The expenditure required to put in place a facility that, when commissioned (normally embedded in a physical asset), is capable of producing goods or services is normally referred to as the capital cost. It could include the purchase of land, purchase of equipment, construction of structures and so on. They may not be limited to initial expenditures, as the purchase of equipment can be seen as a capital expenditure. Typically, capital costs are thought of as one-time costs, although the expenditure to pay for them may be spread over a period of time. Capital costs are fixed and are, therefore, independent of the level of output. The distinction with capital maintenance, which can vary with how hard a capital asset is used and, therefore, should be treated as an operating cost in project appraisal, even where an allowance for depreciation over its asset life is used as a proxy for ongoing maintenance costs.

However, the creation of a physical asset through forms of construction or other processes will be achieved through the employment of human resources, use of machinery and so on – all of which would be seen as operating expenditures on the part of the parties creating the facility. In other words, what for one entity may be seen as a capital cost could be seen as an operating cost by another. It depends on whose perspective you are viewing the expenditures from. Equally, the value of the asset may not reflect the expenditures required to create it, which gets us into the realm of book values, which will be left to the accountants to deal with and are not discussed in this chapter. But as an example, land is generally not depreciated and can sometimes be excluded from project appraisal if its value is not seen to be diminished by its use for the project.

Therefore, what counts as a capital expenditure can be open to interpretation. Note also that capital cost is *not* the same as the cost of capital.

Operational costs

While the capital cost is seen as one-off expenditures, operational costs are regarded as recurrent expenditures required to maintain the flow of goods and services. They may include the cost of employment of personnel, payment for services, insurances, maintenance costs, consumables and much more. Operational costs may be further divided into direct and indirect costs as well as fixed and variable costs.

Fixed and variable costs

Fixed costs include sunk costs, but may be avoided at the end of a project (e.g. from a sale of the asset) – they are sunk, however, until that point. A fixed cost is not sunk where in the long run it may not be fixed – for example, if the plant had an alternative use then at some point it may cease to be fixed. Variable costs that occur in a project are those that are not sunk/fixed – they can be avoided during life of a project if there is no output being produced at a point in time (e.g. no energy is used if the plant does not operate). It is important to distinguish between fixed and variable costs in a project appraisal where volumes of inputs and outputs are likely to vary, particularly if there is a revenue stream associated with its operation. The easiest way of doing this is to identify which costs vary with output (variable) from those costs which do not (fixed). The boundary of the project appraisal is important – overhead costs representing the cost of resources used by an organisation to be able to continue to perform its functions are usually taken to be fixed costs.

Direct and indirect costs

Direct costs are those that can be directly assigned to the production of goods or services, and are very often seen as variable costs. In other words, the level of usage of resources is correlated with the level of production or output. But direct costs and variable costs are not synonymous: the cost of running a machine used in production is not necessarily a direct cost but is likely to be a variable or semi-variable cost. Indirect costs are those not directly attributable to any given function. They may also be fixed or variable. An example of an indirect cost might be the machine used in the production mentioned above, especially if it can be used for more than one purpose. Indirect costs are sometime called overheads.

Working capital

Working capital is usually defined as net current assets, which would include bank balances and cash in hand. In terms of a project appraisal, it would be the additional amount of capital required to run a project created by investment in fixed assets. And it would be capital required until such time as an increase in net cash flow removed the need for the additional financing.

Transfer payments

Transfer payments are one-way payments for which no money, goods or services are received in exchange. They can affect the distribution of costs and benefits among project beneficiaries, as they transfer responsibility for resources from one party to another without altering the availability of the resource as a whole. It has a cost effect. Taxes, duties, royalties and franchise may be considered as forms of transfer payments. On the other hand, if a tax is levied on an output which is included in the market price, this should not be accounted for in a financial analysis, as the producer does not benefit from the tax. Excluding Value Added Tax and other sales taxes is a good example of this.

Opportunity costs

This is the value of the next best alternative, which will be foregone if the proposed project goes ahead. Identifying opportunity costs can be an important aspect of project appraisal when there are alternatives that are mutually exclusive and where there are choices over what projects can be embarked upon. Identifying the opportunity costs involves an assessment of the costs and likely returns for the available alternative options. To an extent, the objective that is to be satisfied must also be considered. In other words, if the objective were to maximise income as opposed to, say, reducing leakage in a distribution system then that will influence the range of opportunities available. Thus, the consideration of opportunity costs can be seen as being part of the financial appraisal of alternative ways to achieve a particular set of goals.

Regulation can play a role in this. Since 2015, Ofwat have set outcome incentives based on a principle that service levels were set at the point where marginal costs equalled marginal benefits, with the benefits informed wherever possible through benefits valuation studies and customer engagement, such as the use of stated preference surveys. Incentive rates for company outperformance and underperformance then set with reference to sharing the marginal benefits for outperformance between companies and customers, and compensating customers based on the net value (benefits less costs) for underperformance. Where regulatory incentives exist and affect revenues, they may be a valid consideration in a financial appraisal, but they should not be considered in a cost benefit analysis.

11.5.1   The importance of efficient and unbiased costs

Whatever cost types are included in project appraisal, the most important factor is that they represent *efficient* costs and the view of cost in the project is *unbiased*. Efficient costs are those that reflect the minimum cost to achieve the outcome from the project over its timeframe. There are many ways of testing what an efficient cost may be. The most common include market testing the cost, including through a tender process. For a repeated activity (assuming this really needs a project appraisal), costs of previous projects may provide a suitable benchmark for what has been an efficient cost, although this risks ignoring inflation, changes in the operating environment and assumes that previous projects have been efficiently delivered. Where market testing is not possible, benchmarking to similar costs and activities can be used. Regulators use tools such as econometrics to set efficient cost benchmarks for company proposals. For instance, Ofwat generally sets cost benchmarks in its price controls based on the upper quartile (top 25%) of the industry – not using the most efficient company recognises that companies are different and efficiency is hard to measure precisely.

The importance of *unbiased costs* provides a balance to the degree to which costs in a project can be identified as efficient. There is a risk that enthusiasm for seeing a project approved encourages overestimates of the benefits and underestimates of the costs. This is known as *optimism bias*. Including contingency and risk allowances based on past projects is one, albeit imperfect, way of trying to avoid optimism bias in project decisions. However, this does not avoid the need for engineers to be conscious of the risks of optimism bias in project appraisal. Are the costs and benefits presented realistic with an appropriate understanding of risk?

11.5.2   Cash flow
Cash flow is what it says it is – the flow of monies associated with a project or enterprise into or out of that project. In the case of a project, the cash flow would be taken over the life of the project. It can be taken as historical cash flows or projected future cash flows. A consideration of net cash flows – that is, the difference between what is coming in and what is going out – can be used to determine if there is an imbalance that would need some form of financing to bridge any deficits.

In the case of water services, the potential income is often highly regulated and set by governments and not by sales in a market. Under these circumstances, the choice of project may have little influence on the income generation. The utility may be operating under a budget constraint, where the challenge is to pick the best projects up to a financial ceiling. Nevertheless, net cash flow is still an important consideration for the reason given above relating to the need for working capital.

Regulation does not necessarily operate on a cash flow basis. Revenues that the utility is allowed to recover may be set on the basis of the average cost of a basket of projects, including the sunk cost in the existing network. For water networks there can be a big difference between the average cost (say of existing water reservoirs) and the cost of the new ones added to the network. Decisions are generally made on the increments being appraised.

11.5.3   Economic efficiency
CBA draws on the idea of obtaining the best possible return for the use of scarce resources. This being the case, the concept of economic efficiency becomes useful as a way of thinking about choices that might be made. As we discuss above, efficiency requires an assumption that the output or outcomes are of equivalent effectiveness in terms of the benefit they delivery. Economic efficiency is something of an umbrella term, as it covers alternative criteria such as

- **Allocative efficiency** – doing the right things – where the available resources are allocated in such a way as to maximise the benefit derived from their use
- **Productive efficiency** – doing things right – when production is achieved at the lowest cost possible
- **Distributive efficiency** – doing right – when the goods and services that are produced are obtained by those who have the greatest need for them
- **Pareto efficiency** – can't do things better – where any change in allocation cannot make one party better off without making at least one other party worse off
- **Dynamic efficiency** – doing things differently – projects that by their nature change what is done, piloting new approaches and innovation that change the cost base.

With the exception of dynamic efficiency, all of these concepts require ceteris paribus assumptions with 'all other things being equal'. As with most of economics, this makes the assumption that there is a market that reflects the actions of an infinite number of rational self-interested economic actors and, thus, what volumes and prices that are observed in a market reflects an equilibrium of supply and demand.

While the various concepts of economic efficiency are important, there is seldom a direct connection made between economic efficiency and CBA, even though how a CBA is carried out can be used to deduce something about the economic efficiency of a project. Furthermore, it can also be noted that economic efficiency can be looked at either from the private or societal perspective. Efficiency can be broken down into two interrelated parts: Should a project that produces goods and services go ahead? If so, what should be the level of provision? The first question only asks if the benefits outweigh the costs for any given project. The second asks how the benefits could be maximised; in other words, what are the various alternatives and which one of these would satisfy the appropriate efficiency condition?

11.6.   Where does the money come from?

When carrying out any form of project appraisal it is always good practice to understand where the project promoter will obtain the financing (the money) to pay for the project. Any form of financing brings its own set of constraints, and may impact on access to further monies. Understanding how a project is to be financed will make dialogues between the project promoter (usually the employer under the contract, or the client for the consulting engineers) and those executing the project much more meaningful. First and foremost, it is important for all parties involved, and especially engineers and contractors, to understand that there are few, if any, clients who have access to unlimited funds. In reality, there never were unlimited funds: it was simply that, in the past, the consequences of cost over-runs were less well publicised than is now the case.

In the water sector, project promoters come in a variety of forms, depending on the country or part of the world being considered. In no particular order, the project promoter could be a government (national or sub-national), a ministry, a department or a statutory corporation; or the project promoter could be a private company that has some form of management contract or franchise for the operation and maintenance of water services (as in France) or that owns the water infrastructure (as in England), or is mutually owned by is customers (as in Wales). Each operates under different constraints when it comes to financing.

The role of investors in society is crucial to concepts of economic efficiency. The Government has a crucial role to play, as the cost that the government can borrow money at from a market is often used as a proxy for the cost of finance that reflects the *risk-free rate*. The Government when it raises debt (unless it fails) can be expected to pay it back, and investors get a low return as a result. In reality, government bonds and debt is not entirely risk free, which can be seen that some governments have a higher borrowing cost than others. The higher the risk that an investor perceives in a project (in terms of whether the value they are entitled to in return for their investment is sufficient for their investment needs), the higher the cost of finance they will require to compensate them for making the investment.

There are generally two types of finance provided by investors. The cost of debt finance, which is often secured on the assets that the investment is being provided to finance, can be observed through the interest payments that are made to the investor until the debt is repaid. Security for debt investors are known as covenants, which specify financial ratios that have to be maintained, maximum levels of debt and trigger events that are circumstances where the debt must be repaid early.

Equity investors on the other hand take a stake in the project or business, and receive dividends or benefit from being able to sell their equity share. If a company fails, debt investors get their money back before equity investors. Therefore, generally, equity investors will have more interest in the outcome delivered by a business or a project as well as whether enough cash is being generated to pay their expectations for dividends, because this should determine the value of their investment if they choose to sell their share. Debt investors generally are interested in whether cash flows are sufficient to pay the interest they are entitled to, and whether they will repay the amount due at the end of the term of their financing.

Equity investors will generally consider a portfolio of investments. They may select a mix of higher and lower risk projects, and will attempt to maximise their overall returns. For instance, investing in one higher risk project that benefits from the price of energy increasing by more than expected and a second higher risk project that benefits from the price of energy decreasing may produce better and safer returns for the investor than investing in two projects less at risk from the price of energy changing. What investors cannot diversify in this way is systemic risk that affects the whole market. For equity investors, utilities and infrastructure are generally lower risk than the average market investment, as long as this is supported by a regulatory framework which fairly recognises that investors should receive a return on their investment commensurate with the risk. Regulation must protect investors as much as customers if they are to invest in water assets that, once in operation, are not allowed to fail or be used for an alternative purpose by law and regulation.

Ofwat has recognised that different investors may have different risk appetites, and it may be possible to finance water infrastructure at a cheaper cost. With Thames Tideway, the project to develop a new super-sewer beneath London, the scale and risks associated with the project meant that a government guarantee was provided to investors, while still maintaining that investors were best placed to carry some of the construction and delivery risk within the returns they received for investing in the project. Ofwat have developed the concept of 'direct procurement for customers' (DPC), where, for major projects such as strategic water resource schemes, the water company competitively tenders both the financing and construction of the project. This is distinct from outsourcing construction by the water company who is financing a project, because the investor will bring innovation and lower whole-life costs by having a greater say in the project design, defining the outcomes and/or operating the project once delivered.

One reason why governments often use private finance to fund infrastructure projects is that the amount that governments can invest without it increasing the cost of government borrowing is limited, particularly by the competing needs of government spending without raising taxes beyond the point that it adversely affects the economy. Infrastructure investment projects often need longer certainty that investment funds are available than governments can provide. A number of alternatives to the short run constraints that relying on government funding can place on a project are available, such as the outsourcing of debt. With private finance intitiatives (PFIs), the government or nationalised industry asks contractors to borrow the money to finance a group of projects, such as a hospital building programme, on its behalf. The government then undertakes to pay an agreed annual charge for the construction and maintenance of the hospital over an agreed period of up to 30 years. Unless there are tax advantages for specific classes of work, this is usually a more expensive way of paying for projects (for a variety of reason), but it has the advantage of keeping the debt off the government accounts. It thus preserves lower government borrowing rates, at least in the short term. The cost is expensive as the government is trying to outsource risk and uncertainty. However, in reality the ability to outsource such risks is limited, requiring long-term contracts that may not be enforceable should risks to the service provider arise in practice, or when the ongoing cost post project delivery proves to be unaffordable.

Despite the long-term concerns and challenges, this model of financing directly from government still predominates in many parts of the world where water services are provided directly by a government department. As far as nationalised industries are concerned, the government will normally require them to borrow money from the government, and, generally, the rate it will charge them will be above the rate that the government itself incurs. In special cases, it may allow this borrowing from government to be 'topped up' with initiatives from the private sector, but seldom through direct borrowing, as this would have to appear on the government accounts. Municipalities and statutory bodies are generally treated in the same way as nationalised industries in the UK, and are restricted to borrowing only from the UK Treasury. However, this is not always the case around the world.

In the USA, for example, it is common for municipalities to raise money by issuing their own bonds, and, as shall be seen in a later section, this can have an impact on the optimal phasing of engineering projects. Under certain circumstances, such bodies (including governments) may also borrow money from banks and other lending agencies, such as international or regional development banks. This borrowing comes with conditions and, often, higher interest rates.

This concept of private investment in public infrastructure is not new. In the case of Bristol Water for instance, in 1846 the UK parliament chose between two rival privately financed initiatives to provide supplies to the city and preferred one that had a social purpose of supplying all the population, rather than the prosperous areas which on paper were the areas likely to be able to pay. The project took 30 years to payback to shareholders having improved the health and wellbeing of the poorer areas so they could afford a piped clean public water supply, ending cholera in the city. Similar private and municipal water projects competed for approval in that era. Many of the great engineering feats of water supply, storage, treatment and distribution from that era remain in use to this day.

This leads to the question of funding for private companies, usually referred to as public companies. What is being referred to are companies that are independent of government ownership (hence private, in one sense of the word), but may be subscribed to by the public (hence public companies), or owned by private individuals (hence private, in a very narrow sense of the word). In some cases they are run as private organisations but the funding comes from government bodies, such as investors that are public sector pension schemes or sovereign wealth funds in other countries infrastructure. In all cases, companies will have shareholders, who have invested their monies in the company in the anticipation of receiving a reward. Sometimes the reward will be an increase in the value of the shares as the company expands, and sometimes it will be a dividend which is paid out of profit. It is unusual for all profit to be paid out as dividend, as normally some will be retained for capital investment. In the case of 'not-for-profit' bodies, any surplus has to be either distributed or re-invested.

It is essential that appraisals contain realistic estimates of cost, as significant cost overruns are unlikely to be greeted kindly by promoters, whether these are public or privately financed projects. There are many examples of publicly quoted companies getting into terminal difficulty because of flawed engineering appraisals, poorly understood risks, and poor stakeholder and reputational management. Economic regulators, such as Ofwat, are particularly concerned as to whether the constraints that financing arrangements place on companies misalign management incentives and focus with delivery for customers and the environment. Once misaligned and the reputational damage done, it can be very difficult to recover, as Southern Water have found on wastewater performance even after extensive leadership changes following failures.

Another example is the demise of Railtrack, which went into the modernisation of the West Coast Mainline on the basis of an estimate of £2.5 billion, and finished up with a project costing almost six times that figure (National Audit Office, 2006). This was a factor in the decision of the incoming Labour government to renationalise the company, with major project cost overruns constraining maintenance expenditure, which ultimately led to disaster and failure. It was the shareholders who lost their investment. More recently, the increasing cost of the publicly financed HS2 did not result in cancellation, but a scaling back of the project that may remove much of the benefits originally envisaged.

Cases such as this illustrate the importance of both project appraisal and a clear understanding of delivery risk when costing the project. Engineers should be conscious of the impact of project appraisal and the potential for their decisions to impact the long-term success of an organisation. Even projects that are successful and keep to time and budget in a construction phase may be delayed during commissioning, as seen most recently with Crossrail. An engineer conducting a project appraisal needs to consider the wider context and be clear in agreeing the assumptions they are expected to consider.

11.7. Techniques and alternatives

The focus of this chapter is on project appraisal, specifically looking at costs and benefits. So far, the elements that might inform an appraisal have been discussed. At some stage, a decision has to be made as to how to operationalise the appraisal. In other words, what technique should be used to inform decision-making? A variety of techniques could be used, depending on the circumstances.

The first characterisation of appraisal techniques is to make the distinction between qualitative and quantitative techniques. This is not as simple as whether financial values are used in appraisal or not. Qualitative analysis involves subjective judgements to analyse value based on human (occasionally expert) judgement. On the other hand quantitative analysis will have a statistical or scientific basis for the values used, for instance energy information from a pump to work out how energy costs vary with volume.

The Accounting for Sustainability (2014) guide to managing future uncertainty usefully characterised different analysis techniques between qualitative assessments of impacts as relatively simple techniques (such as trend analysis) and quantitative analysis as more complex using techniques, such as Monte Carlo simulation, where probability distributions for individual components are combined randomly in order to gain a distribution for the likely outcome. A more innovative approach is to combine qualitative and quantitative techniques into spatial analysis – which may be a series of qualitative trigger points on geography or human behaviour which link together different quantitative modelling. Scenario or spatial analyses may allow the combination of a number of different appraisal techniques. This is useful in situations where discrete events (e.g. a future choice on scale of project) may depend on external factors, such as those involving a third party decision.

Table 11.1 shows some forms of project appraisal, each of which deals with costs but treats consequences in different ways. The more commonly used techniques include least-cost analysis, cost-effectiveness analysis (CEA) and cost-benefit analysis (CBA). The latter is probably the best known and most widely used. Indeed, CBA is a commonly used appraisal tool, and one which the US government, the European Union, the World Bank and other regional and international agencies routinely use to appraise projects. Although the aims of each of these techniques may be different, as evidenced by their descriptive titles, they do share some commonalities. The end point is the achievement of a goal for which there are different alternative options, and a decision is required on how to go about taking into account the costs associated with the achievement of the goal(s).

Table 11.1 Measurement of costs and consequences

Type of appraisal	Measurement or valuation of costs	Identification of consequences	Measurement or valuation of consequences
Least cost analysis	Monetary units	Achieved to the same degree	None, or qualitative
CEA	Monetary units	Single effect of interest, common to alternatives, but achieved to different degrees	Qualitative (e.g. units to which an outcome such as the environment reaches an agreed standard) or quantitative (e.g. modelling of leakage reduction from an intervention)
CBA	Monetary units	Single or multiple effects, not necessarily common to alternatives	Monetary units, often expressed as a benefit–cost ratio
Social return on investment (SROI)	Monetary units	Single or multiple effects, not necessarily common to alternatives	Quantitative, equivalent to CBA but measures social impact of private costs and investment. Can be a qualitative assessment of the social consequences of delivery options to achieve a common outcome.
Scenario analysis	Ranking based on monetary units or qualitative assessment	Running another appraisal type by varying assumptions	Depends on the appraisal technique, but either ranking or monetary units
Adaptive pathways	Qualitative trigger points that may result in a different monetary unit outcome	Identifies an event or trigger point that indicates a series of consequences	Qualitative description of the trigger point or a quantitative monetary or output should an event arise (e.g. intervention if leakage reaches a particular future level)

After Drummond et al. (2006)

In the following section, the techniques mentioned above are briefly discussed. In addition, some other possible approaches to project appraisal are considered that are also used. The additional approaches considered are lifecycle analysis, whole-life costing (WLC) and real options. Each of these approaches to appraisal can and, in some cases, do draw on the techniques already mentioned. They differ in the manner in which they are used, and they may be thought of as alternative ways of identifying the range of costs and consequences associated with alternative courses of action. Each, in turn, is the subject of a growing body of literature detailing their underpinnings and assumptions as well as case studies of their application.

11.7.1 Least-cost analysis

Least-cost analysis focuses on the determination of the project alternative with the least cost, given that the alternatives are mutually exclusive and technically feasible options. To be mutually exclusive, the alternatives must be capable of producing the same outcome or output of a specified standard or quality. If there are differences between alternatives in terms of their output, then these would have to be normalised in order to compare alternatives on the same basis. For projects where the benefits are in the form of a single commodity, such as water supply, and where the potential income stream is not affected by the choice of alternative, then a least-cost approach can be considered.

An example might be the extension of a water supply which could be secured either through leakage reduction or by incorporating new sources. In such a case, if incorporating a new source provides more supply than the leakage reduction, then in order to compare them equally, the relative benefits achieved would have to be normalised. This could be done by valuing the foregone incremental benefits from incorporating the new source and to add it as a cost to the leakage reduction option. In doing so, it would ensure that the two options are essentially equivalent to each other. The alternative with the lowest present value of costs would then be the least-cost option. Note that benefits do not enter the calculation, where they are the same for both alternatives under these assumptions.

11.7.2 Cost-effectiveness analysis

CEA usually considers a single measure of output, such as the level of leakage reduced or the level of dissolved oxygen in a river. The alternatives that are considered may be able to achieve different levels of reduction in the chosen parameter, but at differing costs. The results are expressed in the form of a cost-effectiveness ratio, which is used to compare options. In other words, cost-effectiveness looks for the best way to achieve a given end result. The most cost-effective alternative will be that which achieves the desired outcome at the least cost. There may be constraints on what least cost alternative is acceptable, based on a qualitative assessment. This can be based on value judgements, in which case the analysis may be described as the choice of 'best value' rather than 'least cost'. This is still a cost-effectiveness analysis rather than cost-benefit analysis, because the core outcome (achievement of a particular level of leakage) is not chosen by the analysis, but how it is delivered is considering more than just the least cost.

When the budget is limited, this may be a useful way of deciding which alternative should be adopted where there are a limited range of options. The difference between cost-effectiveness and least-cost methods lies in the way that benefits (consequences outcomes) are handled, as can be seen from Table 11.1. The main difficulty in using CEA is that it can only really handle one measure of output. Where there are multiple outputs, unless some aggregate measure can be derived, the method becomes difficult to apply. Even if such an analysis were to be applied to each of the outcomes, engineers would be faced with the task of deciding which output is the more important and how they should be ranked. This introduces an element of subjectivity, which may not be seen as helpful.

11.7.3   Cost–benefit analysis

In CBA, benefits (consequences) are treated in the same way as the costs by assigning a monetary value to them. The broad objective is to determine whether the outcomes from an alternative justify the costs incurred in realising the benefits. The problem becomes one of ensuring that all the benefits or dis-benefits arising from the proposed project can be identified and costed. In some cases, the difficulty lies in identifying the benefits and the externalities associated with projects alternatives. For benefits, it is important to consider the degree to which they can be attributed to the project, in the sense that they would not otherwise be delivered without the project. As such, CBA has a broader scope than least-cost analysis or CEA. The latter address mainly questions of productive efficiency while CBA helps to inform questions of allocative efficiency to determine if a goal is worth achieving, given alternative uses of the resources.

CBA is used in a wide variety of applications, although it was developed for use in infrastructure projects. It assesses the net impact of a project or alternative project and, thus, is used to inform decisions where budgets are limited. Even if budgets were limited, CBA would determine whether a project was worthwhile as no project should expect to proceed if the benefits are less than the costs. Among the strengths of the techniques is that it is able to consider externalities and other factors that distort prices, allowing market imperfections to be explicitly considered. This allows the techniques to go beyond simple financial effects for the private investor and to consider projects from the point of view of society or of the public. In principle, CBA could be used to evaluate programmes and policies. The aim of CBA is to determine if a project is desirable from the point of view of the private investor and that of social welfare, taking into consideration all the costs and all the benefits associated with the proposed project. It does so by summing the discounted costs and benefits of project alternatives and comparing them. The process of discounting will be discussed below.

The purpose of the project and the nature of the parties responsible for the project will have an influence on determining the extent of the CBA. Where the purpose of a project is to promote local development and is the responsibility of a public body, then the CBA should go beyond just the financial analysis. After all, what is important is the extent to which the project will be able to meet the goal of local development and at what overall cost. A further consideration is the extent to which the carrying out of a project will affect the local 'market' – that is, the availability of goods and prices for goods and services. In regions such as Europe, it would be the exception where a project is of such a scale as to have any noticeable impact on the market and, therefore, such impacts can be ignored.

This, however, will not be the case in other geographic settings. Under these circumstances, aspects such as the impact on wages or displaced consumption would have to be factored in. Many of the international lending agencies, such as the World Bank, Asian Development Bank, the European Development Fund and others, have developed their own guides to carrying out CBA. There are also specialised or standardised methodologies that have been developed by various agencies for use under particular circumstances – for example, the evaluation of flood alleviation or coastal protection schemes by the Department of Environment Food and Rural Affairs in England. Flood schemes generally are only funded in England if they achieve above a particular benefit to cost ratio, not just a ratio that is higher than 1. This in part reflects uncertainty on benefits and risk that costs are higher than estimated (see optimism bias). The rules in flood schemes allow schemes with lower benefit to cost ratios to proceed, but only if they attract private funding, which is a way of revealing that the benefits are more likely to exist if there is a private company willing to partly finance a flood defence scheme. This reflects a form of 'beneficiary pays' rather than just relying on taxpayers in general.

In cases where there is the private provision of services to the public (sometimes referred to as public goods) such as water supply and sanitation, the question of what benefits should be included and whether the appraisal should go beyond that of a financial analysis becomes somewhat blurred. In certain instances, this question will be answered in part by the requirements of government agencies, regulatory authorities as to what should be considered or included.

It might be argued that some of the evaluation or decisions with respect to benefits have already been taken – for example, the disposal of sewage waste by long-sea outfall is no longer an option in Britain; one reason among many being that there is a presumption that the amount the public are willing to pay to improve the marine environment outweighs any additional costs associated with the alternative means of disposal of the waste on land. Thus, any cost–benefit appraisal of a wastewater treatment works can take this as a given and need not include it in the calculation of costs and benefits associated with the various options considered.

Another example may be in leakage appraisals in calculating the impact of traffic disruption. Before the introduction of lane rental schemes which charge utilities for occupying road space, the disbenefit of traffic disruption would be factored into calculating the cost benefit of different leakage reduction approaches. With utilities paying for their roadworks, it can be argued that traffic disruption ceases to be an externality and becomes a direct cost in the appraisal. If traffic disruption was the only factor, CBA would no longer be required and CEA used instead.

The extent to which CBA can incorporate environmental and social costs and benefits is a matter of debate. It is related more to the ability to identify the relevant social and environmental facets to be included and how to go about valuing them than it is to the ability of the methodology to handle them. For some, the very idea of placing a value on the goods and services that the environment supplies is questionable. Leaving that point aside, with environmental impacts considered – and they should be – then not only should the benefits outweigh the costs but they should outweigh the cost plus the environmental cost of the impacts of the project. It might also be said that the environmental cost could also mean external costs, as unpriced environmental damages are externalities associated with the project. The question of how to arrive at environmental costs and, by the same token, environmental benefits is a subject area all of its own. Suffice it to say, this is an area in which techniques such as economic impact assessments (EIAs), life-cycle analysis (see the following section) and environmental economics all have an important role to play. As mentioned earlier, sometimes policy makers will have made these decisions for the appraiser, so an understanding of the context of the law and governance for decision making should always exist before using any project appraisal technique – the reader will notice the number of times we repeat this point.

11.7.4 Social return on investment

A social return on investment (SROI) analysis combines the fixed private outcome from CEA with the valuation of (public) social, economic and environmental benefits from CBA. It is also one way of moving beyond a 'least cost' analysis to 'best value'. Through identifying the social and environmental benefits of a project's outcomes, and valuing them, the wider benefits of costs can be identified. It measures the effectiveness of the costs and resources that a project incurs in achieving wider benefits.

11.7.5 Scenario analysis, adaptive pathways, real options and least-regrets

Scenario analysis is an extension of other appraisal techniques, which examines and evaluates possible events and scenarios that could take place in the future, and predicts the various feasible results or possible outcomes. If probabilities are applied to these scenarios, then an expected value of a project can be calculated. Monte Carlo simulation can also be used to randomise the likelihood of the probabilities that form the scenarios, so a distribution of outcomes can be achieved.

There are limitations with expected values, which Monte Carlo simulation can illustrate. Two projects may have equivalent net values, but may have very different risk profiles. One project may have a very narrow range of outcomes, symmetrically (perhaps normally distributed in statistical terms) around the expected value – in this case there may be a 50% chance of better or lower net value of a project than the average point estimate. A second project may have an 80% chance of somewhat better outcomes than expected, but 20% chance of a disastrous failure that would threaten the financial future of the organisation – the project has asymmetric risks. These projects are clearly not equal and should be treated differently, even if they have the same expected cost benefit.

There are two techniques that can help with these circumstances. An adaptive pathway tries to set out the key factors on a development of a project that can trigger either project interventions or benefit realisation. This can particularly help where there are dynamic efficiencies expected from a project that may only apply in some related scenarios – particularly useful if one project depends on others that are being considered.

Real options appraisal can combine features of both Monte Carlo simulation and adaptive pathways. A real option for a project means that an option to make an investment remains open, but there has yet to be commitment (i.e. the investment has yet to become a sunk cost). Real options appraisal considers the value of committing to a project, compared to the counterfactual of waiting for better information. There may be an opportunity cost to delay, but if more value might be created if uncertainty is reduced, this opportunity cost may be worthwhile.

An alternative to real options appraisal can be to attempt to make least-regrets decisions. Rather than assessing the value of waiting for better information due to uncertainty, this considers what scenarios might apply should a decision not be made. One way to consider this is to minimise the maximum loss that could arise. This is particularly useful when considering low probability/ high consequence impacts that may be outside of past experience, something that water services typically have to consider. For instance, investment could be targeted by investing in short-term maintenance that reduced the immediate risk of supply interruptions. An alternative investment might avoid very high consequence risks (e.g. from a drought), where supplies could not recover from such an extreme event. The least regret option is to avoid catastrophic failure of the water supply system and to identify the lowest cost way of achieving this (e.g. investing in emergency supply facilities). Cyber security of water distribution systems is one modern example of where a least regrets approach may be appropriate in project appraisal, because the potential consequences are outside past experience as to what the cost of failure may be.

11.7.6 Life-cycle analysis

Life-cycle analysis is based on the idea that a detailed examination of the life cycle of a project and all the goods and services that are required to bring it to completion and keep it operating should be carried out. The goal is to compare the full range of social and environmental damages that will arise, and to use this as a means of choosing that which imposes the least burden on the environment. It considers aspects such as the use of raw materials, energy consumption and the amount and type of waste produced. Theoretically, it provides a way of assessing raw material production, manufacture, distribution, use and disposal, and intermediary steps. It has been used to optimise environmental performance, either of a single product or a company, but is much more difficult due to the informational requirements to apply to projects as part of an evaluation. It is important to consider life-cycle analysis if claims of carbon-neutrality or water-neutrality are important to the project objectives.

The procedures of life-cycle assessment have been systematised by the International Organization for Standardization (ISO) through its incorporation into environmental management standards, ISO 14000. Life-cycle analysis is a potentially powerful tool which can assist regulators to formulate environmental legislation, help manufacturers analyse their processes and improve their products, and perhaps enable consumers to make more informed choices.

One of the most identified problems with the application of life-cycle analysis is that of where to draw the boundaries around an analysis. It is usual to ignore second generation impacts, such as the energy required to extract aggregate used in concrete to build the wastewater treatment works to treat the sewage to environmental standards. While life-cycle analysis is a way of finding out about the environmental footprint and converting this into some form of common metric for analysis, comparisons of alternatives are more challenging. As a technique within project appraisal, it is probably best employed as part of an EIA.

11.7.7 Whole-life costing

WLC (Skipworth *et al.*, 2002) is a term that is predominantly, although not exclusively, used in an engineering context with respect to the provision or procurement of service infrastructure. It has been advocated as an alternative decision-making approach that is more appropriate when considering alternative means of achieving a set project objective and the balance between the initial capital investment and subsequent operational costs over the whole life of the project and the associated service infrastructure. A distinctive feature is that, as an approach, it moves away from the perspective of projects as being one-time exercises to considering a stream of potential interventions over a period of time. As such, it is of particular use in the water industry.

The whole life cost of construction in a conventional sense (it) should include such factors as initial construction cost, operating, maintenance and repair charges and an allowance for demolition.

(Chenery, 1984)

In other words, it combines capital and operational costs over the life of the asset infrastructure. This has implications for investment decision-making when compared with alternative capital-budgeting decision-making approaches. The adoption of WLC approaches has been underpinned by developments such as best value in local government, building and constructed assets (ISO/DIS 15686), private finance initiatives and public–private partnerships. The US Environmental Protection Agency (1996) provides a comprehensive treatment of WLC and the range of costs that can be considered for inclusion. These include the initial capital as well as all up-front costs, operational, maintenance, rehabilitation, repair costs and, where relevant, the decommissioning costs. It is not just internal or direct costs that have to be considered but also overhead costs as well. A significant feature of WLC is its attempt to minimise the distinction between capital maintenance and operating costs.

In the context of the provision of water services, WLC has been proposed and adopted as an integral part of decision support applied to the maintenance and rehabilitation of distributed networks such as water mains. In the case of highly regulated privatised WSPs, not only are the costs of service provided important but also the quality aspects.

Generally, the application of WLC to network distribution management aims to achieve the lowest network service provision operating cost, when all costs are considered, while meeting statutory standards and regulatory performance requirements. One of the key differences between WLC application within the utilities sector, as against the built environment, lies in the stage of the service provision cycle. Generally, the service infrastructure already exists and, thus, the focus is not on its provision but on maintaining the serviceability of the assets. Within this context, the physical condition of the service infrastructure and the effects of deterioration on serviceability assume a greater significance. This adds extra importance to the timing dimension of management decisions and interventions, as these impact not just on the net cash flow generated but also on the relative benefits of an intervention, namely its impact on serviceability and performance. This, appropriately, should be the focus of water service regulation.

Thus, in applying a WLC approach to ageing assets, knowledge of the physical performance and the consequences of that performance are important (Box 10.3). This is due to the linkages between such performance and the activities and demand for resources that operating and maintaining the levels of serviceability require.

Box 11.3 Performance modelling

How well a system such as a water distribution system performs depends on a number of factors. If it is already in place, then factors such as age, materials used, usage and so on will have an important influence. If it is a new system, then there will be certain expectations as to how it will perform over its useful life and the level of maintenance it will require. At the same time, how well a system performs and what has to be done to maintain or improve that performance come with a cost. These costs will vary over time, and will depend on the type of interventions made (e.g. repair or replacement), while the type of intervention will also affect future performance as well as running costs. An example is response to leakage – mains repairs (e.g. using encapsulation collars) resolve the leak, but do not improve the underlying condition of the pipe. Mains replacement on the other hand has a longer lead time to fix the leak, but may reduce leakage and the risk of service disruptions.

A growing trend is to couple performance modelling with associated costs, and to optimise overall performance subject to constraints, such as the minimisation of overall costs. The performance model is a mathematical representation of the key interactions and characteristics of the system, and varying the parameters allows the impact of alternative decisions regarding design, management or other interventions to be assessed. Better information from network monitoring and modelling has improved this trend. Modelling can be undertaken surrounding the characteristics of increasingly smaller pipe lengths, targeting burst and supply interruption risk, pressure, leakage, water quality and discolouration outcomes. Characteristics can include water quality, pressure, external factors such as soil and pipe age and material. WLC analysis however needs to consider whether the lowest cost effective approach for a short section of pipe is the same when considering a larger section of network.

11.8. Discounting

Where the time over which a proposed project will come into being and operate is in excess of at least 1 year and where the costs incurred and the benefits that accrue do so at different points in that time horizon, then the concept of discounting becomes a central consideration in the evaluation of alternatives. The vast majority of people would rather have a benefit sooner rather than later. This idea lies at the heart of discounting, and is referred to by economists as a positive rate of time preference, for a variety of reasons. Firstly, because there are varying degrees of uncertainty as to what tomorrow might bring, there tends to be a preference to enjoy benefits today rather than wait until later. Then, there is the general idea that people will be better off in the future than they are today, thus money today is worth more than the same amount in the future because people will be richer – it is hoped. In other words, a given amount of money has more value today than at some time in the future. This idea provides the basis for discounting, and can be applied not just to costs and expenditures but also to other goods and services that are not so easily traded or for which markets exist. Put the other way around, it can be said that an individual would have to be offered some form of incentive to defer consumption until a later point in time: the payment of interest related to the scale of benefit they expect to receive.

If individuals think like this, and society is an aggregation of individuals, then it should follow that society should also make choices about the value of projects or investments on a similar basis. But the question is, should society adopt the same set of preferences as an individual? In other words, should society discount the future to the same degree that private individuals (or entities) do? Many economists argue that there should be a difference between them. One distinction that has been made is that individuals will express a different set of preferences depending on whether they are considering something from their own individual perspective as, say, a customer or from their perspective as a citizen concerned about what may happen in the future. A further argument that has been advanced is that individuals often underestimate the benefits of future consumption and that there is no good reason to carry this short-sightedness over into the social context. It seems, therefore, that there is consensus that there is and should be a difference between the rate at which a private individual views future consumption and the way society views the future. One of the implications of this is that projects that are undertaken on behalf of society (by a government) and which are intended to benefit society should not be discounted at the same rate as projects undertaken by private bodies for their own benefit.

The question once again is: which interest (discount) rate should be used in a particular set of circumstances, especially when carrying out project appraisal? In some cases, interest rates are determined by monetary authorities such as central banks, which may be setting rates in order to achieve objectives other than to do with matching supply with demand – for example, the UK Treasury's Green Book (HM Treasury, latest version 2022), in setting out guidance for the evaluation and appraisal of government investments, proposes that a discount rate of 3.5% be adopted with factors such as risk, optimism bias and the cost of variability dealt with separately and explicitly. The UK Treasury Green Book provides a range of excellent guidance and explanation of the terms involved in public sector project appraisal, and should be a source of reference for anyone conducting project appraisal that includes public goods such as water services.

The STPR excludes the effect of inflation, and this principle of excluding the impact on inflation in cash flows when discounting should also be applied to private investment. It is the 'real' rates rather than 'nominal' rates of interest that should be used. Real rates are nominal rates adjusted for inflation. If there is only one interest rate on offer in the private sector, say 5% per year, and inflation is 2% per year, then the real interest rate would be $5 - 2\% = 3\%$.

For more precision, the Fisher equation should be used to calculate real private discount rates. For this example, this is calculated as $((1+5\%) / (1+2\%)) - 1 = 2.94\%$, reflecting that both interest rates and inflation compound in value over time.

This also has implications for costs and benefits and how they are to be handled. The use of a real interest rate implies that the effects of inflation on costs and benefits should be ignored. All costs and benefits should be accounted for in constant money terms, and the use of any historical cost data should be adjusted accordingly.

Another question that can arise is whether, for long-term projects, those that operate and deliver benefits over an extended period of time, the discount rate employed should be uniform. One of the reasons for asking this question is the impact that the use of a discount rate has on future values (Table 11.2). Table 11.2 illustrates that future values are of increasingly less importance, especially as the discount rate increases. This aspect can lead to some types of projects being disadvantaged when compared with others. Take, for example, two alternatives that achieve the same stream of benefits, but one has a high initial cost but low recurrent costs and the other a lower initial cost but higher recurrent costs. The effect of discounting, other things being equal, will favour the option with the lower initial cost. It is for this reason that in certain instances a decreasing discount rate may be applied to future values. This approach has been adopted by the UK Treasury in its Green Book, where a declining long-term discount rate was set out (Table 11.3). The length of time over which a project is appraised can also have an important influence on the choice between alternatives.

In cases where the time horizon may be long and there are associated environmental impacts as a result of the project, the proper time horizon for the appraisal is the time at which the impacts of the project cease. It is not the time at which the project ceases to serve its original purpose. The examples of nuclear power stations and the effects and mitigations of climate change come immediately to mind. In practice, the choice of discount rate requires the exercise of judgement and the inclusion of sensitivity analysis to explore the effects of different rates on outcomes.

Table 11.2 Present value of 100 at various discount rates

Discount rate %	Time horizon in years				
	5	10	20	50	100
1	95.15	90.53	81.95	60.80	36.97
2	90.57	82.03	67.30	37.15	13.80
4	82.19	67.56	45.64	14.07	1.98
6	74.73	55.84	31.18	5.43	0.29
8	68.06	46.32	21.45	2.13	0.05
10	62.09	38.55	14.86	0.85	0.01

Table 11.3 Declining long-term discount rate

Period of years	0–30	31–75	76–125	126–200	201–300	>301
Discount rate	3.50%	3.00%	2.50%	2.00%	1.50%	1.00%

Even then there should be scope for debate and disagreement. The former senior vice president and chief economist of the World Bank, Joseph Stiglitz, once said that the choice of discount rate

depends on a number of factors, and indeed I have argued that it might vary from project to project depending, for instance, on the distributional consequences of a project. These results may be frustrating for those who seek simple answers, but such are not to be found. The decision on the appropriate rate of discount thus inevitably entails judgments.

(Stiglitz, 1994)

11.9. Evaluation techniques

There are a number of evaluation techniques that can be applied, depending on what objectives a project proposer has. This indicates that there is no 'right' evaluation technique but rather a range of possible options. All of the techniques look at the initial investment required against a project set of returns on that investment and, because they are projections, they will necessarily involve varying degrees of uncertainty. Uncertainty is an additional aspect that has to be dealt with explicitly to inform decision-making. All of the techniques try to assist in answering the question 'Should the project go ahead?' And in some cases they also help to answer the further question 'Is this option better than another one?' There is one technique that ignores the time value of money: the payback approach. Then there are those techniques that are based on the application of the time value of money: discounted cash flow, NPV, the IRR and the cost–benefit ratio. Finally, there are those techniques in which the last-mentioned approaches are combined with other non-monetary factors, referred to as multi-criteria approaches.

11.9.1 Payback period

Payback is probably the simplest appraisal tool. It involves a calculation to estimate the time that would be required for the projected cash flow to cover the initial capital cost or investment required to establish a project. The strengths of this technique are its simplicity and that it gives a quick indication of the ability of a project to pay for itself. It is particularly useful where project time spans are relatively short and where there is a comparison to be made between two or more possible alternatives. Table 11.4 and Figure 11.2 illustrate a simple example of this technique. It can be seen from this simple illustration that the payback period is three years and approximately four months.

The payback method has a major limitation: it just concentrates on cash flow return and ignores other aspects. It pays no attention to what happens after payback, and, more importantly, it ignores the time value of money by assuming that money tomorrow has the same value as money today, implying a zero discount rate.

11.9.2 Discounted cash flow

Discounted cash flow looks at the expected cash flow associated with a project that has been forecast into the future over some finite period of time. The principle of the time value of money (that money today is of more value than money tomorrow or at some other point in the future) is then

Table 11.4 Payback period

Year	Capital Investment	Income	Cumulative Cashflow
0	650	0	−650
1	0	100	−550
2	50	200	−400
3	0	300	−100
4	0	350	250
5	0	400	650

Figure 11.2 Payback period

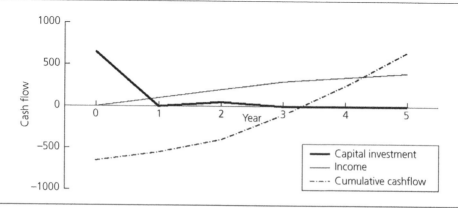

applied to this. Essentially, what is being done is to ensure that the cash flow amounts are equivalent to each other no matter at what point in time they are realised. The way of achieving this is to 'discount' the future amounts in such a way as to ensure that all amounts are equivalent to each other, not in terms of total amount but in terms of the value placed on each of the single units that make up each total amount. In order to do this, there has to be some reference point in time around which the costs are to be compared or related. The usual practice is to take the 'present' as the reference point, but this does not have to be the case.

Whatever reference point in time is chosen, it has to be explicitly stated. The reasons for this are that the present is itself not static: what is the present today will not be the present tomorrow. Discounting also offers a means of comparing the results of similar projects that may have been evaluated at some time in the past, by enabling the results to be compared based on the same values.

Let us assume it has been determined that, for a particular project, the appropriate discount rate is 4% and the cash flow is the same as that given in Table 11.4. But, before that, the way discounting works will be looked at.

Consider a case where there is an amount of money to invest. Given an initial amount invested at the start of the project is A, and an interest rate to borrow this amount of money is fixed at r% per annum (all ignoring inflation as discussed earlier), then at the end of 1 year, the investment would be worth the initial sum (A) plus the interest earned, which would be $A \times r\%$. Thus, in total would be $A + A \times r\%$, *which can be written as*

$$B = A \times (1+r)$$ (11.1)

If this sum were to be left for another year to earn further, then the initial investment would be worth

$$C = A \times (1+r) \times (1+r) \text{ or } C = A \times (1+r)^2$$ (11.2)

Discounting is the reciprocal of this 'compounding'. This can be illustrated by rearranging the above equations. So, the amount at the end of the first year would be $B \times 1/(1+r) =$ the initial amount invested, A. Similarly, for two years the amount at the end of the second year would be $C \times 1/(1+r)^2 =$ the initial amount invested, A. And so on for subsequent years. Applying this to the income stream shown in Table 11.4 gives the results in Table 11.5 and Figure 11.3.

Put another way, this represents the present value of the cash flow.

Table 11.5 Discounted cashflow for r = 4%

Year	Income	Discounted Amount
0	0	0
1	100	96
2	200	185
3	300	267
4	350	299
5	400	329

Figure 11.3 Discounted cash flow

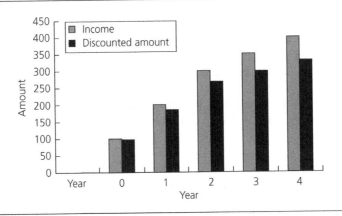

310

11.9.3 Net present value

The discounted cash flow only looks at one part of the proposed project – the benefits arising from the project, represented in this case by the revenue generated from its operations. It does not look at the investment side. The NPV is the present value of the benefits generated by a project less the present value of the investments required to realise the function and operation of the project. Using the amounts in Table 11.4 and discounting them all to year 0, the NPV can be found (Table 11.6).

The NPV of this project over a five-year lifespan is 479 currency units, at a discount rate of 4%. If at the end of the period some of the project components still had a value that could realised, then that would have to be factored in as a benefit: it represents an income for the project. If, on the other hand, no benefit in the form of the value of the project infrastructure can be realised, then the investment made must be regarded as a sunk cost and should not be factored into the calculation. If there is a disposal cost to remove equipment, this should be factored in, particularly if the WLC is being considered.

If the funding for a project involves a loan that is repayable over a period of time, then the cost of servicing the loan and the schedule of payments has to be included in the cash flow.

NPV can be represented in the form of the expression

$$NPV = \sum_{t-0}^{t-T} \frac{benefits - costs}{(1+r)^t} \tag{11.3}$$

where r = the discount rate
t = the life of the project.

The NPV takes the future projected cash flows into account, and its gives absolute values. The value is only relevant to the degree that the discount rate chosen is realistic for those providing the finance. For this reason, an NPV calculation is often better used as an IRR when comparing different projects or projects against an investor's perception of returns they required, which is often described as a 'hurdle rate'.

Table 11.6 Net Present Values

Year	Capital Investment	Present Value of Investments	Income	Present Value of Income	Cumulative Cashflow	Net Discounted Cumulative Cashflow
0	650	650	0	0	−650	−650
1	0	0	100	96	−550	−554
2	50	46	200	185	−400	−415
3	0	0	300	267	−100	−148
4	0	0	350	299	250	151
5	0	0	400	329	650	479
Totals		696		1176		479

11.9.4 Internal rate of return

The calculation of an IRR is an extension of the NPV technique. The IRR is that discount rate which when applied to the costs and the benefits is such that the NPV is zero. The IRR can be determined through trial and error calculation (or using the solver function in a spreadsheet) – for example, in the case of the expenditures presented in Table 11.4, the calculation would show that the required discount rate to set the NPV to zero (i.e. the present value of investments equals the discounted income cash flow) is 22.2%.

The point of calculating the IRR is that it allows a comparison to be made with the cost of capital for the project. In this case, the discount rate (= the cost of capital) has been given as 4%. This project yields an IRR value of 22.2%, which is much greater than the 4%. In other words, the project gives a much higher return than just leaving the money in the bank or investing it in some form of savings; it is, on the basis of the information presented here, a good investment. An investor may take a view of risk as well as the cost of borrowing for the investment in its view of an acceptable hurdle rate, and require a higher IRR than a discount rate based on the borrowing cost. If an investor had defined an acceptable hurdle rate of 15% reflecting their experience of the cost of investment and risk involved in such projects, then the IRR suggests this project has acceptable future cash flows from their perspective.

11.9.5 Cost benefit analysis

The above examples are all financial appraisals, as they consider cash flows. CBA extends NPV to consider costs that may not be immediate cash flows. These could be indirect financial costs and benefits (e.g. overheads) or valuation of natural and social capital impacts of a project.

The process of calculating an NPV has has yielded a single number, the NPV, and, as will be seen below, this can be used for comparative purposes. But the question arises as to whether this is the only metric that can be derived from the figures produced, or the most appropriate one. The most obvious alternative metric, and one quite commonly used, is to look at the ratio of the benefits to the costs. Benefits can be taken as gross benefits or, alternatively, as net benefits after having deducted indirect operational costs from the stream of benefits. Care must be taken not to double account for costs so that, if net benefits are used, the operating costs are not also included in the costs, only the initial investment costs. Changes in indirect operating costs (higher or lower) should be reflected as a benefit. Direct operating costs that arise from a project can be included as either a cost or a benefit, depending on how closely they are linked to the investment decision. If in doubt, it is better to show changes in operating costs as a net benefit, in which case cost reflects the initial investment and its financing.

Using the example in Table 11.6 at a 4% discount rate, the present value of the benefits is 1176 currency units and the present value of the costs is 696 currency units. The figures give a benefit–cost ratio of 1176/696 = 1.69. If values are applied to items that are not financial cash flows, the value of the NPV itself becomes an SOI, and may be better described using a benefit to cost ratio rather than quoting the NPV itself.

More is said about the application of CBA as a means of comparison in the next section. For the moment, it can be noted that essentially CBA is a method of evaluating the comparative economic efficiency of options. Because it seeks to reduce everything to value in monetary terms, it is poor at being able to handle different concepts of value and relative value, what different people might consider as being of importance and therefore more 'valued'. Indeed, much of the criticism of CBA comes down to the reduction of decision-making to just one of monetary value.

Applying discounting in project appraisal that has private investment but public benefits needs to consider how to include the private cost of financing, but discount future public benefits using the STPR. An illustration of this is shown in Table 11.7. A recognised approach to doing this, particularly useful where there is a regulated cost of capital and competing options to consider, is to use the 'Spackman' approach.

1. Convert capital costs into annual costs using the company's cost of capital (in real terms as noted above). This gives a stream of financing costs, which should be included as part of the cost side of the cost benefit analysis.
2. Use the STPR of 3.5% in discounting all costs and benefits, as recommended by the HM Treasury Green Book

This approach to discounting is only necessary where public benefits are being considered. Where there are just private costs and benefits, the cost of capital of the firm providing the capital outlay should be used as the discount rate.

11.9.6 Comparing projects
So far, only single projects have been looked at. All of the techniques outlined above can and are used as a basis for comparing between options.

- In the case of payback, the decision question would be: 'Of the options considered, which one has the shortest payback period?'
- In the case of a discounted cash flow, the question would be: 'Of the options considered, which one yields the highest positive cumulative cash flow?'
- In the case of NPV, the question is similar to that of the discounted cash flow: 'Which yields the highest NPV?'
- And in the case of IRR: 'Which option yields the highest rate of return?'

The answers to all of these questions are of great interest to decision-makers, and can provide the basis for making informed decisions. An example will illustrate the point. Take the case where

Table 11.7 Example of the Spackman approach (Adapted from Joint Regulators Group (2021))

Cash flow	Year 1	Year 1	Year 3	Year 4	Year 5
Capital outlay by the firm	−£1,000				
Firm's cost of financing the above capital outlay, five year bond with 7% interest	−£244	−£244	−£244	−£244	−£244
Consumer benefit	£300	£300	£300	£300	£300
Net social benefit	£56	£56	£56	£56	£56
Net social benefit (at year 0) discounted at the STPR of 3.5%	£54	£52	£51	£49	£47
Net present value at year 0 (sum of the above)	**£253**				

there are two alternative projects that, due to the availability of funds, are mutually exclusive: if one is undertaken, then the other cannot proceed. The case is illustrated in Tables 11.8 and 11.9.

The results are somewhat contradictory: on the basis of NPV, project A would be favoured, while on the basis of IRR and the benefit–cost ratio, project B would be chosen. Where projects are mutually exclusive in terms of their benefits, then the project with the greatest NPV should be chosen. However, if there are constraints over the availability of funding, then it makes sense to invest in the one that would give the highest returns – the highest benefit–cost ratio and, possibly the highest IRR (because in some instances there can be more than one IRR). This can occur if the sign of the annual net benefits changes a number of times.

Given that the results of comparison can be ambiguous and that in trying to determine what the relevant costs and benefits are there will be uncertainty and errors, it makes sense to subject the appraisal to a sensitivity analysis. In this, some of what might be regarded as the key assumptions on which the appraisal is based need to be tested to see how sensitive the results are to their variation. Another way of looking at this is as a way of seeing how robust the assumptions and the outcomes are. If the results suggest that there are one or more parameters that account for variability in the outcomes, then perhaps it would be prudent, if possible, to try to make sure that the most reliable value is determined for that parameter. In other words, it could suggest where further investigation might be required. On the other hand, this may not be as useful as it first sounds, as it might well be that the source of sensitivity was already known, and all that has been done is to demonstrate that fact, which may not be helpful.

Sensitivity analysis might, however, indicate whether the uncertainties are such as top lead to spurious choices. The benefit–cost ratio does give us some indication as to how much things would

Table 11.8 Comparison of projects

Year	Project A		Project B	
	Cashflow	PV of Cashflow	Cashflow	PV of Cashflow
0	−500,000	−500,000	−300,000	−300,000
1	170,000	163,462	50,000	48,077
2	200,000	184,911	100,000	92,456
3	200,000	177,799	150,000	133,349
4	150,000	128,221	200,000	170,961
5	250,000	205,482	250,000	205,482

Table 11.9 Comparison of projects using NPV, IRR and B-C ratio

Technique	Project A	Project B
Net Present Value	359,874	350,325
IRR	21%	25%
Benefit-Cost Ratio	1.72	2.17

have to change so as to make a given alternative marginal. The higher the benefit–cost ratio, the more robust the alternative to uncertainties, unless the uncertainties are mutually reinforcing. As we show earlier, there are forms of sensitivity testing (such as real options analysis) which can be used to test whether a benefit–cost ratio greater than 1 (or a higher value) is sufficiently positive to indicate that a project should proceed.

11.9.7 Multi-criteria analysis

Multi-criteria analysis (MCA) differs from CBA in that provides a way of considering more than one decision criteria. It can be used when some of the impacts or benefits of project alternatives cannot be assigned a monetary value but they might have an importance in the decision making process. Some (including the authors) take the view that all benefits within a project should be assigned a value that allows a CBA to be calculated. However, even if values are assigned to environmental and social factors, MCA still has a role in decision making, through understanding constraints that differ between a project appraisal and the ultimate decision, such as minimum acceptable outcomes, and illustrating the assumptions behind a CBA used in project appraisal.

It is an approach that assumes that different objectives can be expressed with respect to some common denominator by means of trade-offs, so that the gains or losses associated with one objective can be traded off against gains or losses in another objective. The role of decision support is then to facilitate us in the process of evaluation. It may well be that there is no solution that optimises all the criteria and, therefore, the results are used to find compromises between different options. The technique is of use where there is a need to handle a lot of complex information in a consistent way.

MCA is a structured approach to determine what the preferences are for options that satisfy several explicit sets of objectives which have been identified to varying degrees. The objectives that are to be met have to be specified and their attributes identified. It is not necessary that these are in monetary terms; often, they are based on analysis using scoring or weighting of qualitative criteria. It is more important that there are measurable criteria to assess the extent to which the objectives are satisfied. This means that it is possible to combine a number of different criteria or categories within an overall framework – that is, environmental and social criteria as well as costs and benefits.

There is the clear idea that decision-making is informed by more than just costs and incomes. Where it is controversial for decision makers in assigning financial values to environmental and social criteria, then MCA will have a more important role. MCA can be used to identify a preferred alternative, to rank alternatives, to short-list alternatives for further appraisal or to weed out the acceptable from the unacceptable.

One strength of MCA is that it is said to be more representative of decision problems, as it takes into account more than one objective – unlike CBA. However, one of the difficulties with it is how the preferences of different groups or individuals are accommodated within its framework. Different stakeholders may have differing perceptions as to priorities and, thus, it may be difficult to come up with a single solution. Under these circumstances, MCA is a means of uncovering those preferences and a starting point for resolving different understandings, making it part of a negotiation process rather than an end in itself. The value then lies partly in the process of constructing the problem to be analysed within the MCA by indicating why factors are important and how they interact. Hence, used in conjunction with other approaches and undertaken from different perspectives, MCA, together with stakeholder consultations, can inform how differences between individual, social and environmental preferences can be negotiated.

In formulating an MCA the objectives that are to be satisfied have to be identified, and frequently this can be proceeded with in a hierarchical fashion. Broad objectives can be broken down into lower-level objectives, allowing a better chance of being able to assess them. Different forms of multi-criteria evaluation are available, with the most suitable method dependent on the nature of the decision to be taken. A logical option, however, is to undertake a financial appraisal or CBA that reflects the set of options that are acceptable from a governance perspective, which can be elicited through an MCA undertaken as part of a communication and engagement process.

The two simplest MCA methodologies are ranking and rating. Ranking gives each objective a rank depending on its perceived degree of importance relative to the decision that is to be made. These can then be ranked in order of the priority given to them. Rating takes this a step further by giving each objective a score, between zero and 100, say. The scores for all the objectives must add up to 100, meaning that if one objective is scored high, then another must be lower. The next stage that can be applied is weighting: numerical weights are given to each criterion that reflect the relative importance assigned to that criterion. It is the process of deciding what the relative weighting should be that causes problems, and is often where stakeholder input is required.

Using the scoring and the weighting, a simple weighted average of scores is calculated. The use of weighted averages assumes that the preferences are mutually independent of each other, an assumption that sometimes needs to be verified. If preferences are mutually independent, then a linear additive evaluation may be applied in which the scores are multiplied by their weights, and combined to get an overall score. Most MCA approaches use this, and an example is given in Tables 11.10 and 11.11 of a sewerage system with a wastewater treatment works located on an estuary. There are three options to be evaluated, and each option has a collection system, treatment facility and disposal system.

Table 11.10 shows in the first column the criteria, followed by the subcomponents for each of the three main criteria. Each of these has been assigned a weighting on a scale of 1–5, with 5 denoting high importance and 1 denoting low importance. In the following columns, the ratings for each of the subcomponents of the criteria have been assigned. If the ratings for the three collection options are summed across for, say, marine impact, they add up to 100. The higher the rating applied, the higher the beneficial impact associated with it. In Table 11.10, the results of the multiplication of the weights and the ratings are given for each of the criteria and then summed, to produce an overall score for each option. For this hypothetical case, it would appear that option A would be preferred, as it has the best overall performance. Notice that in this example the costs of the various options have not been included. Often in MCA, the costs will be included and given a significant weighting – for example, tender evaluation to comply with utility procurement rules is often undertaken using an equivalent of MCA, and it may also include some absolute pass/fail criteria such as meeting legal requirements and health and safety standards.

MCA procedures differ between each other in terms of how the basic information is processed and the degree of sophistication used. These more complex approaches include analytical hierarchy process, pair-wise comparison, multi-attribute utility theory, outranking methods and fuzzy sets; but a discussion of these is outside the scope of this chapter. Where it is clear that one option dominates the others, the less sensitive decision makers should be to the weighting procedure used.

Table 11.10 Multi-criteria analysis

Criteria	Attributes	Weights	Option A			Option B			Option C		
			Collection	Treatment	Disposal	Collection	Treatment	Disposal	Collection	Treatment	Disposal
Environmental	Marine impacts	5	50	25	80	30	55	10	20	20	10
	Odour	4	40	50	10	40	30	30	20	20	60
	Biodiversity	3	33	50	60	34	40	35	33	10	5
Social	Disruption	5	60	50	10	20	30	30	20	20	60
	Public acceptability	5	20	20	10	40	40	70	40	40	20
	Land use conflicts	4	10	20	10	10	60	45	80	20	45
Operational	Reliability	4	75	5	80	15	5	10	10	90	10
	Flexibility	3	30	35	30	35	35	35	35	30	35
	Complexity	3	65	35	10	15	30	5	20	35	85

Table 11.11 Results of MCA

	Option A	Option B	Option C
Environmental	1604	1202	794
Social	1010	1610	1580
Operational	1255	585	1160
Total Score	3869	3397	3534

Table 11.12 Results of CBA

	Option A	Option B	Option C
Capital Cost	208,850,000	301,500,000	286,200,000
Operation & Maintenance	77,550,000	120,500,000	132,250,000
Benefits	813,600,000	878,700,000	950,850,000
Net Present Value	527,200,000	456,700,000	532,400,000
Benefit-Cost Ratio	2.84	2.08	2.27

Table 11.13 Combined CBA and MCA

	Option A	Option B	Option C
Net Present Value	527,200,000	456,700,000	532,400,000
NPV Ranking	2	3	1
Benefit-Cost Ratio	2.84	2.08	2.27
B-C Ratio Ranking	1	3	2
MCA Score	3869	3397	3534
MCA Ranking	1	3	2
Unweighted score	4	9	5

In the hypothetical example of the sewage system given above, a CBA has been carried out, the results of which are given in Table 11.12. How, then, can both sets of results be used to inform decision-making as, in terms of NPV, option C would be preferred, while option A gives the highest benefit–cost ratio? One way would be to combine the two within an MCA and assign weights to the results from the MCA given in Table 11.11 and the NPV and benefit–cost ratio given in Table 11.12. This is shown in Table 11.13, and, in this example, equal weighting is given to all three criteria. On this basis, option A would seem to be the preferred option.

11.10. Regulation and decision making
11.10.1 Projects in a modern regulated environment
At this point, most of the different methods of project appraisal that engineers will encounter have been presented. The importance of understanding where funding comes from has been discussed, the limitations that various promoters will face have been set out, and 'discounting' has been

explored. While there are a wide range of techniques, most employers or clients restrict themselves to accepting decisions based on simple financial appraisals when more sophisticated techniques are clearly available.

This can simply be a matter of scale – most projects are repeatable examples of projects that have been undertaken previously, and financial appraisal can be sufficient to test what continuous improvements, innovations and efficiency are being realised with one project compared to previous examples. At the other end of the scale are large one-off strategic projects, where an overall project appraisal using techniques highlighted in this chapter may have already been carried out.

Where legal or regulatory standards are involved, the choice of whether to undertake (and sometimes even finance) a project is not relevant. More recently, legislation such as the Water Framework Directive has recognised that where a project is 'disproportionately costly' then a delay in delivery for better information or setting an alternative objective may be appropriate. This can mean that CBA and real options analysis has a role in meeting regulatory standards, particularly where what is disproportionate can be a benefit to cost ratio of 1:10 (i.e. the costs are ten times higher than the measured benefits). As with Environment Agency benefits assessment guidance, this may mean that some environmental factors such as habitats and biodiversity are not assigned financial values as a matter of policy. In many situations, a large part of what would be the economic analysis has already been determined by legislation and regulation. What the engineer has to do is find the most economic way of meeting these standards, and that generally reduces matters to a financial appraisal.

Ofwat's regulatory approach continues to evolve. Whether a project is meeting a legal requirement overseen by the Environment Agency or Drinking Water Inspectorate, or optional ways of improving service improvements, CBA and scenario analysis may well be required. This reflects the maturity of water sector policy and decision making. The Environment Act 2021 sees the setting of long term outcome targets, such as for biodiversity net gain and reductions in leakage, which have legal status overseen by an independent Office of Environmental Protection (OEP).

Converting a process that sets legal outputs and standards that must be met today to one that focuses on measurable outcomes for customers and the environment in the long-term increases the relevance of CBA and advanced forms of scenario testing. While the desired long-term outcome is clear, the trade-offs in the short-term on how this end point is reached become more complex, and there is a real risk that pressures on the affordability on customer bills will result in companies and their regulators postponing necessary investment until it is too late to deliver the outcome. Equally, with compelling CBA analysis and scenario testing, regulators will be concerned that companies will wish to reduce their risk of being able to deliver long-term outcomes through excessive early investments when faced with such uncertainty. Companies are increasingly responsible for identifying the short-term needs rather than regulators, for instance choosing which specific abstraction sites to reduce water taken from the environment in order to meet a long-term abstraction reduction target. With the Environment Act setting targets up to 2050, and water resource management plans (WRMPs) with time horizons as far away as 2085, how long-term investment decisions are made for five year Ofwat price reviews is an emerging challenge.

The first challenge with long-term outcomes is *commitment* – when faced with uncertainty how can accountability for the contribution a particular project makes to the outcome be measured in practice. This increases the importance of post-project appraisal further, but increases in importance the measurement for CBA project decisions of wider social and environmental benefits.

The principle that water companies should deliver wider *public value* when delivering their core activities is becoming increasingly important, reflecting this priority. Ofwat have set out six principles for how greater public value should be delivered. This builds on water companies setting out *social contract* commitments for how they deliver for the long-term for customers and the environment. Individual companies set out their own arrangements, but also agreed collectively some Public Interest Commitments such as to achieve net zero operational carbon emissions by 2030. The principles recognise that companies should *'create further social and environmental value in the course of delivering their core services, beyond the minimum required to meet statutory obligations.'* Ofwat also set out limitations of what water customers should pay for, recognising that, while water companies should show leadership on public value, this should not displace other organisations who are better placed to act, and other sectors such as agriculture to contribute to the cost where they benefit – a recognition that a 'beneficiary pays' principle is gaining ground compared to a 'polluter pays' principle where regulators move to setting long-term outcomes as opposed to short term outputs. This highlights the importance of financial cash flows and the cost of finance – the Spackman discounting approach achieves this by including the cost of finance within a CBA.

As an example, water companies in England are now asked to work together on regional water resource management plans, provisioning water resource planning across sectors and across company boundaries on behalf of society. Regulators are requiring water companies to *provision* the long-term environmental outcome, and the regulation of water companies includes their engagement with agriculture and then the degree to which agriculture pays its fair share towards the water supply solutions water companies provision for their own needs of public water supply, as well as other users. To achieve this, CBA and MCA techniques are being used, with some regional planning groups using sophisticated software modelling techniques in order to achieve what is described as the 'best value' decisions on water resources.

The final change in this regulatory landscape is scenario testing. Companies are now expected to set out a core, least cost and no regrets set of investments for the five-year regulatory plans, in the context of scenario testing against a range of future (exogenous) uncertainties, such as climate change, technology and demand that regulators believe are likely to be outside of management control. Where companies can demonstrate investment for the long-term outcomes in all scenarios requires investment in the next five years, then it will be supported. Even where short-term investment is then not supported, the adaptive pathway as uncertainty reduces, and the point where investment will be triggered, depending on which scenario emerges, will be understood. In this way, Ofwat hopes that *trade-offs* in the short term will allow the most efficient spend without this compromising the ability of the water sector to meet the long-term outcomes government has specified. This increases the importance of both CBA and MCA techniques, but at the same time highlights their limitations without the context and governance of the decisions being understood. It also highlights the importance of valuing social and environmental factors, even where the only issue is the timing of when investment is made and not the ultimate outcome.

11.10.2 Distributional aspects and affordability

Where social and environmental factors and customer willingness to pay have been taken into account in a project's objectives, within MCA or CBA, then there has already been some consideration of distributional impacts. Additional costs are likely to have been included in a project selected on a most cost beneficial, or best value basis (for MCA), rather than based on least cost (or highest financial NPV). These additional costs can be because of social as well as environmental factors, particularly if some social groups (such as those on low incomes) are more vulnerable to service and environmental risks. At its simplest for water and wastewater services, the poorest in society are the least able to afford the alternatives of good public provision, such as bottled water. The issue of inter-generational equity also applies to environmental impacts of water services – not protecting the water environment today means a higher cost of producing and distributing clean water for future generations.

Affordability should be considered as an additional distinct social and distributional impact. This is the affordability of the product or service to consumers, rather than whether investors can afford to make the investment. Affordability is usually a distributional and equity issue. In the case of water services, it may have a direct bearing on the design of how such goods and services are to be paid for. This affordability is distinct from the willingness to pay of consumers in general for the outcome, even where the impact of water bills has been used as a mechanism in its calculation. Consumers in all social groups may be willing to pay for the outcome, but still want the affordability of delivering these outcomes to be mitigated, such as through the phasing of meeting the outcome, or through distributional actions such as social tariffs.

It might be argued that matters of equity and distribution are not the concerns of a service provider but rather of the tax and welfare system. In this regard, the status of the service provider and the general national policies will have a bearing on whether this is a significant issue that should be incorporated into an appraisal or not. In cases where service provision is through some form of public service, then affordability may be of somewhat more importance.

Just as social and environmental benefits can be important, so too are social and environmental costs that might arise through damage to parts of the environment, pollution, loss of biodiversity and other causes. The decision as to whether these are to be included will depend on who is carrying out an evaluation and what the legal and regulatory requirements are.

11.10.3 Commercial projects
The section above implies that even where there are projects that must be delivered, project appraisal can support both choices of how an outcome is delivered and the timing of its delivery. However, there are also *optional projects* where the decision to proceed or not should be driven by commercial factors. Rather than operating within the fixed income of regulated price decisions, optional projects can deliver a new separate or complimentary income stream. An example might be the installation of a small hydroelectric scheme at a raw water reservoir, or a wind turbine on land owned by the utility. Any company will have a number of such opportunities, but it has no obligation to undertake any of them. It will therefore need to be satisfied that the project is worth investing in. To do this, it will normally use the IRR.

One benefit of this rather complex financial analysis is that it lends itself to being an effective filter for such proposals, because the board can set a so-called 'hurdle rate', meaning that if the IRR does not exceed the hurdle rate, the proposal will not be implemented. If the hurdle rate is exceeded, then the highest IRR option is likely to be preferred, but not always. The total capital cost may also play a part – money is a scarce resource. However, if no option beats the hurdle rate, the proposal will fail. In setting the hurdle rate, the board will have to consider a number of issues.

One dominant issue is that it will be using funds that could be applied to other optional projects, and therefore shareholders will expect a good return from any such investment. This alone means that the hurdle rate will usually be in excess of the discount rate adopted by the same entity for the NPV analysis of mandatory schemes. The board may also take the view that it should build in some form of risk premium against things going wrong, such as costs escalating, or the predicted income stream being less than forecast, and, again, this can be dealt with on a generic basis by simply raising the hurdle rate.

11.10.4 Phasing of projects

A further issue to be considered is the phasing and scaling of a project. In other words, what is the design horizon for any project, and can the project be split into a number of phases?

From the standpoint of an economist, the overprovision of facilities ('gold-plating') is a waste of financial resources at both a national level and in terms of an organisation's finances. It is very difficult to predict the future, particularly in terms of demand, and once a decision has been taken, there is only one scenario that will allow it to be proved correct, and any number which will prove it to be in error. Recognising this through adaptive planning sees such issues as managing uncertainty rather than errors with the benefit of hindsight. The longer the planning horizon, the greater the degree of uncertainty.

Ignore, for the moment, the possibility that the assumptions underpinning a proposed project could be incorrect, and consider the simple case of a pumping station which is planned for a 25-year future demand, where the demand is increasing linearly over that period. While pumps could be installed now that would meet the ultimate demand, that would mean they would be running at their theoretical duty just as they reached the end of their design life. Might it not be better, therefore, to install pumps that could meet the forecast in 12 years time? And then, when they are running at maximum capacity, consider what size of pumps would be required to supplement the output. This then leaves the alternatives of building the pumping station with space for extra pumps, or simply building a pumping station that can deal with forecast demand for the next 12 years.

To answer these questions, discounting techniques and, in particular, NPV analysis can be used. By carrying out an NPV analysis of the different options, the future value of money is taken account of and the most attractive option determined. The choice of phasing generally works better for 'above ground' assets such as water and wastewater treatment works than for long-life infrastructure pipe and reservoir assets. This is because for shorter lived assets there can be a genuine choice of life of asset and intensity of use that can be compared to investment size and financing cost. For a pipe and reservoir, the asset is almost always sunk beyond the timeframe for the financing decisions of investors.

11.11. Final words

The appraisal of projects or alternatives is as much about the exercise of judgement as it is about the use and application of appraisal techniques. Each project is, by its very nature, unique and presents new challenges. Those who carry out appraisals should have a blend of technical skills, an understanding of economics and accounting, and an appreciation for the environment and social issues. Increasingly, appraisals are going beyond being an end in themselves that supply answers which decision-makers take up. Our brief description of the evolution of the water regulatory framework in England will give a sense of how the tools of project appraisal are developing in both sophistication and importance. The fundamentals of commercial decision making using NPV and IRR are fundamentals that change little over time. For those who want to explore regulation and decision making further, there are a wide range of references at the end of this chapter.

The process of carrying out an appraisal becomes an integral part of communication and shared understanding that can be used to inform better decision-making. Decisions are not always made on the basis of a formal appraisal, as there are other subjective factors that can influence choices and outcomes. One of the tasks of project appraisal is to try to make explicit what the consequences associated with the available options are.

11.12. Worked examples

The examples apply the discounted cash flow to find NPVs.

11.12.1 Example 1: the effect of operating costs on the choice of project

A client has three alternative options that would each deliver the same level of service.

- Option A would be to construct the works at a cost of £12 million, and the associated operating costs would be £0.250 million per year.
- Option B would be to construct the works at a cost of £7 million, and the associated operating costs would be £0.500 million per year.
- Option C would be to construct the works at a cost of £6 million plus upgrades costing £0.670 million every 5 years, and the associated operating costs would be £0.500 million per year.

Given that the evaluation horizon is 20 years and the discount rate to be used is 8%, which of the three options should be chosen? What would the effect of altering the discount rate to 10%?

See Table 11.14 for the results of the appraisal.

11.12.2 Example 2: the effect of phasing

A client has to build a new sewage treatment works, and is presented with two proposed options.

- Option A: build the whole works now at a cost of £24 million over a 3-year period.
- Option B: build half of the required works now at a cost of £15 million and the remainder in 12 year's time at a cost of £18 million.

The construction period is 2 years for each period and the discount rate is 8%. Which option would you recommend? What would be the effect of reducing the discount rate to 5%?

Table 11.14 Appraisal for Example 1

Example 1

Year	Option A				Option B				Option C			
	Capital Cost	Operating Cost	Discount Factor	NPV	Capital Cost	Operating Cost	Discount Factor	NPV	Capital Cost	Operating Cost	Discount Factor	NPV
0	12.000		1.000	12.000	7.000		1.000	7.000	6.000		1.000	6.000
1		0.25	0.926	0.231		0.5	0.926	0.463		0.5	0.926	0.463
2		0.25	0.857	0.214		0.5	0.857	0.429		0.5	0.857	0.429
3		0.25	0.794	0.198		0.5	0.794	0.397		0.5	0.794	0.397
4		0.25	0.735	0.184		0.5	0.735	0.368		0.5	0.735	0.368
5		0.25	0.681	0.170		0.5	0.681	0.340	0.670	0.5	0.681	0.796
6		0.25	0.630	0.158		0.5	0.630	0.315		0.5	0.630	0.315
7		0.25	0.583	0.146		0.5	0.583	0.292		0.5	0.583	0.292
8		0.25	0.540	0.135		0.5	0.540	0.270		0.5	0.540	0.270
9		0.25	0.500	0.125		0.5	0.500	0.250		0.5	0.500	0.250
10		0.25	0.463	0.116		0.5	0.463	0.232	0.670	0.5	0.463	0.542
11		0.25	0.429	0.107		0.5	0.429	0.214		0.5	0.429	0.214
12		0.25	0.397	0.099		0.5	0.397	0.199		0.5	0.397	0.199
13		0.25	0.368	0.092		0.5	0.368	0.184		0.5	0.368	0.184
14		0.25	0.340	0.085		0.5	0.340	0.170		0.5	0.340	0.170
15		0.25	0.315	0.079		0.5	0.315	0.158	0.670	0.5	0.315	0.369
16		0.25	0.292	0.073		0.5	0.292	0.146		0.5	0.292	0.146
17		0.25	0.270	0.068		0.5	0.270	0.135		0.5	0.270	0.135
18		0.25	0.250	0.063		0.5	0.250	0.125		0.5	0.250	0.125
19		0.25	0.232	0.058		0.5	0.232	0.116		0.5	0.232	0.116
20		0.25	0.215	0.054		0.5	0.215	0.107	0.670	0.5	0.215	0.251
Totals	12.000	5.000		14.455	7.000	10.000	0.329	11.909	8.680	10.000		12.030
Discount rate	8%											

Table 11.15 Appraisal for Example 2

Example 2

Year	Discount Factor	Option 1 Capital Cost	Option 1 NPV	Option 2 Capital Cost	Option 2 NPV
0	1.000	8.000	8.000	7.500	7.500
1	0.926	8.000	7.407	7.500	6.944
2	0.857	8.000	6.859	0.000	0.000
3	0.794	0.000	0.000	0.000	0.000
4	0.735	0.000	0.000	0.000	0.000
5	0.681	0.000	0.000	0.000	0.000
6	0.630	0.000	0.000	0.000	0.000
7	0.583	0.000	0.000	0.000	0.000
8	0.540	0.000	0.000	0.000	0.000
9	0.500	0.000	0.000	0.000	0.000
10	0.463	0.000	0.000	0.000	0.000
11	0.429	0.000	0.000	9.000	3.860
12	0.397	0.000	0.000	9.000	3.574
Total		24.000	22.266	33.000	21.878
Discount rate	8%				

REFERENCES

Accounting for Sustainability (2014) The A4S essential guide to managing future uncertainty. https://www.accountingforsustainability.org/content/dam/a4s/corporate/home/KnowledgeHub/Guide-pdf/The%20A4S%20Essential%20Guide%20to%20Managing%20Future%20Uncertainty.pdf.downloadasset.pdf

Bateman I *et al.* (2002) *Economic Valuation with Stated Preference Techniques: A Manual*. HMSO, Cheltenham, UK.

Bristol Water (2021) *Regulating for Consensus and Trust*. Bristol, UK. https://f.hubspotusercontent30.nct/hubfs/7850638/Regulating%20for%20consensus%20and%20trust.pdf

Chenery C (1984) Whole life cost of construction – informal discussion. *Proceedings of the Institution of Civil Engineers* **76**: 822–825.

Green C (2003) *Handbook of Water Economics Principle and Practice*. Wiley, NY, USA.

HM Treasury (2022) *The Green Book: Appraisal and Evaluation in Central Government. Treasury Guidance*. TSO (the Stationery Office), London, UK. https://www.gov.uk/government/publications/the-green-book-appraisal-and-evaluation-in-central-governent

Innes J and Mitchell F (1995) A survey of activity based costing in the UK's large companies. *Management Accounting Review* **6(2)**: 137–153.

National Audit Office (2006) *The Modernisation of the West Coast Main Line: NAO Report HC 22 2006–2007*. TSO, London, UK.

OECD (2020) Water Governance programme. https://www.oecd.org/water/regional/

McGuffog I (2004) Are consumers willing to pay for environmental economics? *The Business Economist* **36(1)**.

McGuffog T (2016) *Building Effective Value Chains: Value and its Management*. Logan Page, London, UK.

Skipworth P, Engelhardt M, Cashman *et al.* (2002) *Whole Life Cost Accounting for Water Distribution Network Management*. Thomas Telford, London, UK.

Stiglitz JE (1994) Discount rates: the rate of discount for benefit-cost analysis and the theory of second best. In Layard R and Glasiter S (eds), *Cost–benefit Analysis* 2nd edn, pp. 116–159. Cambridge University Press, UK.

Winpenny J (2003) *Financing Water for All. Report of the World Panel on Financing Water Infrastructure*. World Water Council, Maseille, France. www.financingwaterforall.org [accessed 01.09.10].

FURTHER READING

Dixit AK and Pindyck RS (1994) *Investment Under Uncertainty*. Princeton University Press, USA.

Drummond MF, Sculpher MJ, Torrance GW, O'Brian BJ and Stoddart GL (2006) *Methods for the Economic Evaluation of Health Care Programmes*, 3rd edn. Oxford University Press, UK.

Emblemsvag J (2001) Activity-based life cycle costing. *Managerial Auditing Journal* **16(1)**: 17–27.

EPA (1996) *An Introduction to Environmental Accounting as a Business Tool: Key Concepts and Terms*. Environmental Protection Agency, DC, USA.

Figueira J, Greco S and Ehrgott M (2005) *Multiple Criteria Decision Analysis: State of the Art Surveys*. Springer Science and Business Media, NY, USA.

Social Value UK (2012) *A Guide to Social Return on Investment 2012*. The SROI Network. https://socialvalueuk.org/resources/sroi-guide/

Dragan A. Savić and John K. Banyard
ISBN 978-1-83549-847-7
https://doi.org/10.1108/978-1-83549-846-020242014

Chapter 12
Sustainability and climate change

Paul Jowitt
Heriot-Watt University, Edinburgh, UK

Adrian Johnson
Stantec UK, Edinburgh, UK

Kees van Leeuwen
University of Utrecht, The Netherlands

12.1. Introduction
12.1.1 Sustainable water supplies – worth fighting for?

The art of engineering has traditionally been based on a successful combination of the rigorous application of technical rationality and engineering science, its practical application, and the driving force of an economic imperative. That is no longer enough. There are other phenomenological, social and environmental impacts, and factors that need to be anticipated and resolved. This is particularly true for the development and management of water resources and associated water supply systems, where conflicts often arise, and which cannot be resolved by scientific analysis alone. Even where the facts are accepted, their interpretation and consequences are often disputed. The sustainable development of water resources depends on the combination of sound science and effective conflict resolution in the face of divergent objectives and the need to balance costs and benefits between a number of different stakeholders.

This chapter is not concerned with particular techniques of engineering water systems – which are adequately described elsewhere in this book – but with some of the larger global and systems level issues that need to be tackled at the larger scale as well as practical approaches for addressing them. In this chapter the focus is on cities as that is where most of us live, the problems reside, and solutions will emerge (Figure 12.1).

Increasingly, real and emerging water management problems have impacts that either need to account for (non-commensurate) socio-enviro-economic effects, or which accommodate large-scale heterogeneity, and sometimes both. They require transdisciplinary and/or large-scale systems models which cannot be resolved by scientific rationality alone but require the exercise of choice, the early involvement of a diverse set of stakeholders and a much wider understanding of impacts.

The primacy of science, engineering and technology as the key drivers to fulfil an economic imperative is giving way to the primacy of socio-enviro-economic drivers supported by scientific and engineering knowledge of what is possible and, most critically, what might be desirable.

Figure 12.1 Megatrends pose urgent challenges in cities (modified from Koop and van Leeuwen, 2017)

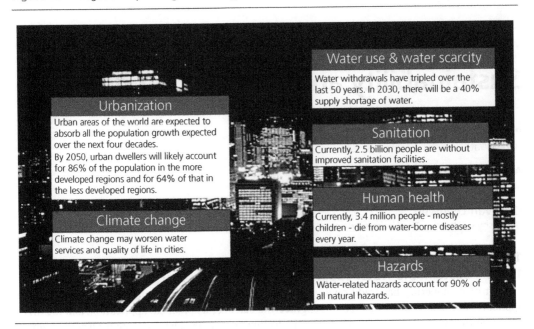

12.1.2   The fundamentals

At the simplest level, a sustainable water supply system is one that supplies anticipated (and reasonable) demands over a sensible time horizon without degradation of the source of the supply or other elements of the system's environment. This requires that the physical boundaries of the water supply system have to be drawn wide enough to ensure that Σ(inputs - outputs) ≥ 0 over time.

At the global scale, the governing activity is the hydrological cycle, in which the supply system is an artificial component, but a component nonetheless. And just as with the hydrological cycle, the water supply system is rarely in a steady state. Water demand varies diurnally and seasonally. The availability of water at the source also varies, and it may be that supply and demand are sometimes out of balance. For the most part, these imbalances can be smoothed out by reservoir storage at the source, storage at the water treatment plant, storage within the distribution systems (service reservoirs), storage at the point of use (cisterns) or transfers from an adjacent system.

Occasionally, the imbalance may lead to water restrictions on some or all users. Inevitably, this causes adverse comment from the users, some of which often finds its way into the media. A letter to the Times on 14th June 1991 (see Figure 12.2) shows one such example.

What the writer of this letter has failed to realise is that elements of the hydrological cycle are rate limited. Aquifers can only be replenished at a certain rate; river channels have a finite capacity; evaporation rates from the oceans are limited. And with the water supply and distribution system itself, the pipe work has a finite capacity – pumps have limited outputs and water treatment plants have a maximum throughput. See also Sabadell, Cape Town, Ahmedabad (Grison *et al.*, 2023) and Melbourne (Van Leeuwen, 2017).

Figure 12.2 Letter printed in the Times, 14th June 1991

Water shortages
From Mr R. Grant Paton

Sir, Amid the constant concern over water shortages, there appears to be a fallacy. Admittedly water is consumed in increasingly large quantities, but surely the world enjoys a closed system. Certainly human consumption of water is quickly returned whence it came. Where therefore does all our water go?

I am, yours faithfully,
R. GRANT PATON,
Odell Cottage,
Queens Lane,
Eynham, Oxford.

A water distribution system can fail in one of three ways

(a) source failure – insufficient water at the point of supply
(b) demand failure – insufficient capacity in the treatment or distribution system to meet transient demands
(c) system failure – a localised failure of a component of the treatment or distribution system, such as a pipe or pump.

In terms of sustainability, these factors extend the boundaries of the water distribution system to include its physical assets, their maintenance and renewal, and, therefore, the need for a sufficient and stable system of financing it, irrespective of the precise nature of ownership.

The boundaries of the water distribution system therefore extend to the socio-economic realm, to include the relationship between consumers (domestic and business) and the water service provider responsible for operating, maintaining and investing in the water supply system. This relationship is usually transacted through water charges, which brings with it the establishment of appropriate and equitable principles of charging. The sustainability issues affecting a water distribution and supply system therefore very quickly encompass the environment, society, economics and finance, and business.

12.1.3 The challenges and need for change
So, the focus of engineering development, including water supply engineering, has changed from technical and cost efficiency to having to address multiple and often divergent social, environmental and economic objectives. Early involvement of stakeholders is key and so is co-creation and the exploration of co-benefits or win-win's (Koop and Van Leeuwen, 2017).

A range of drivers have emerged from socio-political constructs of sustainable development, including the needs for social inclusion and social justice, to avoid environmental degradation and, most recently, to address projected climate change. The latter is now a dominant theme in public discourse and is the subject of increasingly stringent international agreements and national policy. The Paris Agreement, adopted by the parties attending the UN Climate Change Conference in 2015 (COP21), was a landmark in that it is a binding treaty to hold the increase in global average temperature below 2°C above pre-industrial levels and to pursue efforts to limit temperature increases to 1.5°C by the end of the century[1]. Governments have variously declared policies to achieve 'net zero' greenhouse gas (GHG) emissions and have set about decarbonising their economies. Plan-making to adapt to projected climate change has also progressed as collective understanding of the likelihood and magnitude of its impacts, spatially and temporally, and of the vulnerability of human and natural systems to them, have improved. Further progress was made at COP28, but it is still behind the curve of what is required.

The problem is that decision-making, whether at policy level or at the level of individual (water supply) projects, can become mired in an amalgam of conflicting objectives. The continued rise of carbon emissions from the water industry in the face of continued investment to improve service to customers and ensure environmental protection is a classic example of this problem.

Much has been said over the last decade on the need for better decisions, which in turn has led to a proliferation of approaches to achieve them. Notwithstanding all these efforts, the same basic tenet applies: a strong framework for assessing sustainability will assist in agreeing objectives, making choices that are socially acceptable and incorporate effective measurement of progress in achieving agreed objectives.

While setting objectives and success criteria is important, such a framework also needs to recognise and address problems in the existing development paradigms. Some of these issues are explored in this chapter, specifically

- end-of-pipe approaches may deliver certainty of outcome to narrowly defined 'engineering' problems, but this will not deliver an overall affordable, carbon-efficient water service that meets the needs of customers and the environment
- the balance between water as a basic human right and as an economic good does not properly reflect social need or encourage the right customer behaviour
- the dominant use of net present value to determine whole-life costs is inconsistent with achieving future resilience and equity, which must be inherent in a sustainable approach
- given the complex interaction of natural and social factors, insufficient attention is being given to understanding the variety of risks to water supply, with the consequences of more frequent and severe failures and loss of water security.

It is argued that what is needed is the embedding of sustainability principles within a strategic systems approach that encourages innovation through all stages of the asset life cycle, to address the various competing pressures, at local and regional levels, and to deliver a transformed, secure, yet affordable water supply service.

Earlier chapters are largely concerned with techniques applied by individual engineers as they go about their project work; by contrast, this chapter looks at larger system level issues that need to be tackled to ensure sustainable outcomes. While systems thinking (like sustainability) is now widely accepted in principle, there is still too little evidence of its practical application. It is recognised that in many cases it will be beyond the purview of the individual practitioner to influence these issues, but collectively we all have a responsibility to work within our spheres of influence – whether at project or policy level – to help challenge the status quo.

---

[1]https://unfccc.int/process-and-meetings/the-paris-agreement

12.2. Sustainability drivers and the sustainable development goals

Sustainable development has become one of the key axioms of modern times, not just of committed environmentalists but widely adopted across the political, business and social spectra. What are the roots of sustainable development? It is most closely associated with the 1992 Rio Summit (United Nations Conference on Environment and Development (UNCED), 'The Earth Summit') and encapsulated in Gro Harlem Brundtland's definition published in the World Commission on Environment and Development (WCED, 1987) as 'development that meets the needs of the present without compromising the ability of future generations to meet their own needs'.

The roots of the Brundtland definition (WCED, 1987) can be traced back at least to Jeremy Bentham's (1789) dictat that we should seek 'the greatest good for the greatest number'. And even the notion of intergenerational equity – so central to the Brundtland definition – was made explicit by American politician and forester Gifford Pinchot's (1947) addition of the phrase 'for the longest time' to Bentham's imperative.

But in more recent times, the concept of sustainable development has its roots in works such as Garret Hardin's paper 'Tragedy of the commons' (Hardin, 1968), the Club of Rome's Limits to Growth report (Meadows *et al.*, 1972) and the UN's Conference on the Human Environment, held in Stockholm in 1972. It was through the WCED in 1987 and its report Our Common Future (WCED, 1987) that it gathered momentum as the means to bridge the gulf between advocates and critics of progress through economic growth. Since then, 'sustainable development' has been adopted in countless hues as a philosophy and a policy tool by individual nations and agencies.

But it is the Brundtland definition (WCED, 1987) that has come to embody what is, in essence, a very simple idea: ensuring a better quality of life for everyone, now and for generations to come.

The four key objectives which underpin sustainable development are often taken to be

(a) social progress which recognises the needs of everyone
(b) effective protection of the environment
(c) prudent use of natural resources
(d) maintenance of high and stable levels of economic growth and employment.

For convenience, these are often reduced to the three headings of

(a) social inclusion
(b) environmental protection
(c) economic well-being.

But the classification of the dimensions of sustainable development does not prescribe precisely how a particular set of policy or decision options are adjudged. It simply defines the need to adopt a holistic approach and identify the broad criteria that should be used to discriminate between particular choices.

While much has been written on what the terms 'sustainable development' and 'sustainability' really mean, if nothing else, it has become a critique of the status quo (Gibson *et al.*, 2005). The WCED and subsequent UN environment conferences, the reports of the Intergovernmental Panel on Climate Change (2007) and the Millennium Ecosystem Assessment (2005) collectively reinforced the conclusion that present practices of human development are not sustainable and need to change. The need to act has spawned a multiplicity of initiatives, some with a primary focus on a particular dimension in the sustainability lexicon (e.g. natural capital, biodiversity net gain, social value) and reporting methods (e.g. Integrated Reporting according to the six capitals (IIRC, 2013)[2]). There is no single 'one size fits all' model and, to be practical, it must be tailored to suit the level and sector of application.

The Brundtland report also stressed the necessity of international cooperation for sustainable development. This conclusion increased awareness of the importance of sustainable development and led to further research and discussion. This gave rise to the Millennium Development Goals (MDGs) for 2000–2015. The MDGs were focused on the development of the global south and had an anthropogenic focus of development. While the MDGs had some large achievements (United Nations, 2015a), the economic achievements were greater than the environmental goals (Georgeson and Maslin, 2018). With the end of the MDGs in 2015, the Sustainable Development Goals (SDGs) were developed to continue the international agreement to sustainable development, this is known as the 2030 Agenda for Sustainable Development. Here the perception of sustainability has developed to include the requirements of environmental sustainability to achieving social and economic sustainability. The SDGs recognise that 'ending poverty and other deprivations must go hand-in-hand with strategies that improve health and education, reduce inequality, and spur economic growth – all while tackling climate change and working to preserve our oceans and forests'[3]. The approach of the SDGs is people, planet and prosperity to include economic growth, environmental sustainability and social inclusion (United Nations, 2015b).

Water is an important resource with increasing demand but is also a critical requirement to the development of multiple sectors discussed above. As water is so relevant, it was included as an individual Sustainable Development Goal (No. 6) for Agenda 2030 and its implementation matters (United Nations, 2018). The targets set for Goal 6 – *ensure availability and sustainable management of water and sanitation for all* – are focused on achieving measurable improvements within defined timescales for people (particularly the vulnerable), communities and water-related ecosystems across the world.

The SDGs form an internationally recognised set of goals and targets which aim to promote development in the economy, environment and society. There are a total of 17 goals (Figure 12.3) containing 169 targets with a focus on *people, planet, prosperity, peace, and partnerships*. The SDGs aim for a larger, holistic approach than the MDGs.

The SDGs provide a forward-looking vision for governments to consider, anticipate and respond to some global changes and trends that impact and shape the policy environment. Four critical megatrends influencing the achievement of the SDGs in cities and regions are herein identified

(a) demographic changes, in particular urbanisation, ageing and migration
(b) climate change and the need to transition to low-carbon economy
(c) technological changes, such as digitalisation and the emergence of artificial intelligence
(d) the geography of discontent.

---

[2] https://www.integratedreporting.org/what-the-tool-for-better-reporting/get-to-grips-with-the-six-capitals/
[3] https://sdgs.un.org/goals

Figure 12.3 The water-centric 17 Sustainable Development Goals for each sector (United Nations, 2015b; Makarigakis and Jimenez-Cisneros, 2019; with permission)

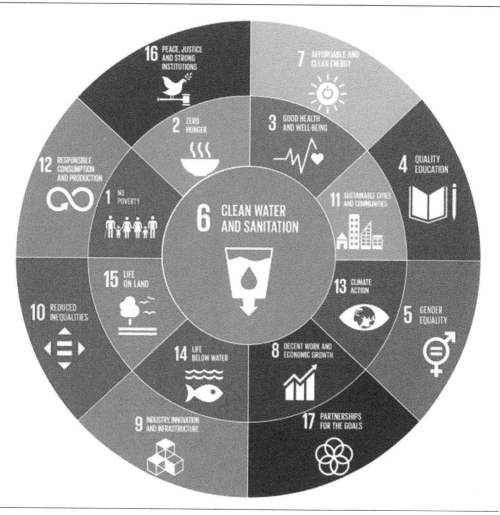

The impact of these four megatrends on people and societies is very much context-specific and therefore requires place-based policies to effectively respond, minimise their potential negative impact on regional disparities and capture the opportunities related to those trends locally.

According to the OECD (2019), scaling up financing for water-related investments is key to achieve the SDGs (Box 1). Water-related investments are key for sustainable development and inclusive growth. Water-related investments have spill-over effects on multiple SDGs, including those on food security, healthy lives, clean energy and marine and terrestrial ecosystems. This reflects the variety of water-related investments and the multitude of different needs these investments can address (e.g. supporting reliable freshwater supply, reducing pollution, providing drinking water, sanitation and wastewater treatment services, irrigation, etc.).

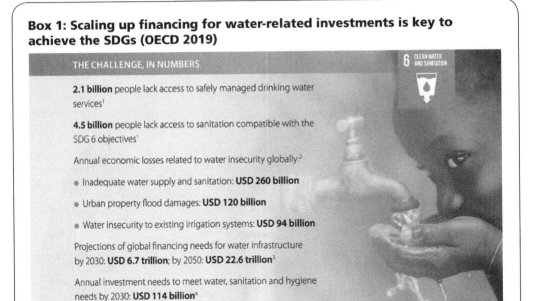

Box 1: Scaling up financing for water-related investments is key to achieve the SDGs (OECD 2019)

THE CHALLENGE, IN NUMBERS

6 CLEAN WATER AND SANITATION

2.1 billion people lack access to safely managed drinking water services[1]

4.5 billion people lack access to sanitation compatible with the SDG 6 objectives[1]

Annual economic losses related to water insecurity globally:[2]

- Inadequate water supply and sanitation: **USD 260 billion**
- Urban property flood damages: **USD 120 billion**
- Water insecurity to existing irrigation systems: **USD 94 billion**

Projections of global financing needs for water infrastructure by 2030: **USD 6.7 trillion**; by 2050: **USD 22.6 trillion**[3]

Annual investment needs to meet water, sanitation and hygiene needs by 2030: **USD 114 billion**[4]

Sources: 1 – WHO-UNICEF, 2017; 2 – Sadoff et al, 2015; 3 – Winpenny (2015); 4 – Hutton and Varughese, 2016

UK data on the UNSDGs for 2023 shows that even in prosperous countries like the UK there is more to be done to ensure that society's water needs are met in ways that sustain a healthy water environment[4].

The EU Water Framework Directive (2000) was a landmark in promoting environmental sustainability in that it was designed to transform the management of water at a catchment level. Rather than prescribing standards, the Directive encourages interested parties to determine objectives for water bodies (typically their restoration to good status or preventing deterioration) within river basin management plans. Achievement of these objectives depends on deploying measures – such as reducing abstractions to restore river flows – that are deemed to be cost-effective both in terms of financial cost and wider environmental cost. Less stringent objectives may be admitted (in the short-term at least) if the costs of achieving them are judged disproportionate. The numerical standards developed by member countries to match the objectives of the WFD have shaped the environmental performance requirements placed on water utilities (among others) over the last 20 years. The carrying over of the WFD into UK domestic law after Brexit demonstrates the perceived efficacy of this framework.

Despite progressive implementation of such legislation, strategies promoting more integrated approaches (e.g. Defra, 2008; EA, 2020) and continued investment by the sector, it has become increasingly clear in recent years that the performance of the UK's water sector is not meeting expectations. In Preparing for a Drier Future (NIC, 2018), the National Infrastructure Commission highlighted that the combined pressures of climate change, an increasing population and the need to better protect the environment means that the UK needs additional supply capacity of 2700–3000 megalitres per day (Ml/d) to ensure continued resilience.

---

[4] https://sdgdata.gov.uk/

The sector is now engaged in planning major new strategic solutions including reservoirs, water recycling plants and inter-regional transfers facilitated by the Regulators' Alliance for Progressing Infrastructure Development (RAPID), tasked with enabling the sector to 'respond to long term water resources challenges while promoting the best interests of water users, society and the environment'[5]. This is a major sustainability challenge, not least because of the extensive wider social and environmental impacts large investment in such physical infrastructure brings.

12.3. Climate change and greenhouse gas emissions

Within the context of Earth's limited natural resources and assimilation capacity, the current environmental footprint of humankind is not sustainable. Assessing land, water, energy, material, and other footprints along supply chains is paramount in understanding the sustainability, efficiency, and equity of resource use from the perspective of producers, consumers, and government (Hoekstra and Wiedman, 2014).

Climate change is one of the most pressing megatrends with impacts, challenges and opportunities varying significantly across territories within and across OECD countries. Some cities and regions are more vulnerable to climate change impacts than others. The global warming at 1.5°C may expose 350 million more people to deadly heat by 2050 (IPCC, 2018), exacerbated by local heat island effects. In Europe, 70% of the largest cities have areas that are less than 10 metres above sea level (OECD, 2010), thus exposed to higher risks of flooding. Moreover, cities concentrate almost two-thirds of global energy demand (IEA, 2016), produce up to 80% of greenhouse gas emissions and generate 50% of global waste (UNEP, 2017). Nevertheless, cities are also part of the solution. Subnational governments are responsible for 57% of all public investment and 64% of all climate-related public investments. Moreover, while transitioning from linear to circular economy, cities contribute to keeping the value of resources at its highest level, while decreasing pollution and increasing the share of recyclable materials.

In recent years, the huge amount of scientific research and modelling, in particular reporting of the Intergovernmental Panel on Climate Change, together with other socio-economic research such as that by Lord Stern on the economics of climate change (Stern, 2006), has meant that climate change and its potential impacts have come to dominate the global environmental sustainability debate. Figure 12.4 is a representation of the Keeling curve showing the increase in atmospheric concentration of carbon dioxide (CO_2) over the last 50 years alongside the increase in population.

According to MacKay (2008), 'The consensus of the best climate models seems to be that doubling the CO_2 concentration would have roughly the same effect as increasing the intensity of the sun by 2%, and would bump up the global mean temperature by something like 3°C.' A 3°C temperature rise will lead to very significant impacts such as the release of methane from permafrost.

By 2023, it appears that the world's climate has progressed further along this trajectory. The message to policymakers from the IPCC in its Sixth Assessment Report (IPCC, 2023) is that 'human activities, principally through emissions of greenhouse gases, have unequivocally caused global warming, with global surface temperature reaching 1.1°C above 1850–1900 in 2011–2020' and that 'Global GHG emissions in 2030 implied by nationally determined contributions (NDCs) announced by October 2021 make it likely that warming will exceed 1.5°C during the 21st century and make it harder to limit warming below 2°C'.

---

[5] https://www.ofwat.gov.uk/regulated-companies/rapid/about-rapid/

Figure 12.4 History of atmospheric CO_2 concentrations and world population growth 1500–2021. Sources: http://clim8.stanford.edu/Sources Population: Historical World Population CO2: Merged ice-core CO2

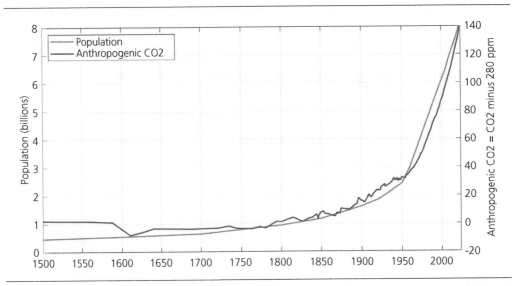

While global consensus on the extent and speed of future climate change and the allocation of responsibility for doing something about it remains elusive, humanity is undeniably increasing the CO_2 concentration in the atmosphere and in the surface oceans through burning of fossil fuels and other activities. The deal to establish a global fund for 'loss and damage', to assist poor nations adversely impacted by climate change, agreed in Egypt in 2022, was not accompanied by a universal commitment to reduce the use of fossil fuels. Despite this, countries across the world are making commitments, devising mechanisms for reducing greenhouse gas (GHG) emissions, have begun decarbonising their economies and making plans to adapt to its impacts.

The Paris Agreement is a legally binding international treaty on climate change. It was adopted by 196 Parties at the UN Climate Change Conference (COP21) in Paris, France, on 12 December 2015. Its overarching goal is to hold 'the increase in the global average temperature to well below 2°C above pre-industrial levels' and pursue efforts 'to limit the temperature increase to 1.5°C above pre-industrial levels'. However, in recent years, world leaders have stressed the need to limit global warming to 1.5°C by the end of this century. That's because the UN's Intergovernmental Panel on Climate Change indicates that crossing the 1.5°C threshold risks unleashing far more severe climate change impacts, including more frequent and severe droughts, heatwaves and rainfall. To limit global warming to 1.5°C, greenhouse gas emissions must peak before 2025 at the latest and decline 43% by 2030. The Paris Agreement is a landmark in the multilateral climate change process because, for the first time, a binding agreement brings all nations together to combat climate change and adapt to its effects. Under the Paris agreement, the EU committed in 2015 to cutting greenhouse gas emissions in the EU by at least 40% below 1990 levels by 2030. In 2021, the target was changed to at least 55% reduction by 2030 and climate neutrality by 2050. Through its Climate Change Act (2008), the UK committed to its own ambitious legally-binding targets for reducing GHG emissions; the overall commitment to achieve a reduction of 80% on 1990 levels by 2050 was replaced in 2019 by a revised legislative target to achieve 'net zero' emissions by 2050[6].

Average global temperatures have risen significantly since the industrial revolution and the last decade (2011–2020) was the warmest decade on record. Of the 20 warmest years, 19 have occurred since 2000. Data from the Copernicus Climate Change Service shows that 2022 was the hottest summer and second warmest year on record. The majority of evidence indicates that this is due to the rise of greenhouse gas emissions (GHG) produced by human activity. The average global temperature is today 0.95 to 1.20°C higher than at the end of the 19th century. Scientists consider an increase of 2°C compared to pre-industrialised levels as a threshold with dangerous and catastrophic consequences for climate and the environment (European Parliament, 2023).

A rise of 3°C in global temperatures above pre-industrial levels by 2100 would be disastrous. Its effects would be felt differently around the world, but nowhere would be immune. Prolonged heatwaves, droughts and extreme weather events could all become increasingly common and severe. Worryingly, slow progress from governments in cutting emissions makes this an uncomfortably plausible scenario.

Agriculture demand accounts for c.92% of the global blue water footprint. The remainder is shared between industrial production and domestic water supply (Hoekstra, 2014; Hoekstra et al., 2012). From land, energy and climate studies, it can be observed that agriculture (e.g. palm oil), and more specifically the livestock sector, plays a substantial role in deforestation, biodiversity loss, water pollution, water scarcity and climate change (Dasgupta, 2021; Hoekstra, 2014; Jalava et al., 2014).

Food systems are responsible for a third of global anthropogenic GHG emissions (Crippa et al., 2021). Large increases in freshwater demand can be expected in the next decades (UNESCO, 2021). These developments are estimated to generate a 40% freshwater supply shortage worldwide by 2030 (2030 Water Resources Group, 2009).

According to Defra (2008), the UK water industry contributes less than 1% to the UK's inventory of GHG emissions, but 'there is still a compelling case for action and … a real risk that this will continue to rise' with increasing water demand, more ambitious standards for water quality and the need for greater resilience in the face of climate change.

A global overview of GHG emissions is presented in Figure 12.5.

Despite efforts by the water sector to reduce energy-related emissions (in part because of grid decarbonisation), there have been continued increases in other areas from substantial investment in new capital assets and rising chemicals consumption, for example. In 2019 water companies in England made a joint commitment to reach net zero 'operational emissions' by 2030 and produced a route map for achieving this[7]. Since then, and in response to the need to achieve net zero *total* emissions by 2050, some are now pursuing science-based targets in line with the requirements of the Science-based Targets Initiative (SBTi). The SBTi's Net Zero Standard requires participating

---

[6]https://www.theccc.org.uk/what-is-climate-change/a-legal-duty-to-act/
[7]https://www.water.org.uk/sites/default/files/2023-08/Water-UK-Net-Zero-2030-Routemap.pdf. Operational emissions in this context include all emissions from the combustion of fossil fuels, direct emissions from treatment processes, the use of grid electricity. GHG emissions from the use of chemicals and capital works are excluded.

Figure 12.5 Global GHG emissions by sector for 2019 (Source: Hale, 2023. ©EconoFact)

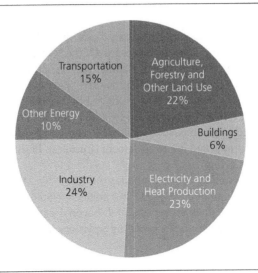

organisations to reduce all their emissions by more than 90% and to permanently remove the remainder (SBTi, 2023). Resorting to the carbon offset market is no longer acceptable.

Emission reductions in the water sector are vital, but reducing emissions is only half of the story: water supply systems are of course particularly vulnerable to the impacts of climate change. Across the world, environment agencies, water suppliers and their regulators have been engaged in assessing the implications of climate change and making plans to ensure future service resilience. For example, in the UK statutory undertakers under the Climate Change Act (2008), including UK water companies, are required to assess risks from climate change to their operations and to report their plans for adaptation. Many are also now expected to demonstrate to their stakeholders that they understand and are planning for potential climate change impacts. The Task Force on Climate-related Financial Disclosures (TCFD) published recommendations for businesses to ensure the effects of climate change – risks and opportunities – become routinely considered in business and investment decisions (TCFD, 2017). These recommendations have been adopted by the UK Government as mandatory reporting for publicly quoted companies, large private companies and LLPs (BEIS, 2022).

In terms of assessing and adapting to such impacts, there remains significant uncertainty both within climate change projections and their application to water resources management for public water supply. The UKCP18 climate projections and hydrological information derived from them, such as the Future Flows Climate dataset[8], provide opportunity to understand climate change risks better. At the highest level, impacts on operations and assets could arise from

- changes in temperature (annual and seasonal)
- change in precipitation patterns and evaporation rates (annual and seasonal)
- changes in the frequency and severity of extreme precipitation events
- changes in the duration, intensity and frequency of droughts
- changes in storm tracks (depressions)
- changes in sea level.

[8] https://www.data.gov.uk/dataset/e48861b5-18f7-4390-87c6-673d2f3a594c/future-flows-climate-data

In some cases, climate change will directly affect water operations and/or assets (e.g. temperature changes affecting the efficiency of treatment processes), whereas in others it will lead to changes in human behaviour or environmental conditions that, in turn, will affect operations and/or assets (e.g. lower flows in rivers affecting the quality of raw water abstracted for treatment, changes in population leading to changes in water demand). So, there is chain of potential first-, second-, third- (and so on) order effects to be considered (UKWIR, 2007; UKWIR, 2013). Second- and third-order effects could include changes in parameters such as

- water demand (domestic and non-domestic)
- river flows and groundwater levels
- river flood levels
- algal blooms and odour nuisance
- changes in soil moisture
- efficiency of biological processes
- water quality
- saline intrusion.

These types of effects have the potential to variously impact water operations and assets in terms of

- operational efficiency
- performance in terms of water quality compliance and flooding frequency/severity
- asset deterioration
- asset failure.

Of course, there may also be interactions between different effects on individual assets or groups of assets, which also need to be systematically evaluated.

The twin challenges of reducing carbon emissions and maintaining resilience to climate change impacts, while meeting society's need for a wholesome and affordable water supply, demand a long-term view involving what future services and the assets to deliver them look like and working out how to get there. In 2020 the Environment Agency in England and Wales published a new National Framework for Water Resources 'Meeting our Future Water Needs' (EA, 2020). This estimates the impact of climate change on national water supply to be around 400 Ml/d between 2025 and 2050 due to lower flows in rivers in summer, changes in reservoir storage and changes in patterns of groundwater recharge. The effect of climate change on resilience varies across the country with southern and eastern regions affected more than the rest. It recognises that the extent of the impact is uncertain and depends on the trajectory of global GHG emissions. As shown in Figure 12.6, nearly 90% of the emissions from the human water cycle arise from the use of water in the home (mainly from water heating for cooking and washing), and are additional to those arising from the operational activities of the water service providers. Thus, if a major shift in the way services are provided could be achieved – specifically to better manage the demand and use of water – this would have the most significant impact.

12.4. Adopting practical, systems-level approaches for more sustainable water supplies

It is clear that the utility engaged in delivering, operating and maintaining water supply infrastructure is faced with addressing not just technological issues but various environmental, social and economic issues too to be sure of delivering sustainable outcomes. While objectives such as the

Figure 12.6 Carbon emissions from the water cycle. Data from Environment Agency (2009)

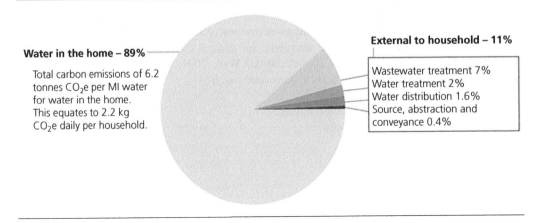

Water in the home – 89%

Total carbon emissions of 6.2 tonnes CO_2e per Ml water for water in the home. This equates to 2.2 kg CO_2e daily per household.

External to household – 11%

Wastewater treatment 7%
Water treatment 2%
Water distribution 1.6%
Source, abstraction and conveyance 0.4%

SDGs and national policies advocate measures for sustainable outcomes, including addressing climate change, the danger is that decision-making becomes mired in an amalgam of competing objectives.

A structured framework for 'sustainability assessment' is needed that practitioners and decision-makers can use to assess alternative courses of action against a range of objective attributes and stakeholder choice criteria. While there are various sustainability assessments, a successful approach will ensure that

- relevant physical, economic, social, cultural, environmental and other factors are identified at an early stage of project development and analysed effectively
- adoption of a set of sustainability objectives that reflect both the primary need for water development and the relevant national/local policies, plans and issues of concern to stakeholders
- a suitable (perhaps radical) range of development alternatives are identified and assessed in terms of their ability to meet the sustainability objectives
- stakeholders, statutory and non-statutory, are engaged from the outset and throughout the process of project development and decision-making
- selected solutions are resilient and adaptable to future environmental and social change
- progress of the chosen solution in meeting the objectives is effectively measured and reported.

Sustainability principles should be applied routinely both at a strategic level to meet policy objectives and legislative requirements and at all levels down to the individual development project. Increasingly, proposals for new water sector investment include assessments of social and environmental impacts – for example, investments under the Water Industry's National Environment Programme (WINEP) in England and Wales are expected to achieve 'wider environmental outcomes' in line with the requirements of the Environment Act (2021). The challenge for the water practitioner is to ensure that such assessments are carried out transparently and so as to deliver solutions that are affordable.

Simultaneous improvement in all three of the usual pillars of sustainability – the economy, the environment and society – is not generally possible, even through the lens of one stakeholder, never mind several, for example, in the context of water supply and distribution is as follows.

- **The environmental dimension** covers the continued drive to meet ever more stringent drinking water quality standards (to safeguard human health) in the face of deteriorations in raw water quality (perhaps as a result of climate change) and contamination, and to do so without risk of interruption (to meet the expectations of twenty-first century society), can result in expensive, high energy, chemically-intensive solutions leading to high greenhouse gas emissions. In locations served by multiple water sources, a more sustainable solution may be achieved by combining treatment upgrades with careful management of catchment sources and refurbishment of existing networks in ways that help reduce the size of downstream works with acceptable risk.

The planned development of major new strategic solutions to increase water supply capacity and resilience has major implications for the environment. New reservoirs, water recycling plants and inter-regional transfers employing large diameter pipelines inevitably require large quantities of material resources in their construction as well as energy to run them. Pumping large quantities of water within and between regions is carbon intensive particularly since they need to be kept in routine operation to ensure their readiness to supply water when required to ensure resilience.

- **The social dimension** encompasses access to and affordability of a sufficient quantity of wholesome water to individuals and communities as well as consumer attitudes about the value of water and the wider environment and how this affects their behaviour. The production and distribution of a wholesome water supply can be considered both a basic human right (a social service) and an economic good. Models have been developed of both. Environmental objectives can theoretically be built into an economic model by incorporating environmental costs and benefits into the valuation. However, this can potentially conflict with the social service model because the resulting tariff changes will tend to adversely affect large households on low incomes. The problem can perhaps be overcome by classifying (household) water use in two ways: *(a)* essential use for drinking, cooking, washing and basic sanitation and *(b)* discretionary use for garden watering, leisure and luxury appliances – and then introducing a systems of increasing block tariffs. This approach will serve to both progressively reduce higher levels of consumption with the consequential environmental benefit and ensure fair access.

In recent years there has been a drive in both the public and private sectors to measure the contribution of projects, notably public infrastructure, to 'social value'. Social value refers to the wider financial and nonfinancial value created by an organisation through its activities in terms of wellbeing, social capital and the environment. In the UK, this has led to the development of a 'National TOMs Framework' structured according to a suite of themes, outcomes and measures[9].

- **The economic dimension.** Investment in water infrastructure is increasingly based on delivering maximum benefits for least whole life cost. The move from decisions based solely on capital cost is vital but whole life costing is too often expressed in terms of the economic concept of 'net present value', based on discount theory that reduces the value of future benefits and costs – in water engineering, this tends to favour low CAPEX and high OPEX solutions. Besides financial impact, high OPEX often means high carbon

---

[9] https://socialvalueportal.com/solutions/national-toms/

emissions and the value of the environment, where it features in the calculation, becomes less valuable year on year.

In recent years, efforts have been made to give more weight to the future environmental and wider non-financial impacts of capital investment – for example, the 2022 revision of the UK HM Treasury's Green Book incorporates a revised valuation of GHG emissions through an updated set of projected carbon values (HM Treasury, 2022). These are projected to rise significantly from now until 2100 and are designed to reflect the monetary value that society places on one tonne of carbon dioxide equivalent. Wider 'natural capital' impacts – such as the UK Government's ENCA (Defra, 2023) – are also increasingly being taken into account in water sector investment planning where this is in the interest of customers.

It is clear that water utilities engaged in delivering, operating and maintaining water supply infrastructure are faced with addressing not just technological issues but various environmental, social and economic issues too to be sure of delivering sustainable outcomes.

A strategic, systems-level approach requires careful thinking and targeted investment but ultimately will be easier and lower cost. The ICE's Systems Approach to infrastructure Delivery (ICE, 2020) shows that the multi-dimensional complexity of improving physical infrastructure (including water supply systems) in ways that meet social needs in an affordable manner, while decarbonising the economy, protecting nature and adapting to climate change, demands nothing less.

Practically, we need to adopt systems-led, collaborative approaches at all levels. One such is PAS 2080: 2023 (BSI, 2023) and supporting guidance document (ICE, 2023). Originally published in 2016, the revised version is a practical framework, which incorporates some key sustainability themes very applicable to water supply systems

- systems thinking and net zero – effective decarbonisation depends on considering asset improvements in the context of the connected networks and wider systems (on the transition to net zero) of which they are part
- a whole-life view – assess and manage the full lifecycle impacts of capital investments (construction, operation, use and end-of-life)
- resilient nature-positive solutions – design to both adapt to environmental change and deliver multiple benefits – natural capital, biodiversity, greater social value, as well as carbon reduction
- collaboration – generate fully integrated value chains with water companies, designers, constructors and product suppliers all working together with regulators and stakeholders.

Importantly, PAS 2080 is not a badge to be achieved on individual projects; it requires organisational change. It's a long-term commitment.

The Value Toolkit[10] and ciriabest[11] are two other examples of practical approaches designed for use by practitioners of infrastructure developments, including those operating in the water sector, to enable them to make value-based decisions and select systems-level solutions that drive better social, environmental and economic outcomes for current and future generations.

---

[10] https://constructioninnovationhub.org.uk/our-projects-and-impact/value-toolkit/

[11] https://www.ciriabest.com/

12.5. Assessing the sustainability of integrated water resources management in cities

International agreements on the need for integrated water resources management (IWRM) have led to major policy initiatives in many countries. IWRM is widely acclaimed by international organisations such as the International Water Management Institute, the Food and Agriculture Organization, the World Bank and various regional authorities. IWRM is defined as a process that promotes the coordinated development and management of water, land and related resources in order to maximise economic and social welfare in an equitable manner without compromising the sustainability of vital ecosystems (UNEP, 2022; United Nations, 2022). The concept and its application is considered by many as pivotal for achieving the water-related UN Sustainable Development Goals (Essex *et al.*, 2020; Pahl-Wostl *et al.*, 2021). As approximately 70% of the population will be living in urban areas by 2050, with the largest growth taking place in cities in Africa and Asia, the pressure for tackling water challenges has shifted to cities (Romano and Akhmouch, 2019). Cities have the responsibility for local resources management, land use and urban infrastructures, and, therefore, can position themselves as arenas for tackling the largest changes (OECD, 2015; Hachaichi and Egieya, 2023).

What is needed for infrastructure projects is a structured framework for 'sustainability assessment' that practitioners and decision-makers can use to assess alternative courses of action against a range of objective attributes and stakeholder choice criteria (Figure 12.7).

12.6. The City Blueprint Approach

There are different structured frameworks to assess the sustainability of cities. In the remainder of this chapter we will focus on the City Blueprint Framework (CBF) that has been applied on more than 125 cities in the world. The aim of the City Blueprint Approach (CBA) is to assess the sustainability of the urban water cycle. The CBA aims to enhance the transition towards water-wise cities by city-to-city learning and sharing best practices (Koop and Van Leeuwen, 2017). Data to support indicator calculations are available on different spatial scales and from a variety of country, river basin and urban sources. The CBA consists of three complementary frameworks (Figure 12.8). The CBA is a diagnostic tool.

Figure 12.7 Typical stages involved in the planning cycle for water according to SWITCH (ICLEI, 2011)

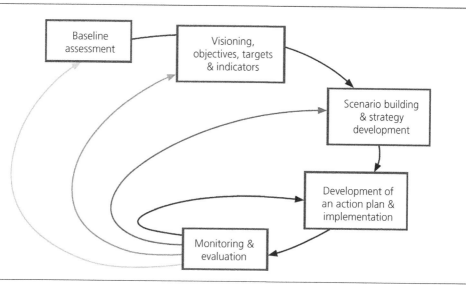

The main challenges of cities are assessed with the Trends and Pressures Framework (TPF) (Koop and Van Leeuwen, 2021a). The water management performances are assessed with the CBF (Koop and Van Leeuwen, 2021b). Where cities can improve their water governance is assessed with the Governance Capacity Framework (GCF) (Koop and Van Leeuwen, 2021c; Koop *et al.*, 2017).

The trends and pressures framework (TPF) assists in setting of city priorities in the social, environmental, financial and governance categories of specific cities (Koop *et al.*, 2022). It is based on a total of 24 indicators. The score of each of these TPF indicators is calculated from zero to ten, where a higher score represents a higher urban pressure or concern. The scores of the 24 indicators are displayed in a spider web diagram. The trends and pressures index (TPI) is the arithmetic mean of the 24 TPF indicators. High TPI scores represent high overall concerns that can form challenges for a city's IWRM. The TPF provides the context in which water managers may gain insights on limitations and windows of opportunity for IWRM.

The City Blueprint (Performance) Framework deals with the adequacy of the city's water management. The CBF provides seven categories subdivided into 24 indicators. Each indicator is scored ranging from zero (poor performance) to ten (good performance) and displayed in a spider web diagram. The CBF is based on 24 indicators divided over seven main categories: *(a)* basic water services, *(b)* water quality, *(c)* wastewater treatment, *(d)* water infrastructure, *(e)* solid waste, *(f)* climate adaptation and *(g)* plans and actions. The Blue City Index (BCI) is the geometric mean of the 24 CBF indicators. High BCI scores represent good performance on IWRM. The TPF and CBF evaluate a total of 48 indicators and the GCF addresses 27 governance questions for water-related challenges (Koop and Van Leeuwen, 2021c; Koop *et al.*, 2017).

The results of the CBA have been summarised in two publications (Koop *et al.*, 2022 and Grison *et al.*, 2023). Reference is made to these papers for all the data, the sources of the data, the calculations and discussions of the results. The main findings will be shortly summarised below. Based on the scoring of TPF and CBF indicators for 125 cities (6000 data points) a simple equation was developed for the estimation of the urban water management performances (BCI*), particularly for cities in data-poor regions. The resulting equation for estimating BCI scores (denoted as BCI*) is shown in Figure 12.9 and the equation below.

Figure 12.8 The City Blueprint Approach divided in its three complementary frameworks (Koop *et al.*, 2022)

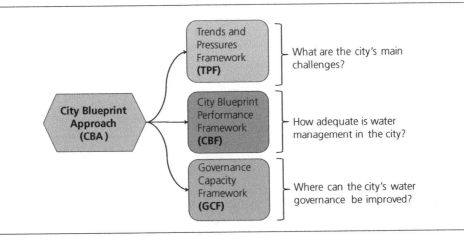

BCI* = *4.25 – 0.396*TPF21 [Government effectiveness] + 0.195*CBF4 [Secondary WWT] + 0.111*CBF8 [Energy recovery]*

The estimation model predicts the BCI* within a range of ± 1.3 (95% prediction interval) from the fully assessed value with a correlation coefficient (R^2) of 0.83. The estimated BCI scores using this model against CBA-assessed BCI scores are shown in Figure 12.9. It shows that the BCI can be predicted in a reliable manner from three parameters: *(a)* government effectiveness, *(b)* secondary waste water treatment and *(c)* energy recovery from waste water.

One of the most important results of the statistical analysis is the relevance of the indicator government effectiveness, one of the governance indicators of the World Bank, in predicting water management performance. Government effectiveness is the most important variable (Multiple R = 0.71 and R Square = 0.50).

This estimation model has been used to estimate the BCI of another 75 cities making the total of 200 cities, representing in total 95% of the world population. The results are shown in Figure 12.10, Table 12.1 and Table 12.2. In more detail, targets regarding drinking water supply have been met in many countries with the exception of some countries in Africa and Asia. Challenges regarding sanitation are still high in countries in Africa, Asia and Latin America. The same holds for management of solid waste, climate adaptation, the percentage of the urban population living in slums and needs for improving government effectiveness, indicating that globally much work remains to meet these targets, especially with regards to urban solid waste management, waste water treatment, air pollution and climate adaptation.

Figure 12.9 Three-variable BCI* estimation model based on CBF and TPF. The plot of the estimated BCI*s against the fully assessed BCIs for the combined 48 CBF and TPF indicators. The solid red line represents a full correspondence of the estimated BCI* and the actual BCI (Y=X; slope = 1) (Grison et al., 2023)

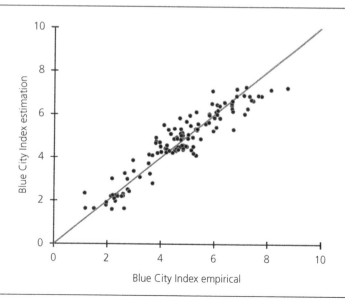

Figure 12.10 Global map of estimated BCI* and fully assessed (empirical) BCI scores for 200 cities. This shows that Latin America, Africa, and parts of Asia generally have BCI scores lower than 4, indicating a great disparity in IWRM. Only Northern Europe shows a distinct cluster of cities scoring higher than 6, whereas Singapore (BCI=8.1) and Amsterdam (BCI=8.7) are the only cities with BCI scores > 8. (Grison et al., 2023)

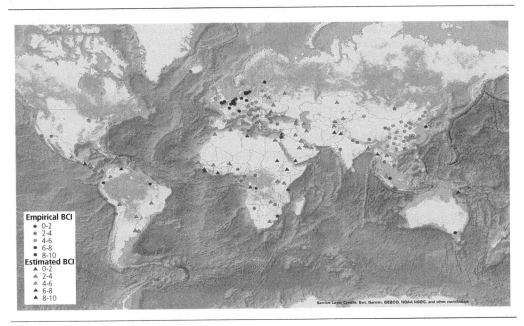

Table 12.1 BCI scores per continent. Regional variation of IWRM in cities among continents as measured by the 125 fully assessed and 75 estimated BCI values (Grison et al., 2023)

Continent	Number of cities	Number of cities with BCI <5	Cities with BCI<5 (%)	Average BCI and standard deviation (in parentheses)
Europe	66	24	36	5.3 (1.5)
Oceania	3	2	67	4.3 (2.3)
Asia	75	65	87	4.0 (1.3)
North America	15	14	93	3.5 (1.1)
Latin America	12	11	93	2.7 (1.1)
Africa	29	29	100	2.4 (1.0)
All	**200**	**145**	**73**	**4.1 (1.7)**

Table 12.2 Distance to targets status for SDG indicators and other relevant targets. For each indicator the total number of people in each country – either the total or urban population – was calculated for which the targets were met (Grison et al., 2023)

No	Indicator	Total or Urban population	People meeting the target		People not meeting the target		Population represented
			million	%	million	%	%
6.1	Achieve universal and equitable access to safe and affordable drinking water for all	Total	6.072	81%	1.380	**19%**	95%
6.2	Access to adequate and equitable sanitation and hygiene for all	Total	4.825	65%	2.628	**35%**	95%
11.1	Urban population (not) living in slums, informal settlements. or inadequate housing	Urban	3.167	75%	1.033	**26%**	77%
11.6.1	Urban solid waste regularly collected and with adequate final discharge out of total urban solid waste generated. by cities	Urban	2.973	42%	4.130	**58%**	91%
11.6.2	Annual mean levels of fine particulate matter ($<10\,\mu g/m^3$)	Total	486	7%	6.967	**93%**	95%
NA	Government effectiveness (>5.0)	Total	4.988	67%	2.465	**33%**	95%
NA	Climate Adaptation - ND-GAIN Readiness (>0.5) (ND-GAIN, 2020)	Total	1.351	18%	6.077	**82%**	95%

The assessments have been used to determine the current status of the implementation of the greater international water and urban agendas (SDGs 6 and 11). We observe that 145 of the 200 cities assessed or estimated have BCIs below five, which means that many cities still have to implement advanced wastewater treatment, energy and resource recovery, and climate adaptation measures. Only two cities have BCI scores > eight. The current state of affairs urges for accelerated improvements: large portions of the global population are far from reaching the SDGs, notably related to water, waste and climate change. This further supports the global assessment performed using the CBA, revealing not only relatively low BCI scores in cities around the world, but also significant regional disparities between Europe and Latin America, Africa and parts of Asia. There is a need to focus on the practical implementation of the SDGs for which global availability and accessibility of data is essential (Essex *et al.*, 2020).

Meeting the UN SDGs is a political choice. Data gaps are preventing adequate implementation of the SDGs. It is not possible to manage a process if progress cannot be monitored, and monitoring of progress is hindered if adequate data is not available (UNEP 2021a, b). To date, funding for SDG 6 targets has been deemed insufficient and the global framework for IWRM shows a poor record of implementation. Unless significant progress is made, it is envisaged that SDG 6 targets will not be met by 2030, which in turn impacts other SDGs (UNEP 2021a, b). The longer it takes to start the actions, the more difficult it will be to overcome challenges of water, wastewater, waste and climate change in cities.

Another way to analyse the data on the 125 cities for which TPF and CBF indicator scores were available is to perform a cluster analysis. This has been described in detail in Koop *et al.* (2023). This hierarchical clustering resulted in five clusters of cities. A summary of the findings is presented in Figure 12.11.

Five clusters of cities can be distinguished which show a pattern of problem-shifting – that is, the shifting of largely preventable problems often in the following sequence: drinking water insecurity,

Figure 12.11 Simplified scheme of a generic path followed by cities to become water-wise based on the hierarchical cluster analysis of 125 urban water management assessments. Problems are often addressed in the following sequence: drinking water security, wastewater treatment and municipal solid waste collection and treatment, climate adaptation and resource recovery (Koop *et al.*, 2022)

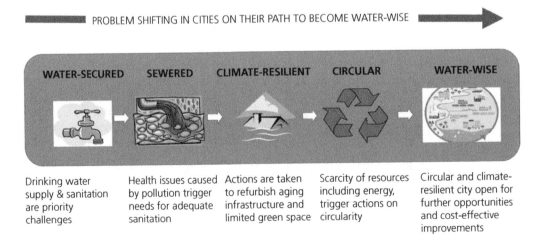

PROBLEM SHIFTING IN CITIES ON THEIR PATH TO BECOME WATER-WISE

WATER-SECURED	SEWERED	CLIMATE-RESILIENT	CIRCULAR	WATER-WISE
Drinking water supply & sanitation are priority challenges	Health issues caused by pollution trigger needs for adequate sanitation	Actions are taken to refurbish aging infrastructure and limited green space	Scarcity of resources including energy, trigger actions on circularity	Circular and climate-resilient city open for further opportunities and cost-effective improvements

pollution caused by inadequate wastewater treatment and solid waste management, inaction on climate change adaptation, and resource depletion. A city that can address and solve all these problems can be classified as water-wise. Currently, only Singapore and Amsterdam can be categorised as water-wise. The analysis shows that currently management decisions on water, waste and climate change often are short-term, reactive and tend to create new problems (Koop, 2019; Peters *et al.*, 2021).

The ability of cities in developing countries to transform to water-wise cities will be critical for human health and environmental sustainability as the world progresses into the twenty-first century. This is most relevant for countries facing fast population growth and high rates of urbanisation. Many developing countries in Africa are facing such developments (Koop and Van Leeuwen, 2017) where water security is broader than the mismatch between supply and demand, and human induced scarcity factors (inequality, poor governance, weak institutions or imbalanced power relations) are equally important (Chitonge, 2020). The five clusters represent challenges in cities but may also provide clusters of solutions – that is, by city-to-city learning (Koop *et al.*, 2022; Dieperink *et al.*, 2023). This would accelerate the transformation to water-wise cities, globally.

Koop *et al.* (2022) provided seven principles to enhance urban water management transformation. These principles are intended to provide general guidance on how city planners, practitioners, politicians, but also the scientific community, can contribute to an accelerated transformation towards water-wise cities. They are summarised in Table 12.3. For further details and references the reader is referred to the original publication of Koop *et al.* (2022).

12.7. Concluding remarks

Water utilities engaged in delivering, operating and maintaining water supply infrastructure are faced with addressing not just technological issues but various environmental, social and economic issues, too, to be sure of delivering sustainable outcomes. Achieving improvements across all such aspects is often difficult because of the multiplicity of stakeholder interests and the complexity of the socio-economic environments within which decisions for making improvements have to be made.

Developing a comprehensive global picture on sustainable water in cities enables us to evaluate the existing gaps in achieving water-related SDGs, in particular SDG 6 (*clean water and sanitation*) and SDG 11 (*sustainable cities and communities*). Unfortunately, most cities do not yet manage their water resources wisely and are far from achieving the SDGs. The main conclusions are:

- The most important parameter in our estimation model of IWRM is government effectiveness, as published by the World Bank, followed by secondary waste water treatment and energy recovery from waste water.
- This supports the statement of the OECD that if you want to 'fix the water pipes, start with the institutions'.
- Only 1 % of the 200 cities assessed can be classified as water-wise (Amsterdam and Singapore).
- Unfortunately, 145 of the 200 cities assessed have low Blue City Index (BCI) scores (BCI < 5), which means that many cities still have to implement advanced wastewater treatment, energy and resource recovery and climate adaptation measures.
- Targets regarding drinking water supply are still a challenge for many cities in Africa and Asia.

- Challenges regarding sanitation are high in cities in Africa, Asia and Latin America.
- As most of people live in cities, achieving the goals of SDG 6 and SDG 11 is crucial for humanity.
- Meeting the SDGs is a political choice. Data gaps are preventing adequate implementation of the SDGs. It is not possible to manage a process if progress cannot be monitored, and monitoring of progress is hindered if adequate data is not available. Smart SDGs are necessary.
- To date, funding for SDG 6 targets is insufficient and the global framework for IWRM shows a poor record of implementation. Unless very significant progress is made, it is envisaged that SDG 6 (and 11) targets will not be met at all by 2030, which in turn impacts other SDGs.
- The seven principles listed in Table 12.3 may provide direction in applying focus and overcome the typical pattern of problem-shifting that we have observed. Such a strategic refocus however requires higher priority and funding for capacity development, implementation and monitoring of the SDGs (European Commission, 2021) across the globe. There is much room for improvement, as only 42% of the 92 environment-related SDG indicators have sufficient data to assess progress made in achieving the targets (UNEP, 2021b). After all, meeting the SDGs, and in particular access to healthy water and sanitation for all (SDG6), underpins a strong improvement in the quality of life and at the same time will also lead to substantial cost-savings in the long run (Koop and Van Leeuwen, 2017). How cities can accelerate these transformations has been addressed in another paper (Dieperink et al., 2023).

Systems approaches that acknowledge the complexity inherent in considering a multiplicity of factors are vital. Practically, we need to adopt systems-led, collaborative approaches at all levels where decisions are made. This is becoming increasingly recognised in both policy and practice but needs to be pressed home in particular by practitioners working at all levels down to individual projects, since it is there that the opportunities to effect lasting change for more sustainable water supply outcomes will be realised.

Table 12.3 Seven principles to enhance urban water management transformation (Koop et al., 2022)

Principle	Description
1	Develop a long-term consistent plan for cities
2	Water as enabler: exploring the co-benefits
3	Water governance is key
4	Solid waste first
5	Waste is an untapped and valuable resource
6	Nature as a building block for cities
7	Change diets: the protein transition

REFERENCES

Bentham J (1789) *Introduction to the Principles of Morals and Legislation*, 1907 reprint of 1823 edn. Clarendon Press, Oxford, UK.

BSI (2023) PAS2080:2023 Carbon management in buildings and infrastructure. https://knowledge.bsigroup.com/products/carbon-management-in-buildings-and-infrastructure?version=standard [Accessed on 17.01.24]

Chitonge H (2020) Urbanisation and the water challenge in Africa: Mapping out orders of water scarcity. *African Studies* 79(**2**): 192–211. https://doi.org/10.1080/00020184.2020.1793662 [accessed 12.12.2023]

Climate Change Act 2008 (chapter 27). The Stationery Office, London, UK.

Crippa M, Solazzo E, Guizzardi, E *et al.* (2021) Food systems are responsible for a third of global anthropogenic GHG emissions. *Nature Food* 2: 198–209. https://doi.org/10.1038/s43016-021-00225-9 [accessed 12.12.23]

Dasgupta P (2021) *The Economics of Biodiversity: The Dasgupta Review*. HM Treasury, London, UK. https://assets.publishing.service.gov.uk/government/uploads/system/uploads/attachment_data/file/962785/The_Economics_of_Biodiversity_The_Dasgupta_Review_Full_Report.pdf [accessed 12.12.23]

Defra (2008) *Future Water: The Government's Water Strategy for England*. TSO, London, UK. www.defra.gov.uk/environment/quality/water/strategy/index.htm [accessed 01.09.10].

Defra (2023) *Enabling a Natural Capital Approach (ENCA) guidance*. https://www.gov.uk/government/publications/enabling-a-natural-capital-approach-enca-guidance/enabling-a-natural-capital-approach-guidance [Accessed 17.01.24].

Department for Business, Energy and Industrial Strategy (BEIS) (2022) *Mandatory climate-related financial disclosures by publicly quoted companies, large private companies and LLPs: Non-binding guidance*. https://assets.publishing.service.gov.uk/media/62138625d3bf7f4f05879a21/mandatory-climate-related-financial-disclosures-publicly-quoted-private-cos-llps.pdf [Accessed 17.01.24].

Dieperink C, Koop S, Witjes M, van Leeuwen K and Driessen P (2023) City-to-city learning to enhance urban water governance: the contribution of the City Blueprint Approach. *Cities* **135**, 104216 https://doi.org/10.1016/j.cities.2023.104216 [accessed 12.12.23]

European Commission (2021) *Evaluating the Impact of Nature-based Solutions: a Handbook for Practitioners*. Publication Office of the European Union, Luxembourg. https://data.europa.eu/doi/10.2777/244577 [accessed 12.12.23]

EA (Environment Agency) (2009) *Water for People and Environment: Water Resources Strategy for England and Wales*. EA, London, UK. www.environment-agency.gov.uk/research/library/publications/40731.aspx [accessed 01.09.10].

EA (2020) *Meeting our future water needs: a national framework for water resources*. EA, London, UK. https://assets.publishing.service.gov.uk/government/uploads/system/uploads/attachment_data/file/872759/National_Framework_for_water_resources_main_report.pdf [Accessed 17.01.24].

Environment Act (2021). https://www.legislation.gov.uk/ukpga/2021/30/contents/enacted.

Essex B, Koop SHA and Van Leeuwen CJ (2020) Proposal for a national blueprint framework to monitor progress on water-related sustainable development goals in Europe. *Environmental Management* 65:1–18. https://doi.org/10.1007/s00267-019-01231-1 [accessed 12.12.23].

European Parliament (2023) EU measures against climate change. https://www.eumonitor.eu/9353000/1/j9vvik7m1c3gyxp/vkqp9dzgriys?ctx=vg9hjjllgxmz [accessed 27.5.2024].

Georgeson L and Maslin M (2018) Putting the United Nations Sustainable Development Goals into practice: a review of implementation, monitoring, and finance. *Geography and Environment* 5(**1**): e00049. https://doi.org/10.1002/geo2.49 [accessed 12.12.23]

Gibson RB, Hassan S, Holtz S, Tansey J and Whitelaw G (2005) *Sustainability Assessment: Criteria and Processes*. Earthscan, London, UK.

Hardin G (1968) The tragedy of the commons. *Science* **162(3859)**: 1243–1248.

Grison C, Koop S, Eisenreich S *et al.* (2023) Integrated water resources management in cities in the world: global challenges. *Water Resources Management* **37**: 2787–2803 https://doi.org/10.1007/s11269-023-03475-3 [accessed 12.12.2023].

Hachaichi M and Egieya J (2023) Water-food-energy nexus in global cities: solving urban challenging interdependencies together. *Water Resources Management* **37**: 1811–1825 https://doi.org/10.21203/rs.3.rs-1956052/v1 [accessed 01.09.23].

Hale G (2023) *The Importance of the food industry for climate change.* https://econofact.org/the-importance-of-the-food-industry-for-climate-change [accessed 01.12.23].

Hoekstra AY (2014) Water for animal products: a blind spot in water policy. *Environmental Research Letters* 9**(9)**, 091003. http://dx.doi.org/10.1088/1748-9326/9/9/091003 [accessed 01.09.16].

Hoekstra AY, and Wiedmann TO (2014) Humanity's unsustainable environmental footprint. *Science* 344 **(6188)**: 1114–1117. https://doi.org/10.1126/science.1248365 [accessed 01.09.17].

Hoekstra AY, Mekonnen MM, Chapagain AK *et al.* (2012) Global monthly water scarcity: blue water footprints versus blue water availability. *PLoS ONE* 7**(2)**, e32688. https://doi.org/10.1371/journal.pone.0032688 [accessed 01.09.18].

HM Treasury (2022) *The Green Book. Central Government Guidance on Appraisal and Valuation.* https://assets.publishing.service.gov.uk/media/623d99f5e90e075f14254676/Green_Book_2022.pdf [Accessed 17.01.24].

ICLEI (2011) *SWITCH training kit. Integrated urban water management in the city of the future.* https://www.gwp.org/globalassets/global/toolbox/references/switch-training-kit-modules-2011.pdf [accessed 01.09.16].

IEA (2016) *Energy Technology Perspectives 2016.* International Energy Agency, OECD Publishing, Paris, France. https://doi.org/10.1787/energy_tech-2016-en [accessed 01.09.16].

IIRC (2013) The International < IR > Reporting Framework. https://www.integratedreporting.org/wp-content/uploads/2013/12/13-12-08-THE-INTERNATIONAL-IR-FRAMEWORK-2-1.pdf [accessed 17.01.24]

ICE (Institution of Civil Engineers) (2020). *A Systems Approach to Infrastructure Delivery.* https://www.ice.org.uk/media/rvnpuajx/ice_systems_report_final.pdf [Accessed 17.01.24].

ICE (2023). *Guidance document for PAS2080.* https://www.ice.org.uk/media/vm0nwehp/2023-03-29-pas_2080_guidance_document_april_2023.pdf [Accessed 17.01.24].

IPCC (2007) *Climate Change 2007: Synthesis Report. Contribution of Working Groups I, II and III to the Fourth Assessment Report of the Intergovernmental Panel on Climate Change IPCC.* Geneva, Switzerland, p.104.

IPCC (Intergovernmental Panel on Climate Change) (2018) *Global Warming of 1.5°C: An IPCC Special Report on the Impacts of Global Warming of 1.5°C above Pre-industrial Levels and Related Global Greenhouse Gas Emission Pathways.* IPPC, Geneva, Switzerland. https://www.ipcc.ch/sr15/download/ [accessed 01.09.19].

IPCC (2023) Summary for Policymakers. In: *Climate Change 2023: Synthesis Report. Contribution of Working Groups I, II and III to the Sixth Assessment Report of the IPCC.* IPCC, Geneva, Switzerland.

Jalava M, Kummu M, Pokka M, Siebert S and Varis O (2014) Diet change - a solution to reduce water use? *Environmental Research Letters* 9**(7)**, 074016. https://doi.org/10.1088/1748-9326/9/7/074016 [accessed 01.09.16].

Koop SHA (2019) *Towards water-wise cities: Global assessment of water management and governance capacities.* PhD Thesis, Utrecht University, the Netherlands.

Koop SHA and Van Leeuwen CJ (2017) The challenges of water, waste and climate change in cities. *Environmental, Development and Sustainability* 19:385–418. https://doi.org/10.1007/s10668-016-9760-4 [accessed 01.09.19].

Koop SHA. and Van Leeuwen CJ (2021a) *Indicators of the trends and pressures framework (TPF).* KWR Water Research Institute, Nieuwegein, the Netherlands. https://library.kwrwater.nl/publication/61396712/. [accessed 27.5.2024].

Koop SHA and Van Leeuwen CJ (2021b) *Indicators of the city blueprint performance framework (CBF).* KWR Water Research Institute, Nieuwegein, the Netherlands. https://library.kwrwater.nl/publication/61397318/.[accessed 27.5.2024].

Koop SHA and Van Leeuwen CJ (2021c) *Indicators of the governance capacity framework (GCF).* KWR Water Research Institute, Nieuwegein, the Netherlands. https://library.kwrwater.nl/publication/61397218/. [accessed 27.5.2024].

Koop SHA, Grison C, Eisenreich SJ, Hofman J and Van Leeuwen K (2022) Integrated water resources management in cities in the world: Global solutions. *Sustainable Cities and Society* **86**, 104137. https://doi.org/10.1016/j.scs.2022.104137 [accessed 01.09.23].

MacKay DJC (2008) *Sustainable Energy – Without the Hot Air.* UIT Cambridge, Cambridge, UK. www.withouthotair.com [accessed 01.09.10].

Makarigakis AK and Jimenez-Cisneros BE (2019) UNESCO's contribution to face global water challenges. *Water* 11(**2**): p.388. https://doi.org/10.3390/w11020388 [accessed 01.09.23].

Meadows DH, Meadows DL, Randers J and Behrens III WW (1972) *The Limits to Growth.* Universe Books, NY, USA.

Millennium Ecosystem Assessment (2005) *Ecosystems and Human Well-Being: Wetlands and Water Synthesis.* World Resources Institute, Washington, DC, USA.

National Infrastructure Commission (2018) *Preparing for a drier future: England's water infrastructure needs.* https://nic.org.uk/app/uploads/NIC-Preparing-for-a-Drier-Future-26-April-2018.pdf [Accessed 17.01.24].

OECD (2010) *Cities and Climate Change.* OECD Publishing, Paris, France. http://dx.doi.org/10.1787/9789264091375-en [accessed 01.09.10].

OECD (2015) *Water and cities: Ensuring sustainable futures. Organisation for Economic Cooperation and Development.* OECD Publishing, Paris, France. https://www.oecd-ilibrary.org/environment/water-and-cities_9789264230149-en [accessed 12.12.23].

OECD (2019) *Making blended finance work for water and sanitation: unlocking commercial finance for SDG 6. Policy Highlights. Organisation for Economic Cooperation and Development.* OECD Publishing, Paris, France. https://www.oecd.org/environment/resources/Making-Blended-Finance-Work-for-Water-and-Sanitation-Policy-Highlights.pdf [accessed 12.12.23].

Pahl-Wostl C, Dombrowsky I and Mirumachi N (2021) Water Governance and Policies. In Bogardi JJ, Gupta J, Nandalal KDW *et al.* (eds.), *Handbook of Water Resources Management: Discourses. Concepts and Examples.* Springer International Publishing, NY, USA, pp. 253–272. https://doi.org/10.1007/978-3-030-60147-8 [accessed 12.12.23].

Peters S, Ouboter M, Van der Lugt K, Koop S and Van Leeuwen K (2021) Retrospective analysis of water management in Amsterdam, The Netherlands. *Water* **13**(8). https://doi.org/10.3390/w13081099 [accessed 12.12.23].

Pinchot G (1947) *Breaking New Ground.* Harcourt, Brace, NY, USA.

Romano O and Akhmouch A (2019) Water governance in cities: current trends and future challenges. *Water* **11**(3). https://doi.org/10.3390/w11030500 [accessed 12.12.23]

SBTi (2023) SBTi Corporate Net Zero Standard. Version 1.1 https://sciencebasedtargets.org/resources/files/Net-Zero-Standard.pdf [Accessed 17.01.24].

Stern, N (2007) *The Economics of Climate Change: The Stern Review.* Cambridge University Press, Cambridge, UK. https://webarchive.nationalarchives.gov.uk/ukgwa/+/http:/www.hm-treasury.gov.uk/sternreview_index.htm [accessed 01.09.23].

TCFD (2017) Recommendations of the Task Force on Climate-related Financial Disclosures. https://assets.bbhub.io/company/sites/60/2021/10/FINAL-2017-TCFD-Report.pdf [Accessed on 17.01.24]

UK Climate Projections (2018) (UKCP18). https://catalogue.ceda.ac.uk/uuid/c700e47ca45d4c4 3b213fe879863d589 [Accessed on 17.01.24]

UKWIR (2007) *Climate Change, the Aquatic Environment and Water Framework Directive*. Water Industry Research, London, UK.

UKWIR (2013) *Updating the UK water industry climate change adaptation framework*, 12/ CL/01/18. Water Industry Research, London, UK. https://ukwir.org/eng/forefront-report-page?object=66618 [accessed 17.01.24].

UNESCO (2021). *The United Nations world water development report 2021. Valuing Water, United Nations Educational, Scientific and Cultural Organization*. https://www.unwater.org/publications/un-world-water-development-report-2021/ [accessed 01.09.10].

United Nations (2015a) *The millennium development report 2015*. https://www.un.org/millenniumgoals/2015_MDG_Report/pdf/MDG%202015%20rev%20(July%201).pdf [accessed 01.09.10].

United Nations (2015b) Transforming our world: the 2030 agenda for sustainable development. *General assembly 70 session* 16301: 1–35. https://www.un.org/en/development/desa/population/migration/generalassembly/docs/globalcompact/A_RES_70_1_E.pdf [accessed 01.09.10].

United Nations (2018) *Sustainable Development Goal 6 synthesis report on water and sanitation*. https://www.unwater.org/publications/sdg-6-synthesis-report-2018-water-and-sanitation [accessed 01.12.23].

UNEP (2017) *Resilience and Resource Efficiency in Cities*, United Nations Environment Program. https://www.unenvironment.org/resources/report/resilience-and-resource-efficiency-cities. [accessed 01.09.10].

UNEP (2021a) *Measuring Progress: Environment and the SDGs*, United Nations Environment Programme. https://www.unep.org/resources/publication/measuring-progress-environment-and-sdgs [accessed 01.09.10].

UNEP (2021b) *Progress on Integrated Water Resources Management. Tracking SDG 6 series: global indicator 6.5.1 updates and acceleration needs*, United Nations Environment Programme. https://www.unwater.org/publications/progress-on-integrated-water-resources-management-651/ [accessed 01.09.10].

UNEP (2022) *What is integrated water resources management?* United Nations Environment Programme. https://www.unep.org/explore-topics/disasters-conflicts/where-we-work/sudan/what-integrated-water-resources-management [accessed 01.09.22]

United Nations (2022) *SDG Indicators. Global indicator framework for the sustainable development goals and targets of the 2030 agenda for sustainable development*. United Nations. https://unstats.un.org/sdgs/indicators/indicators-list/ [accessed 1.2.20]

Van Leeuwen CJ (2017) Water governance and the quality of water services in the city of Melbourne. *Urban Water Journal* **14(3)**: 247–254. https://doi.org/10.1080/1573062X.2015.1086008. [accessed 01.12.10].

Water Framework Directive. Directive 2000/60/EC of the European Parliament and of the Council of 23 October 2000 establishing a framework for Community action in the field of water policy. *Official Journal of the European Communities* **L327**, 22.12.2000, pp. 1–73. Available online from http://eur-lex.europa.eu/en/index.htm

Water Resources Group (2009) *Charting Our Water Future. Economic frameworks to inform decision-making, 2030*. Water Resources Group, MN, USA. https://www.2030wrg.org/wp-content/uploads/2014/07/Charting-Our-Water-Future-Final.pdf [accessed 01.09.10].

WCED (1987) *World Commission on Environment and Development. Our Common Future*. Oxford University Press Oxford, UK.

Dragan A. Savić and John K. Banyard
ISBN 978-1-83549-847-7
https://doi.org/10.1108/978-1-83549-846-020242015

Chapter 13
Digital water supply and distribution

Peter van Thienen
KWR Water Research Institute, The Netherlands

Dragan A. Savić
KWR Water Research Institute, The Netherlands and University of Exeter, UK

Zoran Kapelan
Delft University of Technology, The Netherlands

13.1.   Introduction
13.1.1   Context/motivation

Our world is digitalising rapidly. In many industries and parts of society, including our everyday lives, we experience (to some degree) the benefits of a comprehensive collection of data, of their ubiquitous availability, and, in particular, of the insights that their analysis provides. This improves our ability to make the best decisions (whatever that may mean) based on sufficient information. But there are also other changes in the world. Ongoing population growth and urbanisation, and changes in our climate and the hydrological cycle that probably represent only the start of what is to be expected for the coming decades, put significant stress on water availability and quality, and our water supply infrastructure.

The drinking water industry has started on a journey to digitalise many of its activities – some organisations within the sector aim to digitalise most or all aspects of their activities. Some water utilities have already come a long way and have placed data at the core of most of their processes; others are only just embarking on this journey (Sarni *et al.*, 2019). There are great benefits to be expected, and, indeed, already experienced by the frontrunners, of extensive digitalisation. Digital tools also offer significant help in understanding the stresses that are acting on our water supply systems and choosing how to respond to these. Digital water is here to stay and should be considered in any new system design or existing system expansion, upgrade or rehabilitation.

13.1.2   What-how-why/scope

The digitalisation of water utilities entails *(a)* the creation of digital records of all assets, their history, state and context; *(b)* the continuous collection of data (e.g. through sensors) of operational parameters (technical, environmental, customer-related, etc.); *(c)* the use of models and (machine learning) analysis tooling to evaluate the current state of systems and predict future states; *(d)* decision-making based on the insights generated from these analyses; and *(e)* (partially automated) control of the system based on these decisions. This requires infrastructure for the collection, transfer, storage, making available and analysis of data and control of the system; it also requires interoperability of all components, standards, quality control and, in particular, also a workforce that understands these aspects and is well-trained to function within a digitalised environment.

The expected and experienced benefits of a digitalised water utility are very attractive: through better knowledge and understanding of water supply and distribution systems, including their current state and operational parameters, better decisions can be made on their control and maintenance, but also their design and use under abnormal conditions, potentially resulting in better performance, a higher efficiency, better sustainability and greater resilience. These benefits are often within reach (although they may involve significant re-engineering of existing business processes), but should not be taken for granted, and can and should be evaluated beforehand.

These abstract concepts can be applied to all actors in the water sector and indeed beyond. However, in this chapter, we focus on drinking water utilities and their processes, from source to tap.

13.1.3   Definitions/ontology

In the context of digitalisation, we may talk about smart water and hydroinformatics. In recent years, the focus has been expanded to include water smartness. Water smartness involves recognising and realising the true value of water, and managing all available water resources in such a way that water scarcity and pollution are avoided. Water and resource loops are largely closed to foster a circular economy and optimal resource efficiency, while the water system is resilient against the impact of climate change events (Water Europe, 2016). Digital water provides a set of instruments to achieve water smartness.

Digital water is a term used to describe the penetration and utilisation of data and analysis in all processes within a water sector entity such as a water utility. Closely related are the academic field of hydroinformatics and the practical application of technologies in smart water management. In this chapter, we use the definitions proposed by Van den Broeke et al. (2019). An ontology for these terms is presented in Figure 13.1.

Hydroinformatics focuses on developing tools to support decision-making at all levels, from operations to governance and policy, through to (data) automation, optimisation and control. Examples of hydroinformatics applications include studies, such as river basin analyses, data mining for identification of unknown substances and modelling of water quality and quantity – for example, in the aquatic environment, water production, supply and reuse processes or in water-related health studies.

Smart water management, in turn, is more practice-oriented than hydroinformatics; it focuses on making the most of water resources by leveraging data for action, while drawing on the field of hydroinformatics for its solutions. Smart water management applications aim to increase the efficiency, effectiveness and resilience of water systems – that is, doing more with less (money, time, and/or resources). Examples of applications include process control (real-time, near real-time), resource planning based on measurement data (medium term) and system design (long-term goals).

13.1.4   Chapter objectives

Digital water is very broad, almost all-encompassing. In this chapter, we aim to touch upon all the relevant aspects mentioned in the previous sections. It is impossible to discuss all these thoroughly and comprehensively in a single chapter. It is our intention to provide an overview and a starting point for a more thorough investigation through the references cited in this chapter.

Figure 13.1 Ontology of digital water, which encompasses the academic field of hydroinformatics and its practical application in smart water management (From Van den Broeke *et al.* (2019))

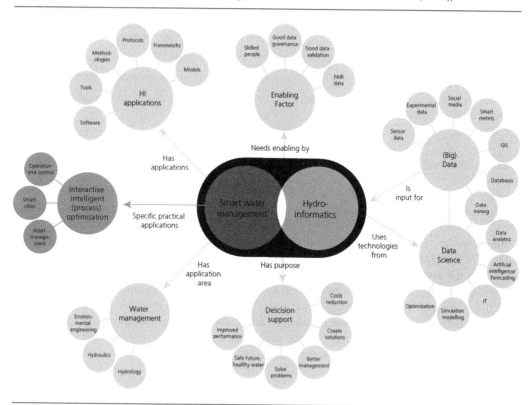

13.1.5   Chapter layout

After this introduction, we continue by briefly sketching the history of digitalisation in drinking water utilities and the current state of this development in Sections 13.2 and 13.3. This is followed by a concise discussion of the elements, technical and non-technical, physical and abstract, that make up smart water systems, in Section 13.4. Moving from theory to practice in Section 13.5, we discuss a number of use cases with examples from practice. The question of why digitalisation may be a good idea and how to answer this, or in other words the construction of financial and other business cases for smart water, is discussed in Section 13.6. The digitalisation of the water sector is not complete yet; Section 13.7 offers a vision of what the industry is moving towards. This chapter ends with a number of guidelines for implementation in Section 13.8 and concluding remarks in Section 13.9.

13.2.   History (brief)

Over a period of more than two thousand years, incremental technological developments have laid the groundwork for modern drinking water treatment and distribution systems (Walski, 2006, gives a comprehensive overview of the latter's history). Mirroring the four industrial revolutions of the past three centuries (Bufler *et al.*, 2017, see Table 13.1), the water industry has adopted many new technologies and applied them in our water supply systems at an increasing pace (although historically often lagging behind the original industrial revolutions).

Table 13.1 Overview of industrial and associated water industry revolutions (Based on Bufler *et al.* (2017))

	Industrial revolution	Water industry revolution
1st	steam/water-powered mechanisation	steel-enabled handling of high pressures
2nd	electricity-powered mass production	pumps and turbines
3d	IT-facilitated automated production	sensoring and modelling of water systems
4th	permanent availability and analysis of data in cyber-physical systems	interfacing of reality and models in cyber-physical systems

The latter two revolutions provide the basis for Smart Water Management, also called Water 4.0.

Starting with the introduction of programmable logic controllers (PLCs) to control processes, paralleled by the application simulation models of hydraulic systems from the 1980s (Walski, 1990), the 1990s and 2000s have shown an increase in the availability, interest and application of online sensors in water treatment plants and distribution networks and SCADA systems to supervise and control them (Jentgen and Wehmeyer, 1994, Ecob *et al.*, 1995, Storey *et al.*, 2011). At the same time, from the 1990s, the academic field of Hydroinformatics, which studies the application of information and communication technology on water systems in practice, rapidly developed. From the 2010s, with the maturation of 'classical' analysis algorithms and machine learning becoming ready for practical application, a comprehensive view of the application of these techniques under the flag of Smart Water Management became popular. Currently, many water utilities worldwide are in varying stages of digitalising their operations (Sarni *et al.*, 2019).

13.3. Current state

Drinking water distribution systems (DWDS) come in different configurations and sizes. Most modern DWDSs are complex systems supplying water to wide ranging customers (residential, commercial, industry, etc.) over a substantial geographical area.

A layered representation of a DWDS is shown in Figure 13.2. The first layer represents the physical assets or infrastructure that comprises water treatment works, supply trunk mains, and large pipe networks with tanks, pumps, valves and other equipment used to distribute water to customers. This layer is, of course, present in all DWDSs although with varying degrees of sophistication. This layer also includes the knowledge of the layout and history of the physical system (to the degree that this is available), such as records or maps of the pipe system. Layers 2–5 represent different elements of smartness added to a conventional DWDS.

Layer 2 denotes the sensing and control aspects of a smart DWDS and layer 3 represents the corresponding data collection and communication system. Sensors and actuators are added to the DWDS to improve its performance (see Section 13.4.3). This started with the introduction of the first supervisory control and data acquisition (SCADA) systems in the 1980s. However, early SCADA systems had only a small number of sensors and actuators installed at most important points in a DWDS, such as water treatment works and pumping stations. Over the years, different sensors, including low-cost pressure loggers, have been implemented on a larger scale and have now made their way into traditionally poorly observed pipe networks. Sensing equipment has improved a lot lately in terms of new and different types of sensors developed, their measurement accuracy, temporal frequency and wireless transmission capabilities.

Figure 13.2 Layered representation of a Smart DWDS (Source: SWAN Forum, https://swan-forum.com)

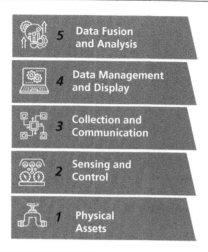

Having said the above, spatial density of pressure and other sensors is still rather low in most DWDSs. In addition, many modern DWDSs still do not make use of smart demand meters (see also Section 13.4.3.1) even though these enable monitoring water consumption, the main driving force of DWDSs, at high spatial and temporal resolutions, and distilling understanding of its components and variability. Notably, water quality sensors are still not widely used and are largely confined to water treatment works (WTW).

Regarding data collection and communication, most modern sensors make use of mobile phone company services (SIM cards) to transmit data regularly to a company control room or wherever else the data is being collected and stored. However, this is not always the case and many DWDSs are still collecting data by other means – for example, manually (data recorded on a device in a field brought into the office manually) or by using vehicles with installed readers that collect data periodically when the vehicle passes by the sensor location (e.g. smart demand metering data collected weekly using automatic meter reading (AMR) technology).

Finally, new smart actuators are being developed too (see Section 13.4.3), but their development and implementation are lagging behind the sensors. As a result, most smart DWDSs have a rather limited level of automation nowadays that is usually focused on addressing some specific issue (e.g. leakage detection or consumption monitoring).

Layers 4 and 5 in Figure 13.2 represent software that accompanies the smart DWDS hardware represented by layers 1–3. The data collected by using new sensors or otherwise is presented to decision makers in a control room or elsewhere by means of software that manages the data and displays it on a screen (layer 4). This capability was introduced with early SCADA systems but the level of sophistication has improved in recent years as the quantity and quality of sensing and other data have increased. Having said this, this type of software still relies mainly on a human operator to analyse the data in order to extract useful information that can be acted upon and this is an increasingly difficult, if not impossible task when data is collected using a large number of sensors.

This is addressed by developing new methods and tools for automated data fusion and analyses (layer 5) which recommend actions to operators. These methods and tools represent the real smartness of a DWDS, when used together with a human operator. Various methods and tools have been developed (Section 13.4.4) but despite some successful commercial stories, many of these seem still confined to the research/academic domain. In addition, the integration of new methods and tools into an existing company IT system has proven to be complex – that is, time consuming and expensive, hence not many DWDSs can say that they have achieved this goal.

In summary, water companies have come a long way in terms of making the operation, maintenance and planning of their systems smarter. However, the level of smartness is wide ranging with different DWDSs being at varying stages of maturity and implementation of smart technologies and solutions. The implementation of different smart aspects is still rather patchy and, in most cases, driven by specific issues and needs.

13.4. Smart DWDS elements

13.4.1 Introduction

This section describes several elements that are most commonly found in a smart DWDS, focusing on the current state of the art and emerging technologies. The aim is to briefly introduce these elements without going into too many details. Interested readers who want to learn more about these are encouraged to look at various references provided in this section.

This section starts with a description of data and its importance in smart DWDS. After this, sensors and actuators (i.e. hardware) are introduced, followed by the introduction of methods and tools (software) that are used to process the data collected. The section concludes with a brief description of the organisational aspects of DWDS.

13.4.2 Data

Data is at the centre of a smart DWDS. Making DWDS smarter is effectively achieved by (a) collecting additional data, especially high frequency and accuracy, temporal and spatial data that was not available before; (b) analysing this data together with existing data to extract useful information; and (c) using the new information to make better operational and planning decisions.

Most DWDS utilities collect a lot of different data already. Examples of more conventional data sources include customer data (bills, complaints), data about company assets (WTW, pipes, tanks, pumps and pumping stations, valves, inspection and other vehicles, etc.), works management data (records of pipe bursts, equipment and other failures, together with corrective interventions undertaken), operational/SCADA system data (near real-time about pressures, flows and water quality plus the status of key control elements such as valves and pumps) and other data (e.g. leakage data, etc.)

However, despite the amount of effort that has been going into data collection, multiple issues still exist. Data is often of insufficient quantity or not collected at all (e.g. smart demand metering data not used to collect high spatial and temporal water consumption data in many utilities). Collected data is largely static in nature and when more dynamic, near real-time data is collected this is often done with a limited number of sensors used. Data completeness and quality issues are present in most data sources. Finally, data collected is often just stored or analysed manually but only in a small number of cases, typically following some failure event. This led to a current DRIP situation (data rich, information poor) in many water utilities.

The above data issues can be addressed in smart DWDSs by collecting new, high frequency, accurate data by using the latest sensors (Section 13.4.3; note, although the costs of these continue beyond their installation and will need to balance the benefits (Section 13.6)) and especially by processing this data using sophisticated methods and tools (Section 13.4.4) to extract useful information. Note that this does come at the const of installation, maintenance and end-of-life replacement.

As mentioned above, the current data situation in smart DWDSs is still far from the ideal one. The same applies to the data and other related standards as well. Different approaches are used to collect, transmit, check, validate, use and store data. Relevant standards for all this are, apart from a few exceptions, largely missing. Most of the emerging products and services for smart DWDSs are a result of individual, uncoordinated initiatives that often end up handling similar data sets in very different ways. This does not help to deal with the above data issues; nor does it encourage, let alone facilitate, data sharing between utilities, technology developers and other stakeholders involved. There is an urgent need to standardise data handling in smart DWDS, for everyone's benefit.

13.4.3 Sensors and actuators

13.4.3.1 Sensors

Sensors have been used in smart DWDSs to collect various data that is subsequently converted into useful information by different methods and tools. As such, sensors are one of the key building components of a smart DWDS. A whole range of sensors have been in use in DWDSs for a long-time now – for example, various pressure loggers/transducers, flow meters and different water quality sensors typically used at WTW (e.g. pH, turbidity and many others). These conventional sensors have a decent measurement accuracy and sampling frequency already but lack the ability to transmit the data collected to company headquarters and/or to do some local data processing. The remainder of this section provides examples of newer, smarter sensors that have these capabilities. Note that these are still only examples shown to illustrate different advanced capabilities of smart sensors – that is, there is no intention to provide a more systematic list of these devices or related information.

13.4.3.1.1 ADVANCED PRESSURE SENSOR (INFRASENSE LABS, 2022)

This is an example of a modern pressure sensor. The new InfraSense TS (Transient Signals) technology enables an automated time synchronisation of data collected at multiple sites which is a challenge with existing sensors, especially when large volumes of continuously acquired high-speed data are collected. When combined with suitable methods this sensor can effectively control and manage the dynamic pressure in a given area resulting in calmer pipe networks with fewer bursts and background leaks.

13.4.3.1.2 SMART DEMAND METERING

These sensors are essentially specialised low flow meters that collect information on water consumption at the individual property level with high temporal resolution (typically minutes). This, in turn, enables the characterisation of water consumption with both high spatial and temporal resolutions, which is a critical input for a number of DWDS decisions to be made at operational, tactical and strategic levels. For comparison, conventional consumption meters record only a volume of water used between two manual readings, usually done every one to six months.

Different smart demand metering technologies exist (Arad Group, 2022). The AMR technology, refered to earlier, collects consumption and other data from property water meters and transfers this data to a central database. The data collection is often done using vehicles fitted with suitable readers that collect data when passing by a property. The newer, advanced metering infrastructure (AMI) technology communicates with metering devices either on request or on a schedule to measure, collect and analyse water consumption data. Whereas the former approach is merely suitable for billing purposes and decision making at tactical and strategic levels, the latter is also useful for day-to-day operational control.

13.4.3.1.3   ACOUSTIC SENSING

These sensors are typically used to detect leaks in a DWDS. The technology behind these sensors is based on fibre optic cables that are installed alongside or inside pipes thus ensuring high spatial coverage. The fibre optic cables continuously monitor vibrations in pipes. When combined with sophisticated AI methods, these sensors can be used to detect and locate pipe bursts and leaks as these leave a specific imprint on monitored acoustic signals.

13.4.3.1.4   WATER QUALITY SENSORS

Water quality analyses, especially the more complex ones, are traditionally conducted by grabbing samples in the field and analysing them afterwards in a laboratory. This, however, is time consuming and expensive, and it does not allow for near real-time water quality data collection that is essential for smart DWDSs – for example, for detecting the presence of contaminants or effectively controlling a level of disinfectant (chlorine) in a pipe network. To address this and other DWDS issues, a number of new water quality sensors have been developed (CETaqua, 2013; US EPA, 2009), including the ones that are capable of detecting a range of contaminants by analysing the light properties of water (Williamson *et al.*, 2014).

13.4.3.1.5   MOBILE SENSORS

Most sensors that are currently used in or developed for the DWDS are based on the concept of monitoring specific parameters of interest at a fixed location in the system. An alternative to this is to build a mobile sensing device – for example, a robotic type device that can move around the pipe network and collect the data using various sensors located at it. KWR in the Netherlands has been working on such a robotic inspection device in collaboration with a Dutch technology development company and seven water utilities in the Netherlands (Van Thienen *et al.*, 2018). The aim is to build a device that can 'live' in the pipe network, move autonomously, and collect and transmit visual and other data. The robot is designed to access (almost) all pipes with diameters in the range of 90–315 mm, which represents two-thirds of the length of the Dutch drinking water networks.

13.4.3.2 Actuators

Collecting data by using sensors is only half of the smart DWDS story. Once the data is collected, it needs to be processed using suitable methods and tools (Section 13.4.4) to extract the useful information that can be acted upon. This is where the actuators come in as remotely controllable devices that can be used to modify the system state and operation based on this new information.

Examples of conventional actuators involve various pumps (fixed and especially variable speed pumps) and valves (e.g. pressure reducing valves – pressure reducing valves (PRVs), time-controlled valves, etc.). These can be controlled manually or automatically/remotely – for example, by way of PLCs that are part of a SCADA or some other control system.

Actuators can be controlled by using some form of central intelligence (e.g. method) or locally, by using simple control rules. An example of the latter is when pumps are switched on and off depending on the water level in the controlling tank.

Smart actuators are more advanced as they have some form of smartness added to them. An example of such a device is the i2OWater's smart PRV for effective pressure management (i20Water, 2022). Unlike conventional PRVs which either have a fixed setting or time-controlled settings (e.g. different for day and night or each hour of the day), the smart PRV is effectively a flow controlled PRV that automatically adjusts its setting based on a number of variables that are observed in near real-time (flows and pressures upstream and downstream of the valve and the pressure at target location). In this way, multiple benefits can be achieved, including reduced future burst rate and related customer interruptions to supply, reduced water loss, optimised pressure in the network resulting in a calmer network, reduced energy and operational costs, and improved customer service. Other smart actuators exist or are being developed.

In summary, actuators are an important building block of a smart DWDS as without these devices it would not be possible to control the DWDS effectively. The area of actuators is relatively underdeveloped when compared to sensors but this is likely to change in the future as true automation of a DWDS is not possible without these devices.

13.4.4 Methods and tools
13.4.4.1 General
These methods are more general in nature and can be used under both normal and anomalous operating conditions in a DWDS.

13.4.4.1.1 REAL-TIME CONTROL (RTC)
One of the goals of a smart DWDS is to automate its operation and maintenance. A number of methods shown in the remainder of this section fall into this category, both for controlling the DWDS under normal (e.g. real-time pump scheduling, e.g. Ulanicki et al, 2007) and anomalous operating conditions (e.g. real-time detection and location of pipe bursts, e.g., Mounce et al., 2002).

Theoretical methods for RTC DWDS have been around for a while (Brdys and Ulanicki, 1994) which coincided with the appearance of the first SCADA systems. These, however, failed to be taken up by the water utilities at a time apart from some local control based approaches where individual devices such as pumps as controlled independently – for example, by using water levels in a nearby tank to decide when to switch on and off a pump. Meanwhile, more complex RTC methods have been developed (see a review in Creaco et al., 2019) but the large scale take up of these methods is still rather patchy, especially when it comes to the adoption of more sophisticated models that make use of predictive model control and across larger geographical areas.

Real-time control of DWDS can be undertaken with different goals in mind – for example, leakage control, pump scheduling, water quality control (e.g. chlorine management within limits) and other reasons. An integral part of RTC control methods is sensor data validation (Branisavljevic et al., 2011).

13.4.4.1.2 DIGITAL TWIN
A digital twin is an important general concept and a tool for smart DWDS. A digital twin is a digital replica of a real DWDS built as a virtual testing ground for the implementation of new smart methods, technologies and solutions.

Most digital twins of DWDS currently exist in a form of a software tool. Typical applications include (see next section for details)

- real-time DWDS state estimation by assimilating sensor measurements with simulation model predictions
- making use of forecasted demands and anomaly detection (e.g. detection of bursts/leaks)
- scenario testing (e.g. to measure the impact of different Covid-19 measures on water consumption (KWR, 2022)).

Digital twins are also becoming increasingly popular in the area of cybersecurity where they can be used to test how a DWDS would respond to different types of cyberattacks (Hassanzadeh *et al.*, 2020).

13.4.4.2 Methods for DWDS operation under normal operating conditions

13.4.4.2.1 PRESSURE MANAGEMENT

This set of methods aims to manage the pressure, especially in DWDS where spatial variations of pressure exist. Most of the methods developed make use of different types of PRVs to reduce the pressure in parts of the network where the pressure is excessively high. This way networks are calmed down resulting in multiple benefits (see Section 13.4.3.2). A number of both academic (Ulanicki *et al.*, 2000) and commercial solutions (i2OWater, 2022) already exist.

13.4.4.2.2 DWDS STATE ESTIMATION

State estimation methods aim to predict system state (pressure, flow, water quality) in near real-time at locations where sensors do not exist. This is done by using methods such as different versions of Kalman and particle filters that are based on the concept of data assimilation (Hutton *et al.*, 2014). In data assimilation, simulation based model predictions are corrected by using incoming observations from the existing sensors. This can be done for both water quantity type variables (Hutton *et al.*, 2014) and water quality variables (Anjana *et al.*, 2019). These way more accurate predictions of system states are made at locations where sensors exist.

13.4.4.3 Methods for DWDS operation under anomalous operating conditions

The focus of the methods presented in this section is on predicting or detecting and locating various anomalous (failure) events, such as pipe bursts and equipment failures, and deciding on how best to respond to these events. The outputs of these methods are often used to inform ongoing asset management and ultimately improve customer service.

13.4.4.3.1 RESPONSE TO FAILURE EVENTS

Once pipe bursts, equipment failures and other similar events are detected and located, a decision needs to be made on how best to limit the damage and subsequently recover from these events. The former requires, often, to act quickly under stressful conditions, hence methods have been developed to assist operators in making relevant decisions – for example, how best to isolate the incident (Jun and Loganathan, 2007) or what is the best set of interventions to apply shortly afterwards to further limit the negative impact (Mahmoud *et al.*, 2018, Nikoloudi *et al.*, 2020).

13.4.4.3.2 DETECTION, LOCATION AND MANAGEMENT OF CONTAMINANTS

A number of methods have been developed so far to detect contamination events in a DWDS, both intentional and accidental ones. Methods developed so far include methods to detect these events, raise the alarms (Perelman *et al.*, 2012), track down the contamination event origin (Sanctis *et al.*, 2009) and decide where best to locate the sensors in the DWDS to ensure timely and effective detection of these events (Weickgenannt *et al.*, 2010).

13.4.4.3.3   METHODS FOR IMPROVED ASSET MANAGEMENT

These methods include methods that make predictions of future pipe burst rates (Kleiner and Rajani, 2001) as a function of different explanatory factors such as pipe characteristics (diameter, material, age), soil characteristics, road conditions and weather. These methods are then used to develop improved evidence-based asset management plans. Another more recent example of a different method involves the visualisation of buried water mains by using augmented reality. This, in turn, enables the technicians involved in the maintenance and repair of pipes to more easily locate relevant assets and hence save time and money.

13.4.5   Organisational aspects

Smart water management relies heavily on technology, but the human element remains essential. From a practical perspective, Sarni *et al.* (2019) conclude, based on multiple case studies, that the effective operation of a smart water solution requires three primary mechanisms to operate. These are engaged leadership, developing dedicated roles and using a bottom-up approach to weave digital projects throughout the utility infrastructure.

These are, to a significant degree, reflected in the experiences of Singapore's Public Utilities Board that have been written down in Ng *et al.* (2020). These authors stress the importance of having 'a pervasive culture that embraces changes', support by its employees both in spirit and execution at all levels of the utility and digitalisation and data analytics training matching individual employees' responsibilities and job scope.

Drawing on experiences from the automotive and aviation industries, Savić (2021) concludes on the relation between automated systems and humans that *(a)* automation requires humans in the loop as an ultimate failsafe; *(b)* human operators need to be trained to understand the capabilities and limitations of systems; and *(c)* automation requires a fallback manual option to provide the possibility of intervention in case an automated system is incapable of dealing with an (unanticipated) situation.

13.5.   Use cases and examples
13.5.1   Introduction

Smart water approaches lead to data-driven applications in water distribution systems for *automatically* extracting value from the data generated by the deployed metering and sensing devices. Research into artificial intelligence (AI) methods, advanced modelling techniques exploiting the topology and physical laws governing water distribution systems, and their hybrids led to a plethora of promising approaches and methods. The number and sophistication of the implementation of those methods in practice are still growing. The following are some of the use cases and example applications.

13.5.2   Operational
13.5.2.1 Demand forecasting

Prediction of water consumption at various temporal and spatial scales is one of the key smart water network applications for drinking water utilities to support operational, tactical and strategic decisions. However, demand forecasting is also one of the most difficult tasks as flow monitoring in distribution systems is still sparse, customer water meters are not read often, sometimes only once per year, and bulk water meters with a higher time resolution (e.g. every 15 minutes) are normally only available at the entry into large distribution areas (e.g. demand management areas (DMAs) and pressure zones). Short-term demand forecasts at 15-minute or hourly resolutions are used to enable the effective operations of a water distribution system, while daily or weekly forecasts are needed for planning the operation of water treatment works to meet system demands (Bakker *et al.*, 2013, Donkor *et al.*, 2014, Romano and Kapelan, 2014, Xenochristou *et al.*, 2020). In addition to standard statistical methods for time series analysis used for short-term demand forecasting, machine learning methods (e.g. Artificial Neural Network and Support Vector Machines) have often been shown to have higher accuracy than time-series approaches (Antunes *et al.*, 2018).

13.5.2.2 State estimation
State estimation refers to inferring the flows and heads in the DWDS at all locations based on measurements collected from limited monitoring locations. State estimation can be formulated as an inverse optimisation problem based on the knowledge of the topology and physical laws (e.g. continuity and energy balance) governing the hydraulic behaviour of the system. The objective of the optimisation is to find the best match between the model predictions and measurements (Andersen and Powell, 2000). However, this approach typically requires more information than is available to achieve the required quality and uniqueness of state estimation. A hybrid approach in which physical laws and a machine learning method based on the topology of the network (graph neural network) are combined is a promising avenue for data assimilation and state estimation in water distribution systems (Xing and Sela, 2022).

13.5.2.3 Anomaly detection
Anomaly or novelty detection refers to the automatic identification of novel or abnormal patterns embedded in large amounts of 'normal' water quality and quantity data (Mounce *et al.*, 2011; Riss *et al.*, 2014). Model-based or physics-based approaches that are often used for anomaly detection rely on highly calibrated hydraulic models, which are often difficult to obtain due to limited field measurements and modelling uncertainties (Hu *et al.*, 2021). Data-driven approaches, however, do not require a hydraulic or a water quality model and have been used extensively to automatically, quickly and accurately identify abnormal patterns in the available data. They often start by forecasting the near-future sensor signal values using machine learning technology. These values are then compared to the incoming observations to assess potential changes in normal patterns and identify an anomaly. An example of the development and application of such anomaly detection is United Utilities' Event Recognition in the Water Network (ERWAN) system for leak and abnormal demand detection (Romano *et al.*, 2020). The ERWAN system uses pressure and flow sensors to measure and monitor the state of water distribution systems. A similar approach can be used for identifying abnormal patterns in water quality measurements caused by intentional or non-intentional contamination of the distribution system (Dogo *et al.*, 2019).

13.5.2.4 Leak localisation
Even in small areas within a water distribution network – for example, DMAs – the typical total pipe length is measured in tens of kilometres. Combined with the fact that most leaks do not immediately appear on the ground, it is difficult to localise a leak to a smaller area for it to be found and repaired. A combination of model-driven and data-driven approaches is still the most promising approach to solving this challenge even with a low density of sensors (Sophocleous *et al.*, 2019; Romano *et al.*, 2013).

13.5.2.5 Water treatment and desalination
Removal of chemicals, particulates and organic materials from the abstracted water is a complex process happening in a water treatment plant that results in safe potable water for distribution. Modelling desalination and water treatment processes and optimising the operations of

the plants have become the subject of model-driven and data-driven research and applications. Model-based methods and feedback control systems are used traditionally to adjust real-time operating parameters based on online water quality monitoring. These models are generally based on strict assumptions that are difficult to adhere to in practice or contain parameters that rely on real-time measurement (Li *et al.*, 2021). Machine learning (artificial neural networks) and nature-based computing (genetic algorithms) have shown promise for a wide range of water treatment and desalination technologies, covering optimisation of ions and pollutant removal, performance and membrane properties as well as cost and efficiency (Al Aani *et al.*, 2019).

13.5.2.6 Energy and water quality management
Water utilities are one of the major users of energy for treatment and distribution of potable water. Minimising operational costs while guaranteeing continuity of water supply and good water quality is one of the key challenges. Various model-based and data-based solutions have been used to optimise operations of pumps and valves in water distributions systems (Mala-Jetmarova *et al.*, 2017). This is often achieved by combining a short-term demand prediction model (e.g. Romano and Kapelan, 2014) with an optimisation model to determine the best pump and valve operation (Abdallah and Kapelan, 2017).

13.5.3 Tactical and strategic
13.5.3.1 Asset management
Asset management models for buried water mains are needed to estimate the optimal renewal policy based on structural deterioration and life cycle costing (Engelhardt *et al.*, 2003). Physical, environmental and operational mechanisms cause deterioration and ultimate failure of these assets. A number of deterioration models rely on machine learning methods, such as artificial neural networks (Harvey *et al.*, 2014) and evolution computing methods (Berardi *et al.*, 2008), to name but a few. They either predict failure behaviour or lifespan of infrastructure. Asset replacement plans balancing investment with expected benefits within a risk-based management framework can then be obtained using multi-objective optimisation tools (Giustolisi and Savic, 2006).

13.5.3.2 Water resource planning and management
Due to increasing pressures on water resources from climate change, population growth and more stringent environmental regulation, risk-based water resource planning is a must to deliver a resilient water supply system in the future (Hall *et al.*, 2020). The approach should deliver both a sequence of investments over a long time horizon and operational policies for shorter-term decision making. While Hall *et al.* (2020) provide a framework for developing the former, an example of the latter using a multi-objective genetic algorithm is given by Morley and Savic (2020).

13.5.3.3 Water distribution network design
Water distribution networks are highly complex, large-scale, long-lasting and capital-intensive assets, which are developed in stages over decadal timescales (Cunha *et al.*, 2019). This implies that there is a high degree of uncertainty (e.g. with long-term demand changes) associated with the design and performance of these systems over the long time horizon. Therefore, the design of such systems has to consider robustness (the ability of a solution to satisfy as many future scenarios as possible) and flexibility (the ability to implement the next-stage solution while keeping a view of the long-term system development) of potential solutions. Nature-inspired computational methods (e.g. simulated annealing, evolutionary computing, ant-colony optimisation) are often used to deal with network design challenges as they naturally deal with discrete variables (e.g. pipe sizes), non-linearity and the multi-objective nature of the problem (Basupi and Kapelan, 2015).

13.5.3.4 Sensor network design

The proliferation of sensor technologies and their increasing use in water distribution networks has led to various methods for the optimal placement of sensors in these systems (Ostfeld *et al.*, 2008). The aim of the sensor network design may be to facilitate the detection of intentional or unintentional contamination of the system or to enable the detection of leaks. Evolutionary computing methods are often used to solve this complex, discrete optimisation problem (Quintiliani *et al.*, 2020).

13.6. Business cases

13.6.1 Introduction

The plethora of technological constructs that fill the smart water toolbox promise to address a range of problems, as described in the previous sections, but they come with a cost of their own, in terms of budget, labour (requiring highly skilled workers and/or potentially displacing workers), materials, energy use and so on. Questions have been raised about whether SMW's focus on technological solutions provides a moral dilemma in the sense that it embodies and promotes a belief in technological fixes that hides the underlying issues of overexploitation of resources and sidesteps discussion on these (e.g. Hartley and Kuecker, 2020). This may be true in some cases, but certainly not in all, as smart water management techniques may actually contribute to recognising and preventing overexploitation (e.g. Morley and Savić, 2020).

It nevertheless seems prudent to evaluate the actual returns from smart water solutions, in terms of social, ecological and economic parameters. These need to be compared to the means invested in realising the solution, but also to alternative spending of the same resources (e.g. rehabilitating a pipe network rather than creating DMAs for leak detection and performing ad-hoc repairs – see Ahopelto and Vahala, 2020).

13.6.2 Approaches and case studies

The financial aspects of non-digital parts of water supply systems are straightforward and need no further discussion in this text. Work has also been done on the sustainability and social aspects of the infrastructure and its operation at various levels of conceptualisation (e.g. Makropoulos and Butler, 2010; Lemos *et al.*, 2013; Behzadian and Kapelan, 2015).

However, up to the present, there seem to be surprisingly few quantitative reports on the returns of smart water management. The International Water Resources Association (IWRA) and K-Water's report (2018) describes a range of case studies in the field (note, however, that the scope of this report is broader than that of this book chapter). The report lists a number of drivers of smart water management, including environmental, economic, social, and technological. Analysing the case studies presented in the report, its authors conclude that 'a financial return on investment is likely to occur in the medium-to long-term', but also argue that it is important to move beyond simple cost-benefit analysis, as there are several additional benefits that are achieved.

Examples of concrete quantifications do exist, however. Monks *et al.* (2021) provide a framework for assessing how digital water meters may contribute to achieving sustainability objectives. This includes an elaborate taxonomy of economic, social and environmental benefits and a stochastic modelling tool for quantitative assessments and monetisation. These authors applied it to a large metropolitan water utility in Australia, illustrating the potential for significant savings in energy, water abstraction and the associated costs.

But the greatest strength of smart water management is not in making the best business case, but to design and operate the best performing system (in terms of reliability, level of service, efficiency, costs, energy consumption, etc.) within the financial, ecological and other boundary conditions that are prescribed by a water utility and/or society (e.g. Morley and Savić, 2020).

As the world's water utilities gain more experience with smart water management and the mid- to long-term time horizons mentioned above are passed, the financial benefits of smart water management will become more clear. The non-financial aspects, such as reduced stress on the water sources and the ecologies that they are part of that may be achieved by smart-metering-fueled water savings and numerically optimised abstraction schemes, provide an immediate return on investment.

13.7. Vision

Many ambitions and technologies have been described in the previous section. But what if these are combined into a comprehensive whole, creating a fully digitalised water utility, without eliminating the human factor, nor indeed by exploiting the capabilities of the employees to the fullest?

Such a water utility would have

- sensors to monitor all relevant aspects of the infrastructure and the medium that is treated and transported by the infrastructure
- systems and procedures to register all changes to the infrastructure that are affected by the water company's employees
- reliable and secure ICT systems to communicate, store, validate, and most importantly, make available the data that is gathered by the sensors to all employees of the water company
- a model or models of the infrastructure that assimilate in real time the data that is collected by the sensors and that is also kept up to date with respect to changes in the infrastructure – let us call this a digital twin – that fills in the blanks between sensors and is capable of predicting the system state in the immediate future for operational purposes and the farther future for a diverse range of activities such as design studies, incident preparedness, and resilience evaluation
- a set of algorithms that run in real time to assist the operators in monitoring and understanding the operation of the system and alert them to anomalies in quality, quantity, and infrastructure state or condition
- algorithms that run in real time to predict the near future state and condition of assets to assist asset managers in infrastructure maintenance
- algorithms that run in real time to detect incidents (e.g. contaminations or bursts) and intervene by controlling actuators in the network, under human supervision
- an infrastructure that is numerically designed and optimised, using the available data and models, to be positioned in the ideal compromise, chosen by the water company, between efficiency, sustainability and resilience
- a workforce that is trained to understand, at least in a qualitative way, the data that is collected, the algorithms that interpret the data and the uncertainties associated with both
- an analytics toolbox that allows all water company employees to perform additional, non-standard analyses on all the data that is available.

In other words, in such a company, operators would be aware of the state and operation of the system to a maximum degree, allowing all decisions to be based and optimised on the appropriate set of information, with humans having the final say. Managers would also be in a position of optimal information availability to make strategic and tactical decisions that contribute to the efficient, sustainable and resilient operation of the water company in the future.

Such a full digitalisation is not a panacea for all the challenges that water companies will face in the coming decades. It does not eliminate the changes in the hydrological cycle caused by climate change, or the degradation of infrastructure by components reaching their end of life. Planet Earth will continue to put bounds on what is possible, and the human factor, including population and behavioural dynamics, policies and politics, will continue to pose and shape challenges, but the digital tooling and full information availability will put the water utility in the best possible position to deal with these.

In such a fully digitalised water company, managers and workers will need to remain alert to the fact that the systems that they rely on, both their internal digital infrastructure but also the services (power, data) and supply chains that they depend on, are not infallible and are, themselves, vulnerable to adverse developments in our environment and society. Fallback options should remain available, including the knowledge in the workforce to apply them.

13.8.   Guidelines for implementation

This chapter presents guidelines for the transition to a digital water future – that is, the implementation of a smart DWDS. The aim is to provide some basic guidance that a water utility should think about when planning a transition to a digital water future. The journey to a digital water future is likely to be an incremental one and different for different water utilities, hence customised roadmaps should be developed. Still, some common principles and considerations exist. Examples of these (not a full list) are shown here below, to aid in creating individual company roadmaps.

Define clear long-term objectives

This is a starting point in the journey to a digital water future. The objectives are identified by analysing the corresponding long-term pressures, drivers and needs. Note that objectives are likely to be different for different water utilities (e.g. leakage reduction, improved water quality, increased resilience to climate change, etc.) as will be the corresponding starting points (i.e. existing levels of digital water maturity – Sarni *et al.*, 2019) and a whole range of other factors affecting the transition (e.g. utility capacity to adopt new technologies, level of funding available, different company setups – private as against public, etc.).

Identify promising technologies and solutions

Once the long-term objectives are identified, water utilities should identify existing and promising new technologies that can help address their needs. The description of the current state of the art and emerging technologies presented in Section 13.4 can assist with this. The chapter not only provides information on both hardware (sensors and actuators in Section 13.4.3) and software (methods and tools, Section 13.4.4) type solutions but also the related data (Section 13.4.2) and organisational aspects (Section 13.4.5). Identifying the right technologies for the given objectives is not an easy task as competing technologies and solutions often exist (e.g. leak detection by using pressure/flow or acoustic sensors, etc.). In addition, identifying solutions that are fit for purpose is often a difficult task that requires specialist knowledge and insight into the latest technologies and trends. The field of smart water technologies is developing fast and what is state-of-the-art today may become obsolete in the future.

Assess new technologies and solutions before large-scale implementation

Most technology developing companies that offer the latest smart water technologies are happy to do pilot projects at a water utility. These projects can be used to manage the risks associated with adopting new smart technologies by exploring and understanding better these technologies in terms of costs and benefits for the utility business and customers, all before full-scale implementation. Pilot projects can also help build momentum for the transition to a smart DWDS through effective demonstration of new technologies.

Put the data at the centre of new technology implementation

Collecting, storing and processing data is central to the implementation of new technologies. Improved data infrastructure and related data management and use can be achieved by a number of different means in a water utility ranging from upgrading existing IT systems to introducing organisational changes – for example, clearly defining responsibilities with respect to data quality, availability and use. The aim is to encourage and empower company staff to use the data daily to improve the operation, planning and general management of a DWDS.

Identify and remove potential technology implementation barriers

These include barriers such as interoperability, regulations, culture and cybersecurity (Sarni *et al.*, 2019). Identifying and working on removing these barriers as early as possible is critical for the implementation of new technologies in smart DWDSs.

Create the right environment that supports the implementation of new technologies and solutions

This can be achieved by a number of different means – for example, by setting the suitable ambition at the chief executive officer and company board level, building a clear business strategy that supports the implementation of new technologies and solutions and/or building an innovation culture that fosters the implementation of new technologies. More information about these organisational aspects can be found in the next section.

13.9.   Concluding remarks

Digitalisation is very much a part of the transition to a water-wise world, as it gives all actors in the water sector the enabling tools. However, there is no single template for that transition, as there are many ways to progress on this journey, and it is probably true that there are as many possible ways as they are water utilities embarking on it. Therefore, this chapter provides the necessary ingredients and considerations but not a single recipe for achieving digital water transition. The ingredients and considerations include basic smart water definitions (Section 13.1), history (Section 13.2), the current state of development (Section 13.3), various elements (Section 13.4), use cases and examples (Section 13.5), business considerations (Section 13.6), the vision (Section 13.7) and guidelines for implementation (Section 13.8).

Digital water is an integrator across utility departments, with technology offering the ability to bring together data and knowledge that are sitting in organisational silos. Leveraging water utility data and publicly available data will result in improved operations, planning and asset management. While the technology is largely available, it is not the greatest challenge for water utilities. The greatest challenge is in the people, the organisations and the culture. We also know that changing an organisation's culture is one of the most difficult challenges and often takes a long time.

REFERENCES

Al Aani S, Bonny T, Hasan SW and Hilal N (2019) Can machine language and artificial intelligence revolutionize process automation for water treatment and desalination? *Desalination* **(458)**: 84–96.

Arad Group (2022) Smart Demand Metering using AMR and AMI Technologies. https://arad.co.il/amr-ami/ last accessed on 15/02/2022.

Abdallah M and Kapelan Z (2017) Iterative extended lexicographic goal programming method for fast and optimal pump scheduling in water distribution networks. *Journal of Water Resources Planning and Management, ASCE (American Society of Civil Engineers)* **143(11)**. https://doi.org/10.1061/(ASCE)WR.1943-5452.0000843.

Ahopelto S and Vahala R (2020) Cost-benefit analysis of leakage reduction methods in water supply networks. *Water* **192**: doi:10.3390/w12010195.

Andersen JH and Powell RS (2000) Implicit state-estimation technique for water network monitoring. *Urban Water* **2(2)**: 123–130.

Anjana GR, Kumar MSM, Amrutur B and Kapelan Z (2019) Real-time water quality modelling with ensemble Kalman filter for state and parameter estimation in water distribution networks. *Journal of Water Resources Planning and Management (ASCE)* **145(11)**. https://doi.org/10.1061/(ASCE)WR.1943-5452.0001118.

Antunes A, Andrade-Campos A, Sardinha-Lourenço A and Oliveira MS (2018) Short-term water demand forecasting using machine learning techniques. *Journal of Hydroinformatics* **20(6)**: 1343–1366.

Bakker M, Vreeburg JHG, van Schagen M and Rietveld LC (2013) A fully adaptive forecasting model for short-term drinking water demand. *Environmental Modelling & Software* **48**: 141–151.

Basupi I and Kapelan Z (2015) Flexible water distribution system design under future demand uncertainty. *Journal of Water Resources Planning and Management* **141(4)**: 04014067.

Berardi L, Giustolisi O, Kapelan Z and Savic DA (2008) Development of pipe deterioration models for water distribution systems using EPR. *Journal of Hydroinformatics* **10(2)**: 113–126.

Behzadian K and Kapelan Z (2015) Advantages of integrated and sustainability based assessment for metabolism based strategic planning of urban water systems. *Science of The Total Environment* **527–528**: 220–231. https://doi.org/10.1016/j.scitotenv.2015.04.097

Branisavljevic, N, Kapelan Z and Prodanovic D (2011) Improved real-time data anomaly detection using context classification. *Journal of Hydroinformatics* **13(3)**: 307–323. https://doi.org/10.2166/hydro.2011.042/.

Brdys MA and Ulanicki B (1994) *Operational control of water systems: structures, algorithms, and applications.* Prentice Hall, NJ, USA.

Bufler R, Clausnitzer V, Vestner R, Werner U and Ziemer C (2017) Water 4.0. German Water Partnership. https://germanwaterpartnership.de/wp-content/uploads/2019/05/GWP_Brochure_Water_4.0.pdf

CETAqua (2013) Review of sensors to monitor water quality. Available at https://publications.jrc.ec.europa.eu/repository/bitstream/JRC85442/lbna26325enn.pdf, last accessed 15/02/2022.

Cunha M, Marques J, Creaco E and Savić D (2019) A dynamic adaptive approach for water distribution network design. *Journal of Water Resources Planning and Management* **145(7)**: 04019026.

Creaco E, Campisano A, Fontana N *et al.* (2019) Real time control of water distribution networks: A state-of-the-art review. *Water Research* **161**: 517–530.

Dogo EM, Nwulu NI, Twala B and Aigbavboa C (2019) A survey of machine learning methods applied to anomaly detection on drinking-water quality data. *Urban Water Journal* **16(3)**: 235–248.

Donkor EA, Mazzuchi TA, Soyer R and Roberson AJ (2014) Urban water demand forecasting: review of methods and models. *Journal of Water Resources Planning and Management* **140(2)**: 146–159.

Ecob D, Williamson J, Hughes G and Davis J (1995) PLCs and SCADA – a water industry experience. In IEEE colloquium on 'Application of PLC Systems with Specific Experiences from Water Treatment'. IEEE, New York.

Engelhardt M, Savic D, Skipworth P *et al.* (2003) Whole life costing: Application to water distribution network. *Water Science and Technology: Water Supply* **3(1-2)**: 87–93.

Giustolisi O, Laucelli D and Savic DA (2006) Development of rehabilitation plans for water mains replacement considering risk and cost-benefit assessment. *Civil Engineering and Environmental Systems* **23(3)**: 175–190.

Hall JW, Mortazavi-Naeini M, Borgomeo E *et al.* (2020) Risk-based water resources planning in practice: a blueprint for the water industry in England. *Water and Environment Journal* **34(3)**: 441–454.

Hartley K and Kuecker G (2020) The moral hazards of smart water management. *Water International* **45(6)**: 693–701. DOI: 10.1080/02508060.2020.1805579.

Harvey R, McBean EA and Gharabaghi B (2014) Predicting the timing of water main failure using artificial neural networks. *Journal of Water Resources Planning and Management* **140(4)**: 425–434.

Hassanzadeh A, Rasekh A, Galelli S *et al.* (2020) A review of cybersecurity incidents in the water sector. *Journal of Environmental Engineering* **146**: 1–15.

Hu Z, Chen B, Chen W, Tan D and Shen D (2021)) Review of model-based and data-driven approaches for leak detection and location in water distribution systems. *Water Supply* **21(7)**: 3282–3306.

Hutton C, Kapelan Z, Vamvakeridou-Lyroudia LS and Savic DA (2014) Dealing with uncertainty in water distribution systems' models: a framework for real-time modelling and data assimilation. *Journal of Water Resources Planning and Management (ASCE)* **140(2)**: 169–183. https://doi.org/10.1061/(ASCE)WR.1943-5452.0000325.

InfraSense Labs (2022) *InfraSense Data Logger and Management System for Monitoring Dynamic Hydraulic Condition*. Available at https://www.infrasense.net/technology.html, last accessed 15/02/2022.

i2OWater (2022) *Advanced Pressure Management*. Available at https://en.i2owater.com/solutions/advanced-pressure-management, last accessed 15/02/2022.

IWRA/K-Water (2018) *Smart Water Management – Case Study Report*. https://www.iwra.org/smart-water-management-case-study-report/

Jentgen LA and Wehmeyer MG (1994) SCADA trends and integration perspectives. *Journal of the AWWA (American Water Works Association)*. https://doi.org/10.1002/j.1551-8833.1994.tb06222.x

Jun H and Loganathan GV (2007) Valve-controlled segments in water distribution systems. *Journal of Water resources Planning and Management* **133(2)**: 145–155. doi:10.1061/(ASCE)0733-9496(2007)133:2(145).

Kleiner Y and Rajani B (2001) Comprehensive review of structural deterioration of water mains: statistical models. *Urban Water* **3(3)**: 131–150.

KWR (2022) *Digital Twin Demonstrator*. Available at https://www.kwrwater.nl/en/projecten/digital-twin-demonstrator, last accessed 15/02/2022.

Lemos D, Dias AC, Gabarrell X and Arroja L (2013) Environmental assessment of an urban water system. *Journal of Cleaner Production* **54**: 157–165. https://doi.org/10.1016/j.jclepro.2013.04.029.

Li L, Rong S, Wang R and Yu S (2021) Recent advances in artificial intelligence and machine learning for nonlinear relationship analysis and process control in drinking water treatment: a review. *Chemical Engineering Journal* **405**: p.126673.

Mahmoud H, Kapelan Z and Savic D (2018) Real-time operational response methodology for reducing failure impacts in water distribution systems. *Journal of Water Resources Planning and Management (ASCE)* **144(7)**. https://doi.org/10.1061/(ASCE)WR.1943-5452.0000956.

Makropoulos CK and Butler D (2010) Distributed water infrastructure for sustainable communities. *Water Resources Management* **24**: 2795–2816.

Mala-Jetmarova H, Sultanova N and Savic D (2017) Lost in optimisation of water distribution systems? A literature review of system operation. *Environmental modelling & software* **93**: 209–254.

Monks I, Stewart RA, Sahin O and Keller RJ (2021) Taxonomy and for valuing the contribution of digital water meters to sustainability objectives. *Journal of Environmental Management* **293**: 112846. https://doi.org/10.1016/j.jenvman.2021.112846

Morley M and Savić D (2020) Water resource systems analysis for water scarcity management: the thames water case study. *Water* **12(6)**: 1761.

Mounce SR, Mounce RB and Boxall JB (2011) Novelty detection for time series data analysis in water distribution systems using support vector machines. *Journal of hydroinformatics* **13(4)**: 672–686.

Mounce SR, Day AJ, Wood *et al.* (2002). A neural network approach to burst detection. *Water Science and technology* **45(4-5)**: 237–246. https://doi.org/10.2166/wst.2002.0595.

Joo Hee Ng, Harry Seah and Chee Meng Pang (2020) Digitalising Water – Sharing Singapore's Experience. IWA Publishing, London, UK. https://www.pub.gov.sg/Documents/Digitalising-Water-Sharing-Singapores-Experience.pdf

Nikoloudi E, Romano M, Memon FA and Kapelan Z (2020) Interactive decision support methodology for near real-time response to failure events in a water distribution network. *Journal of Hydroinformatics* **23(3)**: 483–499. https://doi.org/10.2166/hydro.2020.101.

Ostfeld A, Uber JG, Salomons E *et al.* (2008) The Battle of the Water Sensor Networks (BWSN): a design challenge for engineers and algorithms. *Journal of Water Resources Planning and Management, ASCE* **134(6)**: 556–568.

Perelman L, Arad J, Housh M and Ostfeld A (2012) Event detection in water distribution systems from multivariate water quality time series. *Environmental Science and Technology* **46(15)**: 7927–8522. doi: 10.1021/es3014024.

Quintiliani C, Vertommen I, Laarhoven KV, Vliet JVD and Thienen PV (2020) Optimal pressure sensor locations for leak detection in a Dutch water distribution network. *Environmental Sciences Proceedings* **2(1)**: p.40.

Riss G, Romano M, Memon FA and Kapelan Z (2021) Detection of water quality failure events at treatment works using a hybrid two-stage method with CUSUM and random forest algorithms. *Water Supply* **21(6)**: 3011–3026. https://doi.org/10.2166/ws.2021.062.

Romano M, Kapelan Z and Savić DA (2013) Geostatistical techniques for approximate location of pipe failure events in water distribution systems. *Journal of Hydroinformatics* **15(3)**: 634–651, https://doi.org/10.2166/hydro.2013.094.

Romano M and Kapelan Z (2014) Adaptive water demand forecasting for near real-time management of smart water distribution systems. *Environmental Modelling and Software* **60**: 265–276. https://doi.org/10.1016/j.envsoft.2014.06.016.

Romano M, Kapelan Z and Savic DA (2014) Automated detection of pipe bursts and other events in water distribution systems. *Journal of Water Resources Planning and Management, ASCE* **140(4)**: 457–467. https://doi.org/10.1061/(ASCE)WR.1943-5452.0000339.

Romano M, Boatwright S, Mounce S, Nikoloudi E and Kapelan Z (2020) AI–based event management at United Utilities. *IAHR Hydrolink* **4**: 104–108.

Sanctis ADE, Shang F and Uber JG (2009) Real-time identification of possible contamination sources using network backtracking methods. *Journal of Water Resources Planning and Management* **136**: 444–453. doi: 10.1061/(ASCE)WR.1943-5452.0000050.

Sarni W, White C, Webb R, Cross K and Glozbach R (2019) *Digital Water – Industry leaders chart the transformation journey.* IWA digital water report, available at https://iwa-network.org/publications/digital-water, last accessed 15/02/2022.

Savić D (2021) DIgital water developments and lessons learned from the car and aircraft industries. *Engineering* **9**: 35–41. https://doi.org/10.1016/j.eng.2021.05.013

Sophocleous S, Savić D and Kapelan Z (2019) Leak localization in a real water distribution network based on search-space reduction. *Journal of Water Resources Planning and Management* **145(7)**: 04019024.

Storey MV, van der Gaag B and Burns BP (2011) Advances in on-line drinking water quality monitoring and early warning systems. *Water Research* **45**: 741–747. https://doi.org/10.1016/j.watres.2010.08.049

Ulanicki B, Bounds PLM, Rance JP and Reynolds L (2000) Open and closed loop pressure control for leakage reduction. *Urban Water* **2(2)**: 105–114.

Ulanicki B, Kahler J and See H (2007) Dynamic optimization approach for solving an optimal scheduling problem in water distribution systems. *Journal of Water Resources Planning and Management* **133(1)**: 23–32.

US EPA (2009) *Distribution System Water Quality Monitoring: Sensor Technology Evaluation Methodology and Results - A Guide for Sensor Manufacturers and Water Utilities.* Available at https://www.epa.gov/sites/default/files/2015-06/documents/distribution_system_water_quality_monitoring_sensor_technology_evaluation_methodology_results.pdf, last accessed 15/02/2022.

US EPA (2012) *Canary User's Manual.* Available at https://cfpub.epa.gov/si/si_public_file_download.cfm?p_download_id=513254&Lab=NHSRC, last accessed 15/02/2022.

Van den Broeke J, Pronk T, Vries D and van Thienen P (2019) *Hydroinformatics and Smart Water Management – Current State and Opportunities for the BTO Utilities.* KWR Watercycle Research Institute, report BTO 2019.045. https://library.kwrwater.nl/publication/59835620/

Van Thienen P van, Bergmans B, Diemel R *et al.* (2018) Advances in development and testing of a system of autonomous inspection robots for drinking water distribution systems. *Proceedings of the 1st International WDSA / CCWI Joint Conference,* Kingston, Ontario, Canada.

Walski TM (1990) Sherlock Holmes meets Hardy-Cross: or Model Calibration in Austin, Texas. *Journal of the AWWA* **82(3)**: 34–38. https://doi.org/10.1002/j.1551-8833.1990.tb06933.x

Walski TM (2006), A history of water distribution. *Journal of the AWWA* **98**: 110–121. https://doi.org/10.1002/j.1551-8833.2006.tb07611.x

Water Europe (2016) *The Value of Water – Multiple Waters for Multiple Purposes and Users – Towards a Future-Proof Model for a European Water-Smart Society.* https://watereurope.eu/wp-content/uploads/2020/04/WE-Water-Vision-english_online.pdf

Weickgenannt M, Kapelan Z, Savic DA and Blokker M (2010) Risk-based sensor placement for contaminant detection in water distribution systems. *Journal of Water Resources Planning and Management, ASCE* **136(6)**: 629–636. https://doi.org/10.1061/(ASCE)WR.1943-5452.0000073.

Williamson F, van den Broeke J, Koster T *et al.* (2014) Online water quality monitoring in the distribution network. *Water Practice & Technology* **9(4)**: 575–585. doi: 10.2.166/wpt.2014.064.

Xenochristou M, Hutton C, Hofman J and Kapelan Z (2020) Water demand forecasting accuracy and influencing factors at different spatial scales using a Gradient Boosting Machine. *Water Resources Research* **56(8)**: e2019WR026304. https://doi.org/10.1029/2019WR026304.

Xing L and Sela L (2022) Graph neural networks for state estimation in water distribution systems: application of supervised and semisupervised learning. *Journal of Water Resources Planning and Management* **148(5)**: 04022018.

Dragan A. Savić and John K. Banyard
ISBN 978-1-83549-847-7
https://doi.org/10.1108/978-1-83549-846-020242016
Emerald Publishing Limited: All rights reserved

Index

Printed and bound by CPI Group (UK) Ltd, Croydon, CR0 4YY

19/01/2025

14627899-0001